Hydraulic Transport of Solids in Pipes

HYDROTRANSPORT 10

Proceedings of the 10th International Conference on the Hydraulic Transport of Solids in Pipes - HYDROTRANSPORT 10 - held at Innsbruck, Austria: 29-31 October, 1986. Organised and sponsored by BHRA, The Fluid Engineering Centre, Cranfield, Bedford, MK43 0AJ, England.

ACKNOWLEDGEMENTS

The valuable assistance of the Organising Committee, Corresponding Members and Panel of Referees is gratefully acknowledged.

ORGANISING COMMITTEE

(Chairman)

S. Odrowaz Pieniazek	Seltrust Engineering
D. Brookes	BP International Ltd.
Dr. N.P. Brown	British Petroleum Co. Ltd.
Dr. A.P. Burns	BHRA
T. Richards	ECC International Ltd.
H.F. Ferguson	U.K. Department of Energy
D.E. Jenkinson	British Coal
Dr. M. Streat	Imperial College of Science and Technology
J.F.C.White	The Rugby Portland Cement plc
Dr. P. Wood	IEA Coal Research
Rosemary Pickford Conference Organiser	BHRA

OVERSEAS CORRESPONDING MEMBERS

R. Casagrande	ILF-Laser-Feizlmayr, Austria
Dr. G.G. Duffy	The University of Auckland, New Zealand
Prof. R.R. Faddick	Colorado School of Mines, U.S.A
F. Ferrini	Techfem, Italy
W. Haentjens	Barrett, Haentjens & Co. U.S.A
Prof. T. Kawashima	Tohoku University, Japan
Prof. C.A. Shook	University of Saskatchewan, Canada
P.W.H. Simons	Holthuis BV, The Netherlands
F.M. Want	Alcoa of Australia Ltd., Australia

Hydraulic Transport of Solids in Pipes

HYDROTRANSPORT 10

Editor

A. P. Burns

*BHRA The Fluid Engineering Centre,
Cranfield, Bedford, UK*

Published on behalf of
BHRA THE FLUID ENGINEERING CENTRE
by
ELSEVIER APPLIED SCIENCE PUBLISHERS
LONDON and NEW YORK

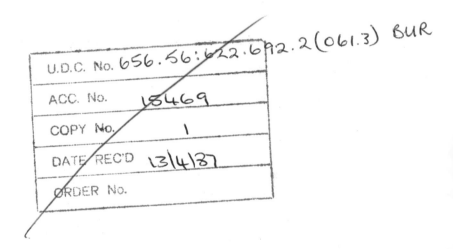

ELSEVIER APPLIED SCIENCE PUBLISHERS LTD
Crown House, Linton Road, Barking, Essex IG11 8JU, England

Sole Distributor in the USA and Canada
ELSEVIER SCIENCE PUBLISHING CO., INC.
52 Vanderbilt Avenue, New York, NY 10017, USA

WITH 59 TABLES AND 283 ILLUSTRATIONS

British Library Cataloguing in Publication Data

Hydrotransport 10 (*Conference: 1986: Innsbruck*)
 Hydraulic transport of solids in pipes.
 1. Hydraulic conveying   2. Bulk solids—
 Transportation   3. Bulk solids handling
 I. Title   II. Burns, A. P.   III. BHRA
 621.8'672   TJ898

 ISBN 1-85166-064-X

Library of Congress CIP data applied for

Printed in Great Britain by Galliard (Printers) Ltd, Great Yarmouth

PREFACE

16 Years of 'Hydrotransport'

Since 1970, BHRA has been organising conferences in the hydraulic transport of solid materials. The first and subsequent conferences have enjoyed strong support from a mixture of academic and industrial delegates. However, in the 16 year period, there have been a large number of changes in technology, economics and environmental considerations which all affect the commercial use of hydraulic transport. This tenth conference in the series continues to reflect a spectrum of interest ranging from fundamental research through to commerical case studies of operating systems; the conferences have always shown this balance.

From the earliest conferences, economic analysis of systems has been a feature and those working in the area have always stressed the importance of this aspect. However, in the period from 1970, there have been wide fluctuations in energy costs and costs of raw materials and significant price changes in both these sectors are still with us. By developing economic models at an early stage, workers in this field of technology have been in a very strong position to adjust their analyses to account for such changes in their basic data. It is encouraging that despite current low energy costs and raw material prices, the economic case for developing hydraulic transport systems is still favourable.

In parallel with economic considerations, political realities have never been neglected when considering the take up of hydrotransport technology and over the conference series, it has been possible to track the political climate in some areas of the world against technical progress of some of the major projects. Again, in the political arena hydraulic transport has not yet reached an end point and changes, especially in America, to legislation will make a further impact in the short term future of major hydraulic transport systems.

The technology itself has fortunately never been pushed into a secondary place by economic or political considerations, and it, along with the other two, has shown a steady progression at all levels of activity. In this, as in most other fields of engineering, the use of increasing computer power has led to the adoption of sophisticated modelling techniques as an aid to understanding both basic and bulk processes in solid-fluid flow. Experimental techniques, instrumentation and field data collection have been able to provide validation for various models as well as providing the insight into the process at a detailed and system level. The content of the tenth conference continues to reflect the trend apparent for the past few years; a trend which has taken us to slurries of high concentration. Two approaches have been very evident. These are the fine homogeneous mixtures of the C.W.M. type using a variety of viscosity reducing additives, or the stabilised approach requiring perhaps less material preparation but offering high concentration transport of a wide range of material sizes.

While it will continue to be the case that the commercialisation of systems will be dictated at the economic and perhaps political level, workers in the area of slurry technology having demonstrated an awareness of these, will be able to bring strong technical resources into the field whenever this has been required. Over the past 16 years, the technology as reflected in the hydrotransport conferences has not been static and there is no reason to believe that over the next 16 years, ultimate solutions will be developed and we can look forward to further innovation in the technical front and as in the past, this conference series will record important developments.

A. P. BURNS

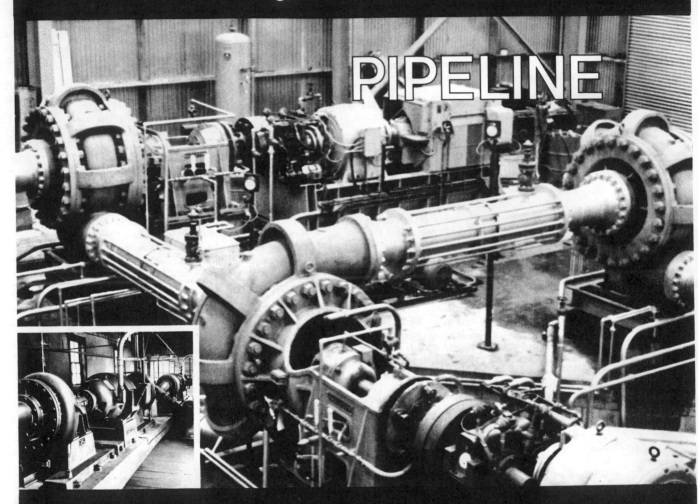

10th International Conference on the
Hydraulic Transport of Solids in Pipes
HYDROTRANSPORT 10
Innsbruck, Austria: 29-31 October, 1986

CONTENTS

CHARLESTOWN ENGINEERING

Process Engineering Technology from the world's largest producer of china clay

Charlestown Engineering is the engineering arm of English China Clays Group.

Charlestown Engineering has wide experience in all aspects of process equipment and technology related to slurries, minerals sizing and liquid/solid separation.

At its manufacturing facilities in Cornwall, Charlestown Engineering produces a wide range of products including: monitors, gravel pumps, filter presses, polyurethane components, control gear and instrumentation.

Charlestown Engineering has designed and installed complex clay process plants in many parts of the world from Brazil to Japan.

Polyurethane lined pipe

Charlestown Engineering pioneered techniques for lining steel pipes with polyurethane. In the late 60's Charlestown Engineering had a requirement to install a 20km buried pipeline to carry highly abrasive micaceous residue. A 20 year life was required.

Charlestown Engineering designed, developed and built a production rig to cast rotationally the polyurethane liner into steel pipes.

A Charlestown Engineering patented process was used to ensure a tolerance of $-0 +2mm$ in liner thickness at bore diameters up to 320mm as standard.

To date, the promised performance of these pipes has more than lived up to the design requirements.

Besides those lines installed in the china clay industry, Charlestown Engineering has supplied polyurethane lined steel pipes to a number of operations, for example, an ore slurry line in Spain and in NE England a coal tailings line which included an offshore section.

Charlestown Engineering also provide a wide range of polyurethane lined steel fittings to match the pipes. A variety of jointing systems are available plus bends, tees, reducers, etc.

Charlestown Engineering
John Keay House, St Austell, Cornwall, PL25 4DJ. England.
Telephone: (0726) 74482. Telex: 45526.

SLURRY METERING

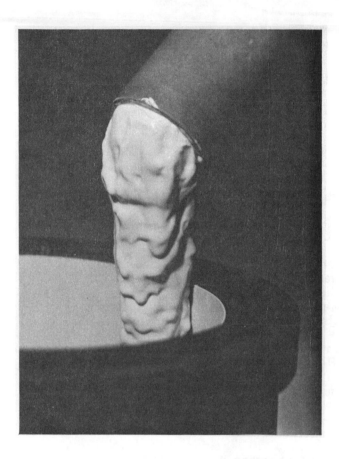

High solids content slurries and pastes and their effect on the performance of flow and density meters is being studied at Warren Spring Laboratory. A pilot plant flow loop is being used to evaluate performance of:

- Electromagnetic flowmeters

- Doppler ultrasonic flowmeters

- Mass flowmeter

- Gamma-ray density meter

- Vibrating tube density meter

This work is being carried out within the Wet Solids Handling Project — a programme sponsored both by industrial member companies and the British Government. The Project is open to new members and the test facility is also available for hire on a fully-confidential contract basis for testing pumps, meters or slurries.

Other Topics covered within the Project include:

- Storage vessel design for cakes and unsaturated wet powders

- Storage vessel design for slurries

- Conveyor selection and design for non-pumpable wet solids

- Measurement techniques for wet solids properties

Further information is available from:

Dr. Nigel Heywood
Warren Spring Laboratory
Gunnels Wood Road, Stevenage, Herts, U.K. SG1 2BX
Telephone: Stevenage (0438) 313388   Telex: 82250

FIT IT AND FORGET IT!

Per-Axel remembers when pipes had to be changed twice a year

P-A works for AB Sydsten in Dalby, Sweden. Ten years ago this company replaced 150 m of steel pipes with Trellex materials handling hose. Slurried diabase and gneiss with particle sizes up to 5 mm are transported in this slurry line.

The 5 mm thick steel pipes previously used were completely worn out after that 35 000—40 000 tonnes of material had passed through them. This corresponds to a life of 6 months, sometimes eight months.

Over the past 10 years about 700 000 tonnes of material has passed through this 152 mm i.d. Trellex slurry line. Wear is negligible and P-A expects the hose to last many years more.

A materials handling hose naturally wears mostly in the bends and a common trick to prolong the life is to rotate the hose half a turn when one side is worn out. At Sydsten they have still not had to employ this trick after 10 years of use.

When you have problems with the transportation of abrasive materials choose Trellex!

Wear costs money! Reduce your costs by installing Trellex materials handling hose. At Sydsten the materials handling hose has so far lasted 20 times longer than the previous steel pipe. It speaks for itself that this must have a favourable effect on the overall economy. Trellex materials handling hose forms part of a complete system for the transportation of abrasive materials. Using three easily replaceable standard components
− the wear-resistant hose
− the externally fitted coupling leaving a completely
  smooth inner tube
− the unique gasket
the system can be tailored to suit the requirements of
any plant.

Apart from the savings due to longer life you get a
maintenance-free system with easier assembly into the
bargain.

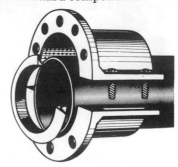

TRELLEBORG
Trellex Products

Trelleborg AB, S-231 81 Trelleborg,

Telex 32928, Tel 0410-51 000 (Int. + 46 410 51 000)

10th International Conference on the
Hydraulic Transport of Solids in Pipes

HYDROTRANSPORT 10

Innsbruck, Austria: 29-31 October, 1986

PAPER A1

HYDRAULIC HANDLING IN THE COAL INDUSTRY

P.A. Wood

IEA Coal Research, UK

SYNOPSIS

The present status of the application of
hydraulic handling and transport systems in the
world coal industry is reviewed and the major
technological problems identified. Topics
discussed include the transport of coal and
mining waste in and around the coal mine or
preparation plant, the long distance transport of
both coarse coal and fine coal, the use of coal
slurry fuels for combustion and the use of
hydraulic handling systems in connection with the
marine transport of coal. There then follows a
discussion of the effects of coal-water interactions on both coal and water quality, and a
statement concerning environmental aspects of
hydraulic handling. It is concluded that there
is a future potential for hydraulic handling
systems in the world coal industry but, in order
to compete with alternative systems that are
available, such systems have to be technically
and economically viable. This will require a
solution to the specific problems identified.

1 INTRODUCTION

IEA Coal Research is interested in advancing the
use of coal worldwide and in promoting international cooperation in research. Part of this
role involves assessing current and developing
technologies that can be, or are being, applied
to the coal industry. Hydraulic handling and
transport is one area that shows continued
promise for various applications in the coal
industry, not only for handling coal but for
handling other materials associated with the coal
industry.

This paper reviews the potential applications for
the use of hydrotransport systems in coal mining
and handling, indicating the major technological
problems as far as the worldwide coal industry is
concerned. This is then followed by a discussion
of the effects of coal-water interactions on
both coal and water quality, and a statement
concerning the environmental aspects of hydraulic
handling.

2 APPLICATIONS

Potential applications for hydraulic handling
and transport in the coal industry include
mining, long distance transport, combustion
and marine transport. Each of these will be
considered in turn.

2.1 Coal mining

Within the coal mining industry hydraulic handling has been used for the transport of both coal
and mine waste, and is a method suitable for the
transport of materials both into and out of a
mine.

2.1.1 Coal handling

Hydraulic handling and transport systems for coal
have been successfully used in conjunction with
the major types of mining equipment; continuous
miners, longwall cutters and hydraulic monitors.
Consequently hydraulic transport is an option for
coal clearance from the face to the surface
preparation plant that can be used for all the
main mining methods; room-and-pillar mining,
longwall mining, shortwall mining and hydraulic
mining. Specific systems will vary in design and
can include sections where slurry is transported
over the floor, in flumes or in pipes.

Floor

Slurry transport over the floor of a coal mine
does not require pumping energy but relies on
gravity flow. Consequently it is necessary for a
suitable gradient to be present in the direction
of flow. Clearly, permitting a slurry to flow
over the floor can result in problems, particularly the difficulty of controlling hydraulic
parameters and the deterioration of floor conditions. In general transport over the floor is
restricted to faces where hydraulic mining is
used, and then only for a limited distance close
to the face. After only a short distance of
transport the slurry will be directed to a flume
or pipeline.

Flumes

Flumes also do not require pumping energy but
again rely on the gravity flow created by a
gradient along the direction of transport. Flow
in flumes is generally turbulent and heterogeneous, and often involves a wide range of particle
sizes - the largest undergoing saltation. Flumes
have been used successfully at several in-mine
installations, generally as an intermediate
transport system between coal mining and subsequent pipeline transport. For example the Sunagawa Mine in Japan uses 3.2 mm thick mild steel
semi-circular flumes which can carry 0.5 Mt of
coal (1.7 Mt of slurry) before needing to be
replaced (Miura and Mase, 1979).

Stewart and Wood (1984) identify several
technological problems associated with flumes
that are related to the optimisation of flume
material and flume cross-sectional shape. Flume

material and thickness must be related to wear rates and life expectancy, while low resistance linings might be necessary for lower gradients. The influence of flume shape seems somewhat confusing and optimisation is a compromise between favourable hydraulic parameters and the problem of installation and subsequent stability. Different cross-sectional geometries result in different hydraulic parameters while the influences of particle shape and size on the performance of hydraulic transport can also depend on flume shape. Installation and stability problems of flumes are dependent not only on their cross-sectional geometry but also on their height. Lateral forces on flumes that are set into the ground can result in buckling or even re-orientation (overturning) of the flumes. Buckling can then encourage blockage while re-orientation can cause spillage.

In general, there is only a limited amount of information available on flume transport and there is a need for more if optimum operating conditions are to be achieved. Specifically, more information is required on: the effect of flume cross-sectional geometry on transport; the controls of wear rates of different con- structional materials and linings; and what conditions of the floor are likely to cause flume deformation.

Pipes

Pipes can also be used for gravity flow but have the added advantage of enabling pumping pressure to be used so that slurry can be transported both horizontally and up inclines. This latter re- quirement is essential in most mining situations if hydraulic handling methods are to be used to raise coal to the surface. Pipeline systems installed in coal mines can consequently be subdivided into two essentially separate systems:

1) horizontal transport systems - operating at lower pressures, eg in the Hansa Hydromine of the Federal Republic of Germany 250 mm diameter pipelines were used to haul up to 12 m^3/min of slurry with minimum operating pressures of 1.2 MPa (Gerstein and Beyer, 1978);
2) vertical transport systems - operating at higher pressures and presenting the greater design and operational problems. The main problems involved are associated with feeding the coal into the system and achieving the sufficiently high pressures needed for vertical transport. Various methods have been used which are dependent on heights of vertical lift required:

   - 260 m lift - centrifugal pumps in series, Loveridge Mine, USA (Alexander and Shaw, 1984);
   - 320 m lift - airlift system, Krasnoarmey- skaya Mine, USSR (National Coal Board, 1977);
   - 518 m lift - hydrohoist, Sunagawa Mine, Japan (Sakamoto and others, 1983);
   - 847 m lift - 3 chambered pipe feeder, Hansa Hydromine, FRG (Gerstein and Beyer, 1978).

Other feeding methods and vertical transport systems are also currently being developed.

Stewart and Wood (1984) identify three major problems for the use of pipelines for coal handling in coal mines. These are:

1) the inadequacy of, and lack of standardisation between, information available on the wear rates of the different constructional materials;
2) optimisation of any vertical hoisting system; and
3) development and optimisation of feeders for coarse coal.

A solution to the first problem will help to build up a data-bank of empirical information which can then be used as a basis for future system design and specification, while a solution to the other two problems will improve efficiency and performance.

2.1.2 Mine waste handling

There are two major aspects where hydraulic transport systems can be used for the handling of waste materials associated with coal mining which can be identified as local disposal and remote disposal.

Local disposal

Hydraulic handling systems have been widely and successfully used for the disposal of washery rejects at many coal preparation plants. These systems are often associated with waste disposal in a slurry pond in the vicinity of the prepar- ation plant, and such ponds can often result in environmental problems.

Remote disposal

Systems that can be classified as remote disposal systems usually involve greater transport distances and ultimate disposal in something other than a slurry pond. The concept can be versatile and four examples include:

1) transport to sea for disposal, eg from Horden Colliery, UK to North Sea;
2) transport to power station for combustion (not direct combustion), eg from Freyming Preparation Plant to the Emile Huchet power station at Carling, France (Roberts, 1985);
3) transport to underground coal mine for stowing, eg at the Ordzhonkidzeugol Coal Mine, USSR (Buchgolc and Zolotariew, 1980); and
4) transport to abandoned underground coal mine for flushing, eg in Scranton, PA, USA (Whaite and Allen, 1975).

It is perhaps the hydraulic handling of waste associated with underground stowing that is the most problematical of these four concepts and such systems should correctly be subdivided into a transport system and a stowing system. Of these two sub-systems it is the transport system that is less problematical and additionally it should be realised that a hydraulic transport system does not necessarily mean that a hydraulic stowing system has to be used. For example, stowing operations in the USSR often use a combination of hydraulic transport with pneumatic stowing. The hydraulic stowing process however encounters many problems although these are generally a problem of the stowing process in general (eg equipment development and optimisat- ion) rather than being specific to hydraulic stowing. However, although the density of a fill that has been placed hydraulically is greater than that produced by any other stowing method, a study by Wood (1983) states that there are still problems associated with a lack of knowledge about the:

1) rheological properties of mine waste slurries;
2) optimum particle size and size distribution;
3) equipment wear and design specification; and
4) characteristics of the waste (sulphur content, weathering etc).

2.2 Long distance transport

The term long distance is obviously somewhat arbitrary and systems are presently operating which are able to transport both coarse (run-of-mine) and fine coal.

2.1.1 Coarse coal transport

Operating systems transporting coarse coal are generally low concentration, use centrifugal pumps and, due to their relatively low efficiency, are confined to shorter distances. The Loveridge Mine of West Virginia, USA for example has an overland transport system conveying about 13 t/min of run-of-mine coal for a distance of 3.8 km to a dewatering and coal preparation plant (Alexander and Shaw, 1984). Other existing systems for coarse coal include a 11 km pipeline for run-of-mine coking coal in the USSR.

In an attempt to improve efficiency alternative systems for the transport of coarse coal at higher concentrations are undergoing considerable research and development effort. Such systems often utilise positive displacement pumps and it is expected that they will be able to transport coarse coal over longer distances and at lower velocities and will therefore have greater efficiencies. These systems often use some mechanism to prevent settling of the coarse particles and include recent developments by:

1) BP and Bechtel - for coal up to 50 mm (Brooks and Dodwell, 1984);
2) Asea-Atom/CSIRO - for coal up to 50 mm (Australian Coal Miner, 1985); and
3) Fibre Dynamics (Duffy and Walmsley, 1985).

As recognised by the Hydrotransport Cooperative Programme of IEA Coal Research the development of coarse coal transport systems would benefit considerably from the development of a universal head-loss model and the further development of coarse coal feeders.

2.1.2 Fine coal transport

Fine coal pipeline systems can be considered as those conveying coal that is smaller than 2 mm and constitute the longest pipelines for coal. These include:

1) Black Mesa Pipeline, USA - 439 km long (Snoek and others, 1976);
2) Donetsk to Black Sea Pipeline, USSR - 200 km long (Eden and others, 1985); and
3) Ohio Pipeline, USA - 174 km long (Halvorsen, 1964).

The technology is perhaps best exemplified by the Black Mesa Pipeline which has transported over 50 Mt of coal since operations started in 1970. Average tonnages of 4.8 Mt/year are transported as a slurry of 50:50 coal to water.

Many other long distance transport systems have been considered in various countries. In the USA, for example, approximately ten routes have been proposed although most of these now seem unlikely to go ahead, primarily as a result of the inability to secure a right-of-way along the pipeline route. Other countries, however, are still actively considering long distance pipelines for fine coal and these include:

1) Canada - 1016 km, Alberta to British Columbia pipeline (Jacques and others, 1982);
2) India - 1000 km, Dhanbad to Bhatinda pipeline (Mining Magazine, 1984); and
3) China - 900 km, Yangtze River coal slurry pipeline (Hill and others, 1985).

In addition to the long pipelines, shorter pipelines for fine coal have also been operated successfully. In France, for example, the Merlebach to St. Avold pipeline has been carrying 1.5 Mt/y over a distance of 10 km since 1952 (Sueur, 1982). Also in France the recently completed La Houve pipeline is expected to carry 0.5 Mt/y over a distance of 7 km (Dulac and Couratin, 1985).

As with coarse coal pipelines, various novel techniques are being developed for the transport of fine coal. These include the oil agglomeration method being developed in Australia (Rigby and Thomas, 1983) and the use of other carrier fluids such as liquid carbon dioxide (Santhanam and others, 1985) or methanol (Keller, 1979).

Two studies (Stewart and Wood, 1984; Wood, 1984) indicate that the major problem with fine coal pipeline transport is the subsequent dewatering of the coal. Despite much research and equipment development adequate dewatering is a difficult and expensive practice. What is needed are improved dewatering capabilities and efficiencies as these would have a significant beneficial impact.

2.3 Combustion and ash removal

In the power generation industry hydraulic handling systems can be used either to handle coal slurry for combustion or as ash removal systems.

2.3.1 Slurry combustion

Slurries containing coal whose purpose is for combustion can be grouped together under the term coal-liquid mixtures (CLM) or coal fuel slurries. There are numerous forms of these:

1) COM - coal-oil mixtures;
2) COW - coal-oil-water mixtures;
3) CWM - coal-water mixtures (also called CWF - coal-water fuels);
4) CMM - coal-methanol mixtures; and
5) CMW - coal-methanol-water mixtures.

Typically a coal-liquid mixture will contain

70-75% fine coal and, in addition to the main
liquid ingredient, may include additives to
reduce viscosity and promote stability (Morrison,
1983). Although both CMM and CMW possess favourable combustion characteristics the high cost of
methanol has discouraged extensive investigation
of these two mixtures. The sequence COM to COW
to CWM represents a successive decrease in the
amount of oil used and much of the present
research interest is in CWM where no oil at all
is needed.

With reference to the combustion process the
important fuel related parameters of a coal-
liquid mixture are:

1) mixture stability;
2) mixture viscosity;
3) atomisation quality;
4) ignition stability;
5) carbon burnout;
6) ash content; and
7) emissions.

As far as handling is concerned other problems,
in addition to stability and viscosity, are
concerned with:

1) preparation of required coal size
   distribution;
2) optimisation of water content. There is a
   need to reduce the water penalty and yet
   still be able to pump the mixture. Some
   mixtures of up to 80 wt% coal have been
   produced (Farthing and others, 1983); and
3) equipment wear. Erosion rates of pumps,
   valves, atomisers and burners are high.

In general, the combustion of COM can be
considered as commercial but the technology of
CWM is now rapidly advancing with several
manufacturers and partnerships investigating
different mixtures (Morrison, 1983). It should
be remembered that coal-liquid mixtures are
generally intended as a substitute for oil - they
would not be used as a substitute for coal.
Also, it is important to realise that slurries
for combustion and slurries for transport are
completely different. It is often stated that an
advantage of hydrotransport of coal is that the
slurry can be burnt directly - this is not
generally true. The size distribution, viscosity
and concentrations of the two types of slurry are
different. There have been however some attempts
to produce a slurry suitable for both purposes
(eg Priggen and others, 1984).

Studies by Morrison (1983) and Siemon (1985) have
identified two tasks that are still to be be
achieved to improve the potential of a CWM:

1) extending burner life; and
2) obtaining prolonged experience of combustion
   in boilers designed for oil.

There are also economic barriers determined by
Siemon (1985) including the availability of the
cheap good quality coal that is needed for fuel
slurries and the fluctuating relative prices of
competing fuels.

2.3.2 Ash removal

Hydraulic handling techniques for ash disposal
from coal-fired power plants and furnaces are

currently being used in several countries,
including the FRG (Pakusch, 1984) and the UK
(Williams and Pitman, 1984).

2.4 Marine transport

The utilisation of hydraulic handling techniques
at marine terminals for the loading and/or
unloading of coal is a somewhat speculative area
that received considerable attention a few years
ago. At present however, interest seems to have
wained and the technology is still effectively
untested as far as coal is concerned (although
the slurry loading of ironsands has been
conducted in New Zealand for many years).

The loss of interest in slurry loading and
unloading may be somewhat premature however as
the present and future increase in world marine
transport of coal will lead to considerable
congestion at terminals where dry handling
occurs. This in turn will result in a need for
additional facilities. Slurry handling would
have the ability to increase considerably the
future handling potential of coal by greatly
enhancing the number of locations where coal can
be handled. The siting of facilities will be
considerably more flexible as:

1) fewer shore facilities are required;
2) ships can be loaded offshore, thus obviating
   the requirement of a near-shore deepwater
   location; and
3) environmental impact is less.

Wood (1984) includes a review of coal slurry
handling at ports. The infrastructure required
to handle coal as a slurry will depend on whether
the terminal is exporting or importing and, in
addition, it will be necessary for coal carrying
ships to be either modified existing tankers or
purpose built vessels. The major problem that
seems to inhibit the use of this technology is
the present lack of necessity to employ an
untested technology rather than an established
one.

3 PROBLEMS

In addition to the technical problems directly
associated with the specific applications already
discussed there are other aspects which might
act to prevent the increased use of hydraulic
handling in the coal industry. Consequently,
whereas the fluid engineer is probably familiar
with the technical problems of hydraulic
handling, it is also desirable that he be aware
of problems that arise by virtue of the very
nature and subsequent use of the coal that is
being handled.

Hydraulic handling of coal requires the immersion
of coal particles in water for extended periods
of time during which coal-water interactions can
be occurring. These interactions can influence
the quality of both the water and the coal thus
inhibiting their usefulness for any subsequent
application (Wood, 1984). Additionally the overall impact of hydraulic handling systems on the
environment should be considered.

3.1 Water quality

As the range of tolerances for the quality of

water for many uses is often very restricted compared to that of coal, it is possible that many coal-water interactions will have significantly greater effects on the quality of the water rather than the quantity of the coal. In general water quality characteristics of recovered water are variable and are a function of:

1) water source or type;
2) coal source or type; and
3) other parameters such as detention time and operating conditions.

In detail however little is known about the influence of hydraulic transport on water quality, despite its importance in predicting the possible effluent quality from a dewatering plant or for determining the design of any subsequent waste water treatment facilities.

The types of water used for hydraulic handling can be fresh water (surface or underground), saline water (underground or sea), sewage plant effluent or other poor quality water, while the range of coals can include all ranks and types. The sulphur content of coal for example can be anything up to 10% and trace element concentrations can vary. Subsequent changes in the concentrations of these elements or compounds in any slurry water are dependent on the:

1) concentration in the coal;
2) concentration of coal in water;
3) solubility of element or compound; and
4) detention time.

In addition to chemical changes, an increased solids concentration of the water can be expected.

Wood (1984) considers that water quality problems are unlikely to be a major problem in general, although some upgrading procedures may be necessary prior to release into the environment. Whatever, it is unlikely that additional treatment will be required beyond that normally used at conventional coal preparation plants or mine sites.

3.2 Coal quality

Changes in coal quality during hydraulic handling can result from either or both of two causes:

1) immersion in water - as a result of coal-water interactions; and
2) handling - as a result of coal passing through pumps, pipes and other equipment where degradation of particles can occur.

Any changes occurring as a result of these processes can be described as changes to the mechanical, chemical, coking or combustion characteristics of the coal.

3.2.1 Mechanical properties

Particle degradation during hydraulic handling is influenced by numerous factors which can be summarised as:

1) equipment specification (eg pipe diameter, pump aperture size);

2) operational characteristics (eg slurry velocity, slurry concentration, distance of transport); and
3) coal characteristics (eg friability, shape, initial size and size distribution, hardness, rank, maceral content).

Although it is possible to make generalisations of the effect of hydraulic handling on the mechanical properties of coal there is a need for more information. Excessive fine coal production for example is an expensive and unwanted result that has adverse consequences on subsequent slurry handling and dewatering and on the treatment of transporting water. It is necessary to collect and analyse degradation data in a standard form so that an empirical data bank can be developed and the degradation process better understood.

3.2.2 Chemical properties

Little information is available on variations in the chemical properties of coal that occur as a result of hydraulic handling. Aspects that would benefit from further investigation include changes in the sulphur, volatile matter and metal contents.

3.2.3 Coking properties

A potential major application for the hydraulic handling of coal would be for the hydraulic pipeline transport of coking coal. However, a study by Wood (1984) confirms that there is controversy over the effects that such handling procedures have on the coking properties of a coal. Some experiments indicate little or no deterioration of caking power while others demonstrate significant deterioration. Furthermore, explanations as to the cause of any deterioration also vary. It might be that the presence of clays, or the effect of particle degradation and subsequent variations in maceral distribution between the various fractions, is the cause. Whatever, it seems likely that any effects on the coking parameters will be specific to particular coals and perhaps even specific to a particular operation.

In the context of coking quality the work with Australian bituminous coals (Rigby and Callcott, 1978; Elkes and others, 1982) is of interest. An oil agglomeration technique has been developed which enables the separation from the coal of refuse materials including the deleterious clays. As a consequence the coal retains its coking properties. A similar method for removing impurities is also being developed in Canada (Energy, Mines and Resources, 1984).

3.2.4 Combustion properties

In addition to the energy penalty associated with the water content of coal-water mixtures there may also be other effects on the combustion properties of the coal itself as a result of prolonged immersion in water. However, very little work has been done to investigate this possibility although one set of experiments (Chugh and Kundu, 1981) suggests that the heating value of coal might increase as a result of immersion (at least during the first eight hours of a twenty four hour test period). This increase in calorific value seems to occur as a result of a reduced ash content.

3.3 Environmental aspects

Wood (1984) considers that in general the environmental impact of hydraulic handling systems for coal can be considered as minimal compared to alternative methods. However some factors that should be borne in mind include:

1) water supply;
2) quality of released water;
3) accidental release or spillage of slurry on land or at sea; and
4) pipeline construction.

4 CONCLUSIONS

Hydraulic handling systems have been designed for many applications in the coal industry, although their use is by no means widespread. The operation of these systems is in general technically successful although there are still numerous problems that need to be evaluated and overcome. This in turn can only serve to improve the economics of such systems.

Some specific tasks that need to be achieved in order to improve hydraulic handling technology include:

1) optimisation of material selection procedures for pipes and flumes;
2) collection and analysis of wear data for pipes and flumes;
3) development and optimisation of feeders for coarse coal;
4) collection and analysis of degradation data for coarse coal;
5) optimisation of dewatering procedures for fine coal; and
6) understanding the effects of hydraulic transport on the coking properties of coal.

As for the future, there is potential for an increase in the use of hydraulic handling and transport systems for coal and for other materials associated with the coal industry. However it must always be remembered that such systems constitute just one of the options available to the designer/planner. Any hydraulic handling system has to compete with alternative systems that might well be better tried and tested. Thus hydraulic handling systems must be technically viable and economically sound. What a forum such as the BHRA Hydrotransport Conference is doing is helping to show that these systems are indeed viable and sound. Such conferences are:

1) investigating, and helping to eliminate, the problems and so improve the understanding;

2) providing information so that design specifications can be determined; and

3) building up the evidence to show when, where and in what circumstances a hydraulic handling system has the technical and economic advantage over other systems.

5 ACKNOWLEDGEMENTS

The permission of the Executive Committee of IEA Coal Research to publish this paper is gratefully acknowledged. The views expressed are those of the author and not necessarily those of the Executive Committee or IEA Coal Research.

6 REFERENCES

ALEXANDER, D.W.; SHAW, R.L. (1984) Coarse coal slurry transport at Loveridge Mine. In: *Liquid-solid flows and erosion wear in industrial equipment*, ASME Fluid Engineering Division Symposium, New Orleans, LA, USA, 11-17 Feb 1984. New York, NY, USA, American Society of Mechanical Engineers, pp 95-100 (1984)

BROOKES, D.A.; DODWELL, C.H. (1984) The economic and technical evaluation of slurry pipeline transport techniques in the international coal trade. In: *Hydrotransport 9*, Proceedings of Ninth International Conference on Hydraulic Transport of Solids in Pipes, Rome, Italy, 17-19 Oct 1984. Cranfield, UK, BHRA The Fluid Engineering Centre, pp 1-31 (1984)

BUCHGOLC, W.P.; ZOLOTARIEW, G.M. (1980) Mixed hydraulic-pneumatic transportation system of stowing material. *Przeglad Gorniczy*; 36 (5); 255-258 (May 1980)

CHUGH, Y.P.; KUNDU, A.K. (1981) Coal quality changes during slurry transportation of coal. *Mining Engineer*; 140 (233); 541-545 (Feb 1981)

DUFFY, G.G.; WALMSLEY, M.M. (1985) The pipeline transport of solids using a novel stabilizing suspending medium. In: *Proceedings of Tenth International Conference on Slurry Technology*, Lake Tahoe, NV, USA, 26-28 Mar 1985. Washington, DC, USA, Slurry Technology Association, pp 115-122 (1985)

DULAC, B.; COURATIN, P. (1985) France's La Houve Coal Pipeline. In: *Proceedings of Tenth International Conference on Slurry Technology*, Lake Tahoe, NV, USA, 26-28 Mar 1985. Washington, DC, USA, Slurry Technology Association, pp 63-66 (1985)

EDEN, R.; EVANS, N.; CATTELL, R. (1985) *World coal: an aide-memoire*. Energy Discussion Paper 33, Cambridge, UK, University of Cambridge, 86 pp (1985)

ELKES, G.J.; RIGBY, G.R.; SIMSON, H.A.; MAINWARING, D.E. (1982) Integrated coal upgrading and slurry transport. *Chemical Engineering in Australia*; ChE-7 (3); 39-43 (1982)

ENERGY, MINES AND RESOURCES (1984) *Alberta/Canada Energy Resources Research Fund - Program Summary and Eighth Annual Report*. Edmonton, Alberta, Canada, Alberta Energy and Natural Resources, 39 pp (1985) (ISBN 0-86499-217-3)

FARTHING, G.A.; DALEY, R.D.; VECCI, S.J.; MICHAUD, E.R.; MANFRED, R. (1983) Properties and performance characteristics of coal-water fuels. Paper presented at the *Fifth International Symposium on Coal Slurry Combustion and Technology*, Tampa, FL, USA, 25-27 Apr 1983. Pittsburgh, PA, USA, US Department of Energy, pp 1204 (1983)

GERSTEIN, L.; BEYER, E. (1978) The hydraulic transport systems at the Hansa Hydromine – planning, construction and operation. In: *Hydrotransport 5*, Fifth International Conference on the Hydraulic Transport of Solids in Pipes, Hannover, Federal Republic of Germany, 8–11 May 1978. Cranfield, UK, BHRA Fluid Engineering, pp J7.87–J7.100 (1978)

HALVORSEN, W.J. (1964) Operating experience of the Ohio coal pipeline. *Coal - Today and Tomorrow*; 18–22 (Jun 1964)

HILL, R.A.; MILLER, C.S.; WANG, J.-W. (1985) Yangtze River coal slurry pipeline. In: *Proceedings of Tenth International Conference on Slurry Technology*, Lake Tahoe, NV, USA, 26–28 Mar 1985. Washington, DC, USA, Slurry Technology Association, pp 57–62 (1985)

JACQUES, R.B.; AUDE, T.C.; RICKS, B.L.; RABB, D. (1982) Alberta coal supply pipeline study - 1981. In: *Proceedings of the Seventh International Technical Conference on Slurry Transportation*, Lake Tahoe, NV, USA, 23–26 Mar 1982. Washington, DC, USA, Slurry Technology Association, pp 377–388 (1982)

KELLER, L.J. (1979) Methacoal technologies: viable new routes to energy sufficiency and economic stabilization. In: *Impact of the National Energy Act on Utilities and Industries due to the Conversion to Coal*, Houston, TX, USA, 4–6 Dec 1978. CONF-781253 Silver Spring, MD, USA, Information Transfer Inc, pp 128–134 (1979)

LEE, H.M. (1980) *The future economics of coal transport*. ICEAS/D2 London, UK, IEA Coal Research, 85 pp (1980)

MINING MAGAZINE (1984) Indian coal pipeline? *Mining Magazine*; 150 (6); 525 (Jun 1984)

MIURA, H.; MASE, S. (1979) Operation and maintenance of slurry transportation system at hydraulic coal mine. In: *Proceedings of the Fourth International Technical Conference on Slurry Transportation*, Las Vegas, NV, USA, 28–30 Mar 1979. Washington, DC, USA, Slurry Transport Association, pp 43–50 (1979)

MORRISON, G.F. (1983) *Combustion of coal liquid mixtures*. ICTIS/TR24 London, UK, IEA Coal Research, 51 pp (Nov 1983)

NATIONAL COAL BOARD (1977) *Report of a visit to study hydraulic mining in the USSR, July 1977*. Stanhope, Bretby, Burton-on-Trent, UK, National Coal Board, Mining Research and Development Establishment, 63 pp (1977)

PAKUSCH, G. (1984) Hydraulic ash disposal systems in coal-fired power plants. *Rohre, Rohrleitungsbau Rohrleitungstransport*; 23 (9); 395–398 (Sep 1984) (In German)

PRIGGEN, K.S.; SCHEFFEE, R.S.; MCHALE, E.T. (1984) Pipelining of high density coal-water slurry. In: *Proceedings of the Ninth International Technical Conference on Slurry Transportation*, Lake Tahoe, NV, USA, 21–22 Mar 1984. Washington, DC, USA, Slurry Technology Association, pp 141–146 (1984)

RIGBY, G.R.; CALLCOTT, T.G. (1978) Slurry beneficiation and transportation system offers advantages for handling coking coals. *Australian Mining*; 70 (2); 18–20 (Feb 1978)

RIGBY, G.R.; THOMAS, A.D. (1983) Slurry handling and transportation developments. In: *International Conference on Bulk Materials Storage, Handling and Transportation*, Newcastle, NSW, Australia, 22–24 Aug 1983. National Conference Publication No 83/7, Barton, ACT, Australia, Institution of Engineers, 5 pp (1983)

ROBERTS, B. (1985) Hydraulic transport of coal at Lorraine Collieries. *Industrie Minerale*; 67 (2); 79–84 (Feb 1985) (In French)

SAKAMOTO, M.; INOUE, H.; UCHIDA, K.; KAMINO, Y.; SAITO, M. (1983) Progress of hydrohoist for coarse coal slurry transportation. In: *Proceedings of the Eighth International Technical Conference on Slurry Transportation*, San Francisco, CA, USA, 15–18 Mar 1983. Washington, DC, USA, Slurry Transport Association, pp 297–304 (1983)

SANTHANAM, C.J.; DALE, S.E.; PEIRSON, J.F.; BURKE, W.J.; HANKS, R.W. (1985) Use of LCO2 to transport low rank coals. In: *Proceedings of Tenth International Conference on Slurry Technology*, Lake Tahoe, NV, USA, 26–28 Mar 1985. Washington, DC, USA, Slurry Technology Association, pp 291–298 (1985)

SIEMON, J.R. (1985) *Economic potential of coal-water mixtures*. ICEAS/E8 London, UK, IEA Coal Research, 100 pp (1985)

SNOEK, P.E.; AUDE, T.C.; THOMPSON, T.L. (1976) Utilisation of pipeline delivered coal. In: *Hydrotransport 4*, Fourth International Conference on the Hydraulic Transport of Solids in Pipes, Banff, Canada, 18–21 May 1976. Cranfield, UK, BHRA Fluid Engineering, pp E1.1–E1.12 (1976)

STEWART, D.; WOOD, P.A. (1984) *Equipment for hydraulic handling of coal*. ICTIS/TR 27 London, UK, IEA Coal Research, 76 pp (1984)

SUEUR, J.M. (1982) Hydraulic transport of slurry to supply a power station. In: *Ninth International Coal Preparation Congress*, New Delhi, India, 29 Nov–4 Dec 1982. Calcutta, India, Coal India Ltd, pp H4.1–H4.12 (1982)

WHAITE, R.H.; ALLEN, A.S. (1975) *Pumped slurry backfilling inaccessible mine workings for subsidence control*. Information Circular 8667, Denver, CO, USA, US Bureau of Mines, 83 pp (1975)

WILLIAMS, J.M.; PITMAN, J.S. (1984) Polyurethane lined ductile iron pipe for the hydraulic handling of furnace bottom ash. In: *AshTech '84 - Conference Proceedings*, Second International Conference on Ash Technology and Marketing, London, UK, 16–21 Sep 1984. London, UK, Central Electricity Generating Board, Ash Marketing Branch, pp 239–248 (1984)

WOOD, P.A. (1983) *Underground stowing of mine waste*. ICTIS/TR23 London, UK, IEA Coal Research, 67 pp (1983)

WOOD, P.A. (1984) *Applications for the hydraulic handling of coal*. ICTIS/TR28 London, UK, IEA Coal Research, 99 pp (1984)

10th International Conference on the
Hydraulic Transport of Solids in Pipes

HYDROTRANSPORT 10

Innsbruck, Austria: 29-31 October, 1986

PAPER A2

COAL-WATER SLURRY

OXIDATION AND RHEOLOGY

E. Carniani

Snamprogetti S.p.A., Italy

G. Gabrielli

P. Baglioni

Department of Chemistry

University of Florence, Italy

SYNOPSIS

It is common knowledge that the preparation of
fluid and stable concentrated coal-water
suspensions depends on several factors.
The most important are nature of the coal,
particle size distribution of the ground
material, presence of fluidizing and stabilizing
additives and preparation process of the
mixture.
The degree of surface oxidation of coal is also a
very important factor since the fluidizing
process strictly depends on the properties of the
solid-liquid interface.
In order to find out the relations between
oxidation of coals and their rheological
properties in concentrated coal-water
suspensions, 4 coal samples of different origin
and with the same particle size distribution were
considered.
The most important rheological parameters of the
slurries at different coal oxidation times were
evaluated trying to determine relations between
the above parameters and the oxidation degree.
The influence of oxidation on the tendency of the
coal particles to agglomerate when the slurry is
subjected to mixing or pumping was also
evaluated.
The results obtained showed a considerable
dependence of the rheological properties of the
slurry on the degree of oxidation of the coal.
This dependence can mainly be interpreted as an
increase of the hydrophilicity of the coal
surface and consequent water absorption.

NOMENCLATURE

τ = shear stress (N/m^2)
K = consistency index
$\dot{\gamma}$ = shear rate $(1/s)$
n = power law index

INTRODUCTION

It is well known that to obtain coal-water
mixtures homogeneous, fluid, pumpable over long
distances via pipelines and with extended
stability in storage tanks, it is necessary to
act on two main parameters:

- particle size distribution

- addition of suitable fluidizing and stabilizing
  additives

With an optimal particle size distribution
(Ref. 1,2) the void fraction is minimized and
therefore as much water as possible is available
for the lubricating function.
The addition of suitable additives acting at the
solid/liquid interface (Ref. 2) allows the
dispersion of the solid phase both because of
electrostatic repulsion exerted by the surface
charges of the adsorbed additive and because of
the steric hindrance caused by the film of
molecular species adsorbed which prevents direct
contact among the particles.
However, it is known that when the particle size
distributions and the type of additive are
constant, the rheological properties of coal
slurries strictly depend on the type of coal used
and in particular on its surface characteristics
(Ref. 3,4).
Therefore, if this technology is to be extended
to most coals, it is necessary to know the
relation between surface characteristics of coals
and properties of slurries.
The surface feature of coal which principally
affects the characteristics of slurry is the
balance between hydrophilic and hydrophobic
properties.
In fact at its surface, coal, due to its
essentially hydrophobic nature, can have mineral
substances and functional groups containing
oxygen (mainly carboxyl groups (COOH), carbonyl
groups (CO) and hydroxyl groups (OH)), which
greatly increase the concentration of the
hydrophilic sites capable of binding with water;
these species therefore subtract water from the
fluidification process and modify the adsorption
mechanism of the additives utilized.

Since all coals are characterized by a different degree of surface oxidation and, except for anthracite, they are easily oxidable even when simply exposed to the air at room temperature, a study was performed to correlate the said oxidation condition with the rheology of coal-water slurries both in static and dynamic conditions.

The study is remarkably interesting for two fundamental reasons: on one hand it is possible to use coals with different degrees of oxidation in the preparation of slurries and on the other hand it is possible to correlate, by a controlled oxidation, the surface hydrophilicity with the rheological behaviour.

EXPERIMENTAL PROCEDURE

Four coals coming from different seams and with suitable particle size distribution were used (Figure 1).

The chemical characteristics of these coals are given in Table 1.

The oxidation was carried out on approximately 1,000 g of each sample subdivided into 4 crystallization vessels, 200 mm diam. each, placed in a forced ventilation oven at 100 °C. The samples were mixed at intervals to obtain a homogeneous oxidation.

The duration of the whole process was 8 hours; after 4 hours, half of the samples were taken out of the oven.

The following analyses were carried out on the oxidized coals for comparison with the initial coals:

- Ultimate analysis

- Surface oxidation

- Wettability

- Moisture

- pH

- Zeta potential with and without the fluidizing additive

- Adsorption isotherms of the fluidizing additive

Moreover the following analyses were carried out on the slurries prepared with the same coals:

- Viscosity

- Static stability

- Rheological agglomeration parameters

Analytical methodologies

The proximate analyses and the determination of total sulfur were carried out according to ASTM methodologies.

A Perkin - Elmer "Elemental Analyzer" mod. 240 C was used for the analysis of C, H, N, calculating the percentage of oxygen by difference.

To evaluate surface oxidation, the total acidity and acid groups were analyzed according to the methodologies proposed by Schafer (Ref. 6, 7).

The total acidity was calculated from the decrease in concentration of a barium hydroxide solution used to exchange the acid groups (which are present on the surface of the coal), and the carboxyl groups were determined by exchange with a barium acetate solution according to the following reaction:

$$2 \, (-COOH) + Ba \, (CH_3COO)_2 \rightleftharpoons \begin{matrix} -COO \\ -COO \end{matrix} Ba + 2CH_3COOH$$

After separation of the coal by filtration the acetic acid was titrated with a standard alkali solution.

The wettability of coals was determined according to Esumi et al. (Ref. 8) by measuring the sinking time of a small quantity of coal suspended on the surface of a surface-active agent solution of fixed type and concentration.

The moisture was determined by comparison with the non-oxidized coals leaving the sample exposed to air for 24 hours and drying it in an oven for 1 hour at 105 °C according to ASTM method D 3173-73.

The pH of coals was measured on 50% solid - 50% water suspension utilizing a pH-meter with combined electrode.

The zeta potential was determined with a PEN KEM analyzer mod. 500 on a very diluted suspension of coal in water.

The adsorption isotherms of the fluidizing additive before and after the oxidation were carried out by stirring solutions of additive having different concentrations and containing the same quantity of coal for 30 minutes and by spectrophotometrically measuring the concentration of the additive in the filtrate.

The viscosities of suspensions were determined with a rotational viscometer Contraves Rheomat 30 using the couple MS-C. The shearing stress values measured at the various shear rates enabled us to determine the rheological behaviour of the sample and to calculate the value of the effective viscosity.

The value of the viscosity was obtained from the ascending portion of the curve of the diagram $\tau - \dot{\gamma}$ after submitting the sample to intensive shear treatment to eliminate the time dependency of the fluid (Ref. 9).

To measure the static stability a gravimetric method was standardized.

This method mainly consisted in allowing the suspension to settle in a plexiglass cylinder for a standard time. The cylinder was then introduced into a freezer to solidify the sample.

The solidified sample was removed from the cylinder and the solid content was determined in the two end sections, having the same thickness. The per cent ratio between the solid content of the upper section and the one of the lower section allowed the calculation of the per cent stability which ranges between 0 and 100%.

The stability in dynamic conditions, that is stirring the slurry by vertical mixers, was evaluated by measuring the trend of significant rheological parameters at suitable time intervals.

The tests were carried out on about 1000 g of slurry introduced into a 1 liter capacity vessel and subjected to stirring at 250 r.p.m. using a vertical mixer IKA-WERK mod. RW 20 DZM and a stainless steel blade stirrer.

During the tests the vessels were kept in a thermostatic bath to maintain the slurries at 20°C.

RESULTS AND DISCUSSION

Table 2 illustrates the analysis of C, H, N, O and the O/C ratio for the original coals and for the same coals after oxidation.

For all coals except the Polish coal there was a per cent increase both in oxygen and in the O/C ratio.

For the Polish coal these two parameters remained almost constant before and after the oxidation and that was probably due to both the lower oxidability of this coal and to the different type of oxidation which, in this case, can be considered mostly superficial.

Table 3, which indicates the values of total acidity and acid groups for the coals before and after 4-8 hours of oxidation, confirmed the previous results which showed that after oven heating there was a considerable increase in the superficial groups containing oxygen.

Once the oxidation had been checked, the aim of the analysis was to find out the main effects of the different degree of oxidation on the characteristics of coals; Table 4, in particular, indicates the percentage of moisture, the pH and the wettability of the four coals before and after oxidation.

It can easily be observed that the moisture increases as the oxidation degree increased, indicating a large concentration of hydrophilic groups, the pH decreased due to the higher concentration of acid groups, and the wettability time decreased because of a higher percentage of hydrophilic groups. Two fundamental parameters, characteristic of the solid-liquid interface, were determined to evaluate the effects of oxidation on the superficial properties, which are particularly important for the utilization of coals in the preparation of slurries.

These parameters are the zeta potential and the adsorption isotherm of the fluidizing additive whose characteristics were described in previous patents (Ref. 2).

The comparison, in Table 5, between the values of zeta potential measured on coals before and after oven treatment, showed a slight decrease as the oxidation increased. An explanation can be found in the superficial increase of the negatively charged groups such as the partially dissociated carboxylic acid and phenolic groups.

Figures 2, 3, 4 and 5 represent the adsorption isotherms of the additive at 20°C for the 4 coals before and after oxidation.

As can be easily observed, the form of the isotherms was approximately of the Langumir type (Ref. 10), and remained constant when the oxidation increased whereas the quantity of additive adsorbed in the saturation varied; in fact it decreased as the oxidation degree increased.

This indirectly confirms the presence of a higher concentration of hydrophilic groups at the surface, since the adsorption of the additive at the coal surface was interpreted to be essentially due to hydrophobic interactions and therefore hindered by the presence of hydrophilic groups.

Since the coals were examined in order to utilize them for the preparation of coal-water mixtures a study was also performed on the influence of the oxidation degree on the rheological properties of slurries.

This study mainly consisted in the determination of the significant rheological parameters both on the slurries prepared according to the procedures indicated in the experimental section and on the slurries stirred in laboratory to simulate the effect of pumping and circulation in a pipe.

The possibility of comparing the pumping conditions in the pipe and the stirring conditions in laboratory, at least as far as the rheological behaviour was concerned, was cautiously confirmed by several laboratory tests carried out on different types of coals.

Figure 6 shows the typical rheograms of slurries having the same concentration prepared with original coals and with the same coals subjected to oxidation for 4 and 8 hours.

As can be noted, the three slurries showed, in the range of shear rates examined, a rather pseudoplastic behaviour, power-law type, which can be represented by the equation:

$$\tau(\dot{\gamma}) = K \dot{\gamma}^n$$

where:

τ = shear stress (N/m^2)
K = consistency index
$\dot{\gamma}$ = shear rate $(1/s)$
n = power law index

and agrees with the analyses of other authors (Ref. 11,12).

Therefore, the rheological behaviour can be essentially characterized by the two parameters, K and n; their values for the various tests are indicated in Table 6 together with the values of static stability.

It can be observed that K increased considerably as the oxidation increased thus indicating a reduction of the fluidity of suspensions depending on the oxidation degree.

That could be due both to a higher water

absorption, and therefore to a higher concentration by volume of the solid phase, and to a lower dispersion of the coal particles partly confirmed by the decrease of the n value. The increase of viscosity as the oxidation degree increased was accompanied by more stability.

As for the rheological behaviour of slurries when stirred, Figure 7 shows the typical rheograms of slurry subjected to stirring using the procedures illustrated in the experimental section.

Figure 7 shows that extension of the stirring time not only caused an increase of viscosity but also a more evident pseudoplastic behaviour with a reduction of the n index which indicates a decrease of the dispersion degree and consequently the onset of an aggregation process. In confirmation, Figures 8, 9, 10 and 11 indicate the variations of the K and n indexes which, as demonstrated in the previous figure, represented in a rather precise way the rheological behaviour of slurries as a function of the stirring times for the 4 coals before and after oxidation.

The trend of the n and K parameters can be compared for the 4 coals as illustrated:

a) Non oxidized coals

An initial slight reduction of the K index followed by a gradual increase till a certain stirring time, which varies for the 4 coals, caused a sharp increase of K accompanied by a contemporaneous decrease of n.

The initial slight reduction of viscosity with the agitation can be ascribed to a better homogeneization and dispersion of coal particles.

The sharp increase of K and the contemporaneous reduction of n corresponded to an irreversible agglomeration process, the mechanism of which, although not completely clarified, seemed to correspond to a stronger adhesion among the particles due to a partial or total decrease of the double layer charge or to its variation in thickness and to the formation of aggregated structures possibly favored by the hydration water.

b) Oxydized coals

The general trend of K and n values as a function of the stirring time can be compared with that previously observed for the same coals non oxidized but where the time required for the agglomeration was much shorter.

That was due both to a lower adsorption and stability of the additive at the surface because of the presence of hydrophilic groups according to the behaviour of the adsorption isotherms and to the fact that the hydrophilic groups enhance the absorption of water and therefore the formation of hydrated structures particularly favourable to the production of coal particles agglomerates.

CONCLUSIONS

The experimental results previously discussed enabled us to draw the following conclusions:

1. Coals can be subjected to controlled air oxidation according to very simple experimental procedures. Therefore the effects of oxidation on the characteristics of coals can be defined in very short times.

2. The main effect of oxidation on the coal surface was the increased hydrophilicity, therefore the oxidation represented a method to evaluate the influence of hydrophilicity on the adsorption of the additives utilized for the preparation of slurries.

3. The studies performed in this direction permitted the clarification and confirmation of the adsorption mechanism of the additives used for the preparation of the slurries and reported in previous patents (Ref. 2).

4. From the experimental tests carried out it was possible to deduce that the rheological behaviour, which is almost similar to the pseudoplastic behaviour, did not vary as the oxidation degree of coals changed.

5. The most oxidation sensitive rheological parameter was the K index which increased as the oxidation degree increased and thus confirmed the important role of superficial hydrophilicity.

6. The agglomeration time of slurries subjected to stirring (mechanical stresses) was considerably reduced when the slurries were prepared with oxidized coals.

These results suggested the experimental activity to be performed to test the slurriability of coals and allowed the prevision of the rheological behaviour of slurries prepared with coals having different oxidation degrees.

REFERENCES

1. Ferrini, F., V. Battarra, E. Donati, C. Piccinini: "Optimization of Particle Grading for High Concentration Coal Slurry". Hydrotransport 9, Conference on Hydraulic Transport of Solids in Pipes, Paper B2, October 1984.

2. Ferroni, E., G. Gabrielli, P. Baglioni, F. Ferrini, E. Carniani: "Aqueous Coal Suspension". Italian Patents no. 21885 A/81, no. 20261 A/82.

3. Kaji, R., Y. Muranaka, K. Otsuka, Y. Hishinuma, T. Kawamura, M. Murata, Y. Takahashi, Y. Arikawa, H. Kikkawa, T. Igarashi, H. Higuchi: "Effects of Coal Type, Surfactant, and Coal Cleaning on the Rheological Properties of Coal Water Mixture". Proc. Fifth Intern. Symp. on Coal Slurry Combustion and Technology, Tampa, Florida, 1983, Pittsburgh, U.S. Department of Energy, pp. 151-175.

4. Igarashi, J., N. Kiso, Y. Hayasaki, M. Yamamoto, T. Ogata, K. Fukuhara, S. Yamazaki: "Effects of Weathering of Coals on Slurriabilities". Proc. of Sixth Intern. Symp. on Coal Slurry Combustion and Technology, Orlando, Florida, 1984, Pittsburgh, U.S. Department of Energy, pp. 283-303.

5. ASTM, Annual Book of ASTM Standards, Part 26, 1979.

6. Schafer, H.N.S.: "Determination of the Total Acidity of Low-rank Coals". Fuel, Lond., 1970, 49, 271-280.

7. Schafer, H.N.S.: "Carboxyl Groups and Ion Exchange in Low-rank Coals". Fuel, Lond., 1970, 49, 197 - 213.

8. Esumi, K., K. Meguro, H. Honda: "Wetting of Coal by Nonionic Surfactant and Mixtures of Nonionic and Anionic Surfactants". Bull. Chem. Soc. Jpn., 56, 3169-3170, 1983.

9. Phulgaonkar, S.R.: "Time and Shear Dependent Rheology of Coal Water Dispersions".
Proc. of Seventh Intern. Symp. on Coal Slurry Technology, New Orleans Louisiana, 1985, Pittsburgh, U.S. Department of Energy.

10. Giles, C.H., Mc Evan J.H., Nakahwa S.W., Smith D., J. Chem. Soc. 3973, 1960.

11. Al Taweel, A.M., O. Fadaly, J.C.T. Kwak: "Effect of Coal Properties on the Rheology and Stability of CWS Fuels". Proc. of Seventh Intern. Symp. on Coal Slurry Technology, New Orleans, Louisiana, 1985, Pittsburgh U.S. Department of Energy.

12. Tadros, Th. F., A. Gybels: "Principles of Preparation of Stable Coal-Water Suspensions and Some of Their Applications". Atlas Chemical Industries N.V. Everberg, Belgium.

TABLE 1 - ANALYSIS OF COALS TESTED

Coal source	Poland	USSR	South Africa	U.S.A.
% wt., air dry				
Moisture	2.0	5.4	2.5	1.9
% wt., dry basis				
Ash	11.6	15.7	15.9	7.9
Volatile matter	30.4	38.8	26.3	36.9
Fixed carbon	58.0	45.5	57.8	55.2
Total sulfur	0.86	0.28	0.70	0.96

TABLE 2 - <u>ULTIMATE ANALYSIS OF COALS BEFORE AND AFTER 8 HOUR AIR</u>
<u>OXIDATION</u>

Coal source		<u>% wt., dry basis</u>				
		C%	H%	N%	O%	O/C
Poland	original	74.3	4.0	1.8	7.4	0.100
	oxidized	75.0	3.5	1.6	7.4	0.99
USSR	original	67.6	3.9	2.8	9.7	0.143
	oxidized	65.8	4.0	2.0	12.2	0.185
South Africa	original	70.0	3.2	2.3	7.9	0.113
	oxidized	62.9	1.7	2.4	16.4	0.261
U.S.A.	original	75.7	5.0	2.1	8.4	0.111
	oxidized	67.0	4.8	1.8	17.6	0.263

TABLE 3 - <u>CHANGES OF FUNCTIONAL GROUPS CONTAINING OXYGEN</u>

Coal Source		Total Acidity meq/g	Carboxilic Acid meq/g
	n. ox.	<0.01	<0.02
Poland	4 h ox.	0.09	0.06
	8 h ox.	0.31	0.08
	n. ox.	0.58	<0.02
U S S R	4 h ox.	0.61	<0.02
	8 h ox.	1.14	0.09
	n. ox.	0.02	<0.02
South Africa	4 h ox.	0.22	0.12
	8 h ox.	0.56	0.12
	n. ox.	<0.01	0.05
U.S.A.	4 h ox.	0.20	0.12
	8 h ox.	0.20	0.12

TABLE 4 - CHANGES OF MOISTURE, pH, WETTABILITY

Coal Source		Moisture wt. %	pH	Wettability min.
	n. ox.	2.0	7.28	5.0
Poland	4 h ox.	2.1	7.12	2.5
	8 h ox.	3.7	7.07	1.0
	n. ox.	5.4	7.71	7.5
USSR	4 h ox.	6.7	7.10	4.5
	8 h ox.	7.0	6.95	3.5
	n. ox.	2.5	6.65	1.8
South Africa	4 h ox.	5.3	6.40	1.0
	8 h ox.	5.4	6.35	1.0
	n. ox.	1.9	4.65	>60
U.S.A.	4 h ox.	3.2	4.42	>60
	8 h ox.	3.5	4.31	>60

TABLE 5 - ANALYSIS OF ZETA POTENTIAL

Coal Source		Original mV, 20°C	+0,5% Additive mV, 20°C
	n. ox.	-14	-56
Poland	4 h ox.	-21	-63
	8 h ox.	-20	-63
	n. ox.	-19	-47
USSR	4 h ox.	-21	-51
	8 h ox.	-20	-53
	n. ox.	-18	-54
South Africa	4 h ox.	-18	-56
	8 h ox.	-18	-59
	n. ox.	-15	-47
U.S.A.	4 h ox.	-22	-51
	8 h ox.	-21	-56

TABLE 6 - <u>VISCOSITY AND STABILITY OF CWM</u>

Coal Source		Cw % wt.	K	n	Stability %
	n. ox.	68	22	0.88	81.5
Poland	4 h ox.	68	30	0.88	90.1
	8 h ox.	68	40	0.90	91.3
	n. ox.	64	32	0.72	97.2
USSR	4 h ox.	64	35	0.88	98.1
	8 h ox.	64	68	0.66	97.8
	n. ox.	65	22	0.85	41.5
South Africa	4 h ox.	65	49	0.82	82.4
	8 h ox.	65	46	0.82	84.8
	n. ox.	64	20	0.94	54.9
U.S.A.	4 h ox.	64	25	0.89	66.1
	8 h ox.	64	32	0.68	65.4

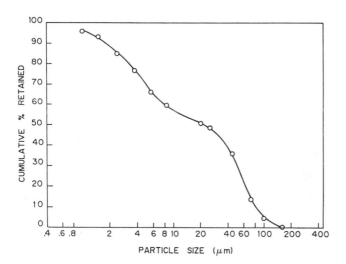

Fig. 1. *Typical particle size distribution of the coals tested.*

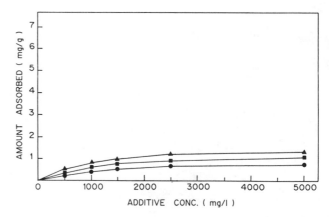

Fig. 3. *Adsorption isotherms of the additive on Russian coal before and after oxidation:* ▲ *original,* ■ *after 4 h oxidation,* ● *after 8 h oxidation.*

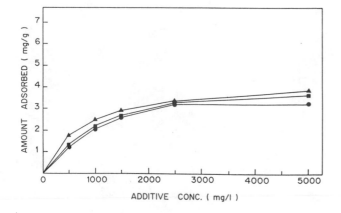

Fig. 2. *Adsorption isotherms of the additive on Polish coal before and after oxidation:* ▲ *original,* ■ *after 4 h oxidation,* ● *after 8 h oxidation.*

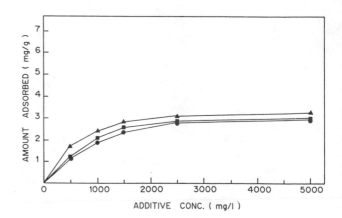

Fig. 4. *Adsorption isotherms of the additive on South African coal before and after oxidation:* ▲ *original,* ■ *after 4 h oxidation,* ● *after 8 h oxidation.*

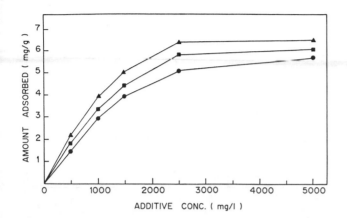

Fig. 5. *Adsorption isotherms of the additive on American coal before and after oxidation:* ▲ *original,* ■ *after 4 h oxidation,* ● *after 8 h oxidation.*

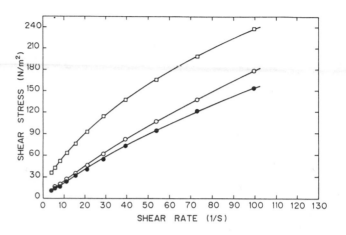

Fig. 7. *Typical flow curves of slurries at different stirring times:* ○ *0 time,* ● *after 3 h,* □ *after 8 h.*

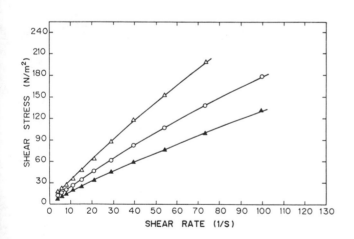

Fig. 6. *Typical flow curves of slurries prepared with coal at different degrees of oxidation:* ▲ *original,* ○ *after 4 h oxidation,* △ *after 8 h oxidation.*

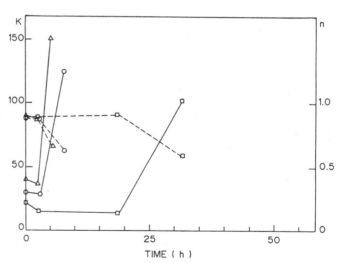

Fig. 8. *Effect of long term stirring on rheological parameters of slurries of Polish coal at different degrees of oxidation:* □ *original,* ○ *after 4 h ox.,* △ *after 8 h ox.* —— *K,* – – – – *n.*

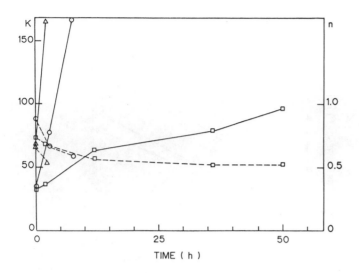

Fig. 9. Effect of long term stirring on rheological parameters of slurries of Russian coal at different degrees of oxidation: □ original, ○ after 4 h ox., △ after 8 h ox. ——— K, – – – – n.

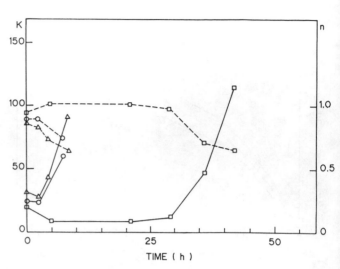

Fig. 11. Effect of long term stirring on rheological parameters of slurries of American coal at different degrees of oxidation: □ original, ○ after 4 h ox., △ after 8 h ox. ——— K, – – – – n.

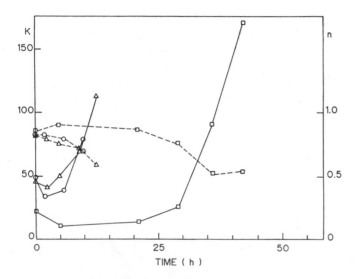

Fig. 10. Effect of long term stirring on rheological parameters of slurries of South African coal at different degrees of oxidation: □ original, ○ after 4 h ox., △ after 8 h ox. ——— K, – – – – n.

10th International Conference on the
Hydraulic Transport of Solids in Pipes

HYDROTRANSPORT 10

Innsbruck, Austria: 29-31 October, 1986

PAPER A3

PRODUCTION PLANTS AND PIPELINE SYSTEMS FOR SNAMPROGETTI'S COAL WATER
SLURRIES. RECENT EXPERIENCE AND CURRENT PROJECTS IN ITALY AND USSR.

D. ERCOLANI

Snamprogetti S.p.A., Fano, Italy

Summary

The paper will describe the world's first large scale coal-water slurry
fuel project, the Belovo-Novosibirsk coal slurry pipeline system, which
will be implemented by Snamprogetti in USSR.
The project includes:

- a production plant which will be built in Siberia near a coal field;

- a 256 km long, 20 in. pipeline to transport the coal-water mixture to a
  power plant.

The research and development work, forming the back-ground of the project,
together with the specific experimental activity performed to define the
main design data are described.
Finally the paper will summarize the main characteristics of the project
regarding the slurry preparation plant, pipeline and pumping stations,
boiler modifications of the power plant.

1.   INTRODUCTION

In the year 1985 Snamprogetti has been awarded a contract by
Techmashimport of the U.S.S.R. for the implementation of the first
world's large-scale coal-water slurry fuel project, the Pilot-
Commercial Coal Slurry Pipeline Belovo-Novosibirsk.

The project includes:

- a high concentration coal-water slurry preparation plant, which
  will be built at Belovo (Siberia) near a coal field, having a
  capacity of 3 MTA of dry coal;

- a 256 km long, 20 in. pipeline to transport the slurry fuel to a
  power station at Novosibirsk, where it will be burned directly
  without further modification.

The project will be performed in joint cooperation between Soviet
organizations and Snamprogetti, which will supply licence and know-
how, basic design, main equipment for slurry production plant,
pumping stations, pipeline terminal and the combustion system.
Snamprogetti has developed its own CWM technology always aiming at
the following objectives:

- to produce an alternative fuel to heavy fuel oils;

- to develop an alternate mode of handling, trasporting, storing and
  burning coal.

The Russian project really will permit to apply these basic concepts:
in fact the Belovo-Novosibirsk slurry system represents the first
industrial application of the whole CWM technology in the world,
since it consists of an integrated system for the production,
transportation by pipeline, direct combustion of CWM in a thermal
power station, without dewatering or any other treatment of the coal
before combustion.
In particular, the technical aspect of the transport of CWM by
pipeline must be strongly enhanced.
In fact, slurry production plants, even if they are smaller than the
Belovo plant, are presently in operation in Italy and in other
countries, producing slurry fuels for pilot or industrial boilers;
therefore, a certain experience in slurry production and combustion
is already existing. But the transportation of CWM by pipeline is
faced for the first time in the world just in this project.
The coal-water slurry process, based on technology developed and
patended by Snamprogetti, will be applied to this large system.
The technology has been developed and tested in pilot scale, followed
by subsequent application in semi-industrial plants (Laviosa, Livorno
and Enichem-Anic, Porto Torres) each having a capacity of 100,000 t/y
of slurry.
Both plants were designed by Snamprogetti and are presently in
operation.
Aim of the present paper is to describe the development work performed on Snamprogetti's technology, together with the experimental
activity carried out to provide the technical basis for the
implementation of the project.
Furthermore, the paper will describe the main characteristics of the
Russian project, including the slurry preparation plant, the pipeline
and pumping stations and boiler modifications of the power plant.

2.   DEVELOPMENT OF THE SNAMPROGETTI'S TECHNOLOGY

In 1980, Snamprogetti started the R & D activity on high coal
concentration slurries.
These slurries were developed to make a liquid fuel available as an

alternative to heavy fuel oil but with the physical characteristics of fluidity, homogeneity and extended stability to make the transportation via pipeline particularly attractive, even for long distances.

This technology really offers meaningful advantages over the conventional 50/50 slurry technology, therefore representing a new mode of handling, transporting, storing and burning coal.

The main advantages offered by this new high coal concentration slurry technology over the conventional method can be summarized as follows:

. All the coal processing is confined to the initial slurry production plant, therefore no additional processing or modification is required at the delivery terminal before combustion.

. Coal dewatering and the following treatment of effluent water are eliminated.

. Coal-water mixtures are non-settling suspensions and behave as a homogeneous fluid inside pipelines. There are no limitations as to critical velocity and critical slope of the line. Additionally, long shut down periods of the pumping system can be tolerated, with no difficulties in restart.

The main objectives appointed by Snamprogetti for its research and development program have been achieved in three basic steps.

The work in laboratory scale was the first step.

Different types of coals were analyzed to investigate the chemical and physical parameters which affect the final properties of the slurries. Different types of chemical additives (both fluidizing and stabilizing agents) were tested to evaluate their performance and to understand the mechanism of their action.

Studies were carried out to optimize the grain size distribution of the coal in the slurry and the bimodal grain size distribution was found to be the optimum curve for the preparation of concentrated slurries. The bimodal distribution permits to reach the best results in terms of concentration of the mixture, fluidity and stability, which, in most cases, can be obtained without the use of stabilizing additives.

The second step was performed in pilot plant scale, by using the test facilities available at the Pipeline Engineering Centre of Snamprogetti in Fano, Italy: a slurry production plant, with a capacity of 250 kg/h, was set up to check different process solutions and to demonstrate the reliability of the Snamprogetti's process, based on the bimodal grain size distribution concept.

At the same time, extended pumping tests in 4 to 10 in. diameter pipes were performed to study the rheology of the mixture in pipe flow, to measure friction losses and to check the reliability of the correlations used for their predictions. This second experimental activity was concluded by the combustion tests, carried out with the purpose to evaluate the combustion properties of the coal-water mixtures and to obtain an useful feed-back to improve their overall characteristics.

The final result of this research program was the definition of the Snamprogetti's process, suitable for preparing coal-water mixtures fluid, stable and pumpable over long distances.

Particular attention was devoted to the development of the combustion technology. Since 1982 Snamprogetti has performed important combustion trials with the main objectives the check the performance of different mixtures in semi-industrial scale and to develop and select combustion systems suitable for industrial application.

These combustion trials were performed in close cooperation with important users and burners manufacturers and by using their test facilities, whose main features are summarized on TABLE 1.

These tests allowed Snamprogetti to verify the high atomizing properties of its mixtures and the possibility to achieve high combustion efficiency, up to 99%.

The third important step for development of the Snamprogetti's technology was the design and construction of demonstrative and industrial slurry production plants (TABLE 2).

In 1984, the first Italian CWM production plant was awarded to Snamprogetti by Industrie Chimiche C. Laviosa.

This plant is located at Livorno, has a design capacity of 100,000 t/y of slurry and is in operation since March 1985. Main purpose of this plant is to provide slurry for long term combustion tests in large capacity boilers and to serve small industrial consumers located within the regional market area.

In 1984/85 another Italian plant, designed by Snamprogetti, was set up. This is the Enichem-Anic pilot plant, located in the petrochemical complex of Enichem-Anic at Porto Torres, in the Sardinia Island. The nominal capacity of this plant is 100,000 t/y of slurry. Its operating activity started on December 1985 with the production of mixtures prepared by using petroleum coke as feedstock.

It has to be pointed out that this plant simulates entirely the process scheme adopted for the slurry production plant at Belovo.

The entire production of this first Anic plant is committed to internal use, mainly for the performance of long term combustion tests on an oil designed boiler with a 300 t/h steam capacity, situated in the Porto Torres chemical plant.

Almost simultaneously with the Russian project, another important contract has been awarded to Snamprogetti by Enichem-Anic. In fact the pilot plant at Porto Torres will be followed by a new, larger slurry production plant, whose basic design is presently underway at Snamprogetti.

This plant will have a design capacity of 500,000 t/y of petroleum coke-water slurries or coal-water slurries; construction is scheduled for 1986/87 and the commissioning is expected by the second half of 1987. Purpose of this future industrial plant is to supply slurry fuels to feed all the oil-designed boilers existing in the petrochemical complex of Porto Torres.

3.   THE U.S.S.R. PROJECT

3.1   Specific experimental work

The implementation of the Russian project is based on a wide experimental work, required to define the main design data and the duty specifications for the main equipment.

The main experimental activities were:

- complete characterization of the coal, whose main characteristics are shown in TABLE 3 and analysis of its superficial properties.

- optimization of the slurry characteristics to define the best formulation (selection of the grain size distribution, able to meet the requirements for combustion and to assure the best compromise amongst maximum achievable concentration, maximum allowable viscosity, amount of additive and energy consumption);

- specification of the grinding equipment. The mills adopted for the production plant are conventional (ball and rod mills) but, considering their particular application, it was necessary an accurate work of optimization and the study of their internal arrangement. This required the definition of the optimum L/D ratio, the most efficient rotating speed, filling factor, type and size of the grinding media, the analysis of the slurry flow along the mills. This activity was performed by using two pilot scale mills

having overall dimensions of 0.6 m x 2.5 m and 1 m x 1.6 m.
For scaling-up purposes, these experimental results were compared
to the operating results coming from the mills installed on Laviosa
and Anic slurry plants.

- Pumping tests. Since the transport of CWM via long distance
  pipeline represents the most innovative technical aspect of the
  Russian project, the pumpability of the slurry was thoroughly
  investigated.

- Combustion tests in small and industrial scale burners.

The investigation on the slurry pumpability is worthy of a better
description.
Over 30 tons of slurry were prepared by using the specified Russian
coal to perform the pumping tests. These tests were carried out in a
200 m. long, 8" diameter test loop, fully jacketed and provided with
a chiller unit to keep constant the temperature of the slurry.
The procedure adopted for this experience, closely simulating the
operating conditions expected in the Belovo-Novosibirsk pipeline,
was:

. continuous pumping of the mixture for 200 h at constant velocity.
  (Transport simulation, having the main purpose to identify possible
  modifications of the slurry after extended pumping);

. shut-down of the loop for 400 h at the temperature of 1°C, which is
  the minimum operating temperature expected (shut-down simulation);

. restart of the flow and again continuous pumping for 200 h.

The more important results of these experiments can be summarized as
follows:

. good agreement between measured friction losses and the values
  predicted by using rheologycal data coming from rotational
  viscometers (Figure 1 gives an example of this comparison);

. the rheology of the mixture was practically constant throughout
  the continuous pumping test.
  In particular, the effective viscosity measured on slurry samples
  taken out from the loop at different time intervals was constant,
  with variations less than + 5% with respect to the average value at
  750 cP.

. the grain size distribution remained practically constant;

. the slurry flow was restarted easily after those 400 h of shut-down
  at the temperature of 1°C, and no settling of the solid particles
  was identified. This confirmed the good characteristics of static
  stability of the mixture.

These results give a clear evidence that a slurry suitably formulated
for pipelining can be transported without risk of propertiees
modifying after extended pumping and confirm the reliability of the
CWM transportation technology via pipeline developed by Snamprogetti.
The final result of the specific experimental work performed on the
Russian coal was the formulation of a coal-water mixture capable to
meet all the requirements for transportation, storage and combustion,
with the following characteristics:

- design concentration   : max 65%
- additive   : 0.5% by weight
- top size   : 350 microns

- viscosity   : less than 800 cP at operating
                                 conditions
- stability   : over 1 month (without stabilizing
                                 chemicals)
- combustion efficiency   : 97-99% (achieved during combustion
                                 tests).

In addition to the shut-down/restart tests, a good demonstration of
stability has been offered by the mixture during the latest
combustion trials.

At the end of January 1986, 160 t of slurry made with coal provided
by USSR were produced by the Laviosa plant, at Livorno, and 800 drums
of that slurry were shipped to USA for combustion tests.

The drums were opened in the second week of March, 50 days after
preparation, shipment, storage and the slurry was found in good
conditions, stable and fluid.

3.2 Description of the project

For a better comprehension of the design criteria and the technical
solutions adopted for the Russian project, it is important to recall
that this integrated slurry system is defined pilot and commercial at
the same time:

- the plant is considered commercial or industrial from the point of
  view of the size, the capacity and the continuous and reliable
  service which has to be assured;

- the plant is also a pilot plant, since it will be used as a huge
  test rig to check the performance of important types of equipment,
  in view of future projects of larger slurry pipeline systems.

The main characteristics of the Belovo-Novosibirsk coal slurry
project are summarized in TABLE 4.

The slurry production plant at Belovo has a nominal capacity of 3 MTA
of dry coal. The slurry will be produced by using a complete wet
process, designed to obtain slurries with the bimodal grain size
distribution curve.

The plant will consist of 7 production lines, each having a nominal
capacity of 60 t/h of dry coal.

The process scheme adopted for each production line is described on
FIG. 2.

Dry ground coal is available at the battery limits of the plant, with
a top size of 6-8 mm, stored in buffer storage bins. This coal is
divided into two streams.

The first stream is conveyed to the ball mill, where it is mixed with
water and the fluidizing chemical. The product from the ball mill,
which is a very fine slurry forming the fine fraction of the bimodal
curve, is pumped to the rod mill, where it is mixed with the second
stream of the dry ground coal. In this second grinding stage, whose
purpose is the production of the coarse fraction of the curve, the
specified grain size distribution and the final concentration are
reached.

Then the slurry passes through mixers to achieve a better
homogenization and to improve its rheological properties and is
finally stored in two tanks, having a capacity of 5000 m^3/each, which
will provide the buffer storage to feed the main pipeline.

Before reaching the main pipeline, the slurry can be pumped through
the test loop, 1 km long and with the same diameter of the main
pipeline. The purpose is to monitor the main parameters of the
mixture (flow rate, density, viscosity, friction losses), so as to
act on the pipeline control system in case the characteristics of the
slurry deviate from the established operating parameters. The
pipeline linking the Belovo plant to the power station at Novosibirsk
will have a length of 256 km and a diameter of 20 in. (Fig. 3). This
size has been specified by the Client in view of future increase of

the throughput. Considering the severe climatic conditions existing in Siberia, the pipeline will be buried at a depth of 2 m, in order to maintain the slurry at temperatures always above 1°C and to prevent freezing.

Three main pumping stations are required: the first one is located at the initial end of the pipeline, close to the production plant, while the second and third ones are located after 90 km and 170 km, respectively.

Each pumping station consists of 2 main slurry pumps, reciprocating, single acting, triplex, driven by variable speed electric motors.

Design capacity and design discharge pressure of each pump are 250 m³/h and 100 bars, respectively, while the installed power of each electric motor is 900 kW.

Additional reciprocating pumps, operating as stand-by units, will be supplied by the Client.

In case of failures in the intermediate pumping stations, it is always possible to assure the transport of the slurry to Novosibirsk at reduced flow rates, by-passing the second pumping station or the third one or both intermediate pumping stations (Fig. 4)

The pipeline terminal (Fig. 5) consists of two slurry storage tanks, with a capacity of 20,000 m³ each. A heater/chiller unit is provided to maintain the slurry at the desired, constant temperature. Side entering mixers, working in discontinuous operation, are provided to have a uniform temperature distribution inside the tanks.

The thermal power station at Novosibirsk consists of six units, each having the capacity of 220 MW. The boilers were originally designed to burn dry pulverized coal and were provided with burners tangentially firing.

For the conversion to the use of coal-water mixtures, each boiler will adopt a new fuel feeding system, a new arrangement for the burners at the corners, a compressed air production system for the atomization of the mixtures.

Since it has been considered useful to maintain the capability of burning coal-water fuels or dry pulverized coal, in each corners of the boiler burners for CWM and burners for dry pulverized coal will be inserted in the same windbox and placed at alternate elevations (Fig. 6). Switching from one type of fuel to the second one will be performed automatically. Likewise, it will be possible to switch over from pulverized coal or coal-water slurry firing to oil.

Finally, with regard to the schedule of the project, the basic and detailed design will be performed during 1986. Delivery of the main equipment and civil works will be completed by 1987. The erection of the plants and the commissioning of the Belovo-Novosibirsk coal slurry pipeline system are expected to be completed on 1988.

4. CONCLUSIONS

The main target of the CWM technology developed by Snamprogetti(production of slurries suitable for transportation via pipeline and for direct burning in electric power plants) has been finally achieved.

After intensive theoretical and experimental activities, the Snamprogetti's technology has been checked in pilot scale and has found the first application with the realization of semi-industrial plants in Italy.

The viability of the basic concepts followed during the technical development and the good level of reliability reached have permitted the achievement of the important contract in U.S.S.R., where the CWM technology will find an ideal case of application through the construction of the Belovo-Novosibirsk coal slurry project.

This project really represents the first industrial application of the whole CWM technology in the world, since it consists of an integrated system for production, transport by pipeline, direct combustion of coal-water mixtures in a thermal power station.

The location of the plant permits to stress another important point.

The natural scenario of the USSR project is the South-Western Siberia, which is an enormously wide area, with many mineral resources, mainly coal, but very far from the utilization areas. This situation confirms the validity of the option offered by the slurry pipeline transportation technology, which has been actually adopted as the most suitable solution to link the coal mines to the end user. Therefore, the Belovo-Novosibirsk coal slurry pipeline system will mark an important milestone for the development of both the CWM technology and the hydraulic transport of coal by pipeline and can provide a meaningful incitement for the adoption of these technologies in future and more important projects in USSR and in other countries.

REFERENCES

1. Laganà, V., Ercolani, D., Romani, G.: "Production plants and pipeline systems for Snamprogetti's Reocarb: technical and economic considerations", Proceedings: 10th International Conference on Slurry Technology, Lake Tahoe, USA, March 26-28, 1985, Washington, D.C., Slurry Technology Association, 1985.

2. Ercolani, D., Grinzi, F., Romani, G.: "Snamprogetti's Reocarb: from the production plants into the boilers", Pre-print for 8th International Symposium on Coal Slurry Fuels Preparation and Utilization, Orlando, FL, May 27-30, 1986.

TABLE 1

TESTS PERFORMED FOR COMBUSTION TECHNOLOGY AND BURNERS DEVELOPMENT

TEST FACILITIES	BURNER CAPACITY (kg/h of CWM)
ENEL (ITALY)	400-3800
DUMAG (AUSTRIA)	100
PEABODY-HOLMES (U.K.)	500-2000
COMBUSTION ENGINEERING (USA)	1300-5200
BABCOCK-WILCOX (USA)	250
ENICHEM-ANIC (ITALY)	3000

TABLE 2

RECENT EXPERIENCE AND CURRENT PROJECTS
ACHIEVED BY SNAMPROGETTI IN ITALY

PLANT/CLIENT	LOCATION	CAPACITY (t/y of CWM)	START
I.C. LAVIOSA	LIVORNO	100,000	March '85
ENICHEM–ANIC	PORTO TORRES	100,000	December '85
ENICHEM–ANIC	PORTO TORRES	500,000	Basic Design Starting

TABLE 3

MAIN CHARACTERISTICS OF THE RUSSIAN COAL

ASH	10%
VOLATILE MATTERS	40%
FIXED CARBON	50%
HIGH HEATING VALUE	6600 kcal/kg
GRINDABILITY	40–45 HGI

TABLE 4

MAIN CHARACTERISTICS OF THE BELOVO–NOVOSIBIRSK
COAL SLURRY PROJECT

PLANT CAPACITY	3 MTA (Dry Coal)
LOCATION	BELOVO
PROCESS	Wet Milling Process
PRODUCTION LINES	7
CAPACITY OF EACH PRODUCTION LINE	60 t/h (Dry coal)
PIPELINE	20 in. x 256 km
PUMPING STATIONS	3
POWER STATION	6 units x 220 MW/each
LOCATION	NOVOSIBIRSK

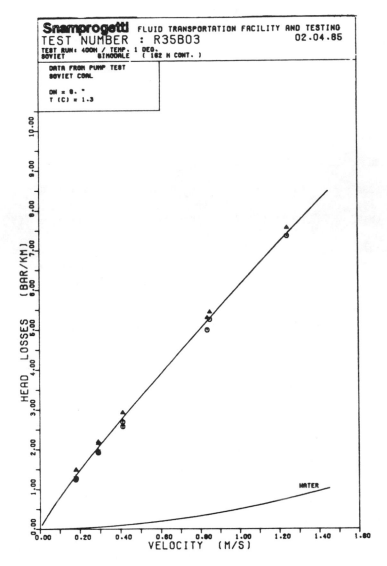

Fig. 1. Example of comparison between measured friction losses and predicted values.

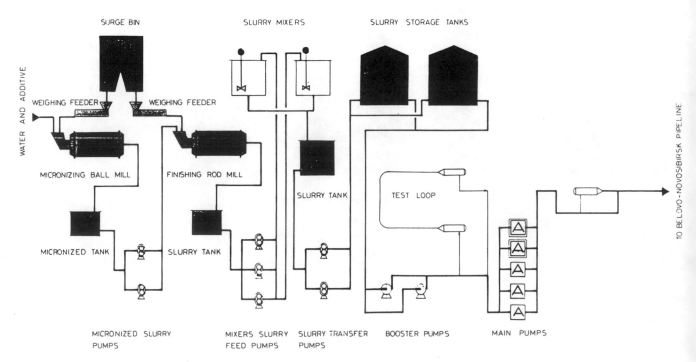

Fig. 2. Slurry preparation plant and initial pumping station at Belovo.

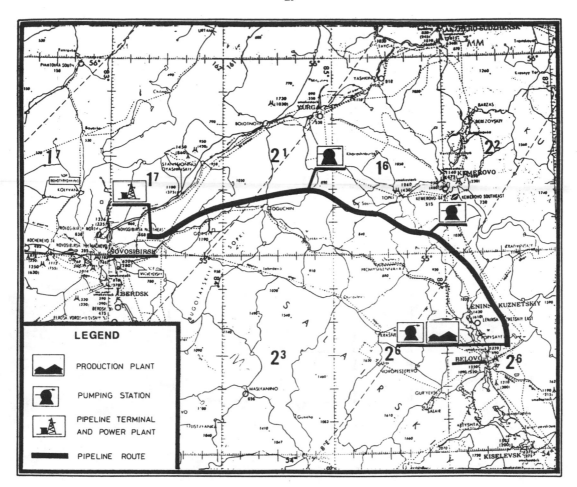

Fig. 3.   Belovo–Novosibirsk coal slurry pipeline.

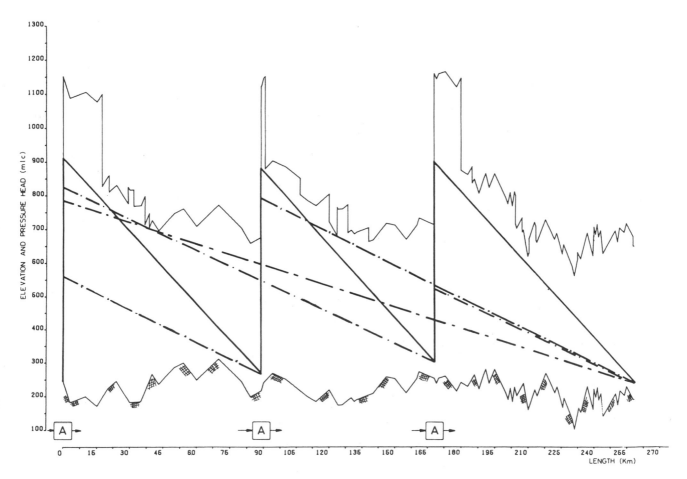

Fig. 4.   Coal slurry pipeline Belovo–Novosibirsk.

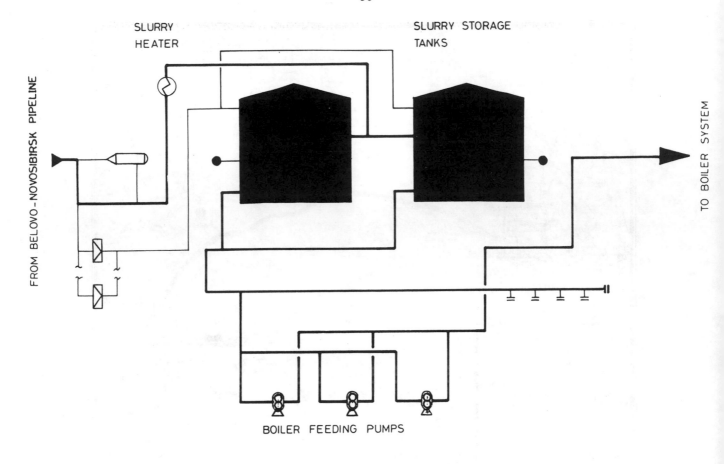

Fig. 5. *Pipeline terminal at Novosibirsk.*

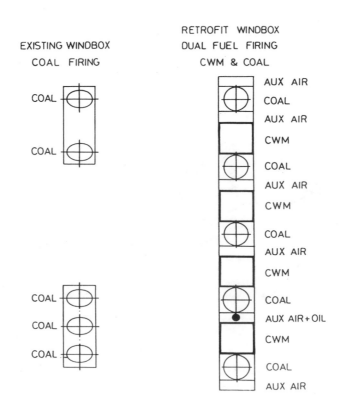

Fig. 6. *Conversion of the boiler from coal to Reocarb.*

10th International Conference on the
Hydraulic Transport of Solids in Pipes

HYDROTRANSPORT 10

Innsbruck, Austria: 29-31 October, 1986

PAPER A4

INVESTIGATION INTO THE TRANSPORT BEHAVIOUR OF HIGHLY CONCENTRATED FINE-GRAINED COAL-WATER SLURRIES (DENSECOAL) USING A 470 M³/H TRIPLEX PISTON PUMP

U. Brandis, R. Klose (Salzgitter Industriebau GmbH / FRG)
P. W. H. Simons (GEHO Pompen Holthuis B.V. / NL)

Abstract

The hydraulic transport of highly concentrated coal-water mixtures over long distances cannot yet be considered the state of engineering.

Numerous pipeline projects and the cost advatage of a closed chain of transport for Coal-Water Fuels (CWF) from the mine up to the ultimate consumer have given rise to world-wide research activities, especially in the transport sector.

Salzgitter has extensive laboratories and testing facilities for investigating the relationships involved in solids pipelining.

The effect of the pipeline diameter and the influence of the solids concentration on the pressure drops were examined in test loops of DN 200, DN 300 and DN 400 diameters. At this the slurries were pumped using a 470 m³/h triplex piston pump.

Also the transferability of rheological data between rotational viscometer readings and measurements made in the pipeline were investigated.

Nomenclature

A, B	Coefficient	(--)
c_w	Concentration by weight	(--)
D	Diameter	(m)
K	Fluid consistency index	(--)
ΔL	Length	(m, km)
n	Flow behaviour index	(--)
ΔP	Pressure loss	(Pa)
Q (x)	Distribution index	(--)
Q	Flow rate	(m³/s)
r	Radius	(m)
T	Temperature	(°C)
$v, \bar{v}$	Velocity	(m/s)
x	Particle size	(μm)
η, η_B	Viscosity	(Pas)
T_r	Shear Stress	(N/m²)
T_0	Yield Stress	(N/m²)
T_w	Wall Shear Stress	(N/m²)

1. Introduction

This report describes the results of two large-scale tests with highly concentrated fine-grained coal-water slurries. The aim of the tests was to examine the flow behaviour of such suspensions. For this purpose pipes with differing diameters on a working scale were used as tube viscometers, whereby pressure drop curves were recorded for each pipe dimension. A comparison of the results should agree with theoretically established approaches /1/ thereby supporting the application of such techniques for larger pipe diameters.

Additionally and in parallel to the individual measuring series of the second test (No. 141509), dynamic samples were taken from the pipe. These were analysed in the laboratory with a rotational viscometer of the type Haake RV 12 (Measuring head 150, profiled measuring cylinder); the total measuring and evaluation procedure was carried out using a computer. By comparing the rheological data obtained by these two different viscometers one should be able if it is possible to calculate in advance the pressure drops in pipes by using results obtained under laboratory conditions.

2. Description of the pilot plant

2.1 The test loop and related instrumentation equipment

The experiments, using highly concentrated fine-grained slurries, were carried out in a pilot plant on a semi-industrial scale, as shown in Fig. (2.1). This consisted of an open circuit with a storage tank of approx. 120 m³, an agitating unit and three loops with DN 200, 300 and 400 each with a length of 222 m, 248 m and 274 m respectively. Each pipe possessed its own measuring systems to record data such as temperature, flow rate, density and pressure differential. The suspensions were pumped using a three-cylinder piston pump which is normally used in long-distance transport and which is dealt with in more detail in the following paragraph (2.2). The control of the pump, the recording of the data and the processing of the measured values were also carried out using a computer.

2.2 The pump

In order to maintain the rheological slurry characteristics and as long distance slurry pipelines have to be equipped with positive displacement pumps to generate the required pressure, a new generation coal slurry pipeline pump, designed and developped by GEHO PUMPS - HOLTHUIS of the Netherlands, was installed.

GEHO have already several years' experience with high pressure pumping of fine coal slurries with concentrations of 64 to 70 per cent by weight and viscosities up to 3,000 cP.

New slurry pipelines require more, powerful and extremely reliable pumps. GEHO started 1979 with the developement of a new generation of coal slurry pipeline pumps for this duty. A prototype was completed in 1982, after which extensive testing began on a unique water/slurry test facilities built at the GEHO factory. It includes an installed motor rating of 2,000 horsepower in order to allow full, continuous performance testing as on slurry pipelines.

The pump is a horizontel, triplex single-acting piston pump, model TZP. The size is ideal for coping with proposed slurry pipeline parameters and the design is extremely attractive when considering pump operating and maintenance cost and demonstrates lowest cost per ton pumped coal.

The pump power end is of a fabricated construction. It has an internal gearing and crankshaft mechanism with crossheads and connecting rods. All bearings are roller bearings, designed for a B-10 life of 100,000 hours at rated load.

The pump end fluid has a modular design and allows for easy access. Quick inspection or replacement of expendables is ensured by means of a hydraulic tightening and release system for cylinder and valve covers and a hydraulic release system for valve seats.

The pump has a minimum number of wear parts and because of the single-acting design there are no stuffing boxes. Piston rods are no longer wear parts.

Newly designed valves ensure extremely long valve life and permit operation with particle size passages of up to 8 mm. With its largest piston diameter the pump can handle flow rates of up to 850 m³/h. Average pressure attainable is in the region of 130 bar.

The pump has performed very satisfactory for the test carried out at SALZGITTER test facility and has contributed further to GEHO's experience with the pumping of coal slurry.

3. Background to the arithmetic model used for evaluating the tests

If the solid concentration c_w exeeds 60 % highly concentrated fine-grained coal-water mixtures can be regarded as homogeneous and stable, i.e. the particles are distributed equally over the pipe cross-section during transport. As no solid deposits occur, the transport can be carried out in the laminar flow area. The latter is favourable from an energy consumption point of view.

Generally the mixtures described above demonstrate a shearing behaviour which is independent of time and which can be characterized by the most general form of the classical rheological equation for a yield pseudo plastic:

$$(3.1) \qquad T_r = T_0 + K \left(-\frac{dv}{dr}\right)^n$$

T_r = shear stress
T_0 = yield stress
K = fluid consistency index
n = flow behaviour index
$-(dv/dr)$ = shear gradient, velocity gradient

From this the known laws according to Newton, Bingham and Ostwald can be derived as special cases.

The viscosity η of a mixture can be calculated from the slope, i.e. the first derivation of the flow curve shown in Eq. (3.1).

$$(3.2) \qquad \eta = \frac{d\,T_r}{d\,(-dv/dr)} = K \cdot n \left(-\frac{dv}{dr}\right)^{n-1}$$

For the Newton and Bingham fluids (n = 1) the flow curve is linear, i.e. the viscosity is independent of the shear gradient and equal to the fluid consistency index K.

Eq. (3.1) can now be stated as follows:

$$(3.3) \qquad -\frac{dv}{dr} = \left(\frac{T_r - T_0}{K}\right)^{\frac{1}{n}} = f(T_r)$$

Considering the equilibrium of forces in a cylindrical pipe section the following equation is normally applied:

$$(3.4) \qquad T(r) = \frac{r}{2} \cdot \frac{\Delta P}{\Delta L}$$

At the pipe wall, with r equal to D/2, the following is valid

$$(3.5) \qquad T_w = \frac{D}{4} \cdot \frac{\Delta P}{\Delta L}$$

The flow rate Q is given to:

$$(3.6) \qquad \dot{Q} = 2\pi \int_0^{D/2} r \cdot v(r)\, dr$$

Taking both Eq. (3.3) and Eq. (3.4) into account we achieve after several manipulations a relationship between the flow rate and the shear stress:

$$(3.7) \qquad \dot{Q} = \frac{\pi}{8} \cdot \frac{D^3}{T_w^3} \int_{T_0}^{T_w} T_r^2 \cdot f(T_r) \cdot d\,T_r$$

The lower integration level in this case is the yield stress T_0 as

$$(3.8) \qquad f(T_r) = 0 \quad \text{for} \quad 0 \le T_r \le T_0$$

By solving the integral we have the following relationship:

$$(3.9) \qquad \frac{\dot{Q} \cdot 8}{\pi \cdot D^3} = n \left(\frac{T_w - T_0}{K}\right)^{\frac{1}{n}} \cdot \frac{1}{T_w^3} \left\{ \frac{(T_w - T_0)^3}{3n+1} + \frac{2 T_0 (T_w - T_0)^2}{2n+1} + \frac{T_0^2 (T_w - T_0)}{n+1} \right\}$$

From the above Eq. (3.9) and Eq. (3.5) the Hagen-Poisseuille equation can be derived as a special case of Newton flow behaviour ($T_0 = 0$, n = 1, k = η):

$$(3.10) \qquad \dot{Q} = \frac{\pi \cdot D^4}{128 \cdot \eta} \cdot \frac{\Delta P}{\Delta L}$$

Also, from the same Eq. (3.9) for the Bingham flow behaviour (n = 1) we have the so-called Buckingham equation:

$$(3.11) \qquad \frac{\dot{Q} \cdot 8}{\pi \cdot D^3} = \frac{T_w - T_0}{K} \left\{ \frac{1}{2} \left(\frac{T_0}{T_w}\right)^2 \left(1 - \frac{T_0}{T_w}\right) + \frac{2}{3} \frac{T_0}{T_w} \left(1 - \frac{T_0}{T_w}\right)^2 + \frac{1}{4} \left(1 - \frac{T_0}{T_w}\right)^3 \right\}$$

Using Eq. (3.5), Eq. (3.11) now is changed to:

$$(3.12) \qquad \frac{\Delta P}{\Delta L} \cdot \frac{D}{4} = \frac{8 \cdot K}{D} \cdot \bar{v} + T_0 \left\{ \frac{4}{3} - \frac{1}{3} \left(\frac{T_0}{\frac{\Delta P}{\Delta L} \cdot \frac{D}{4}}\right)^3 \right\}$$

with the average velocity $\bar{v}$ defined as

$$(3.13) \qquad \dot{Q} = \frac{\pi}{4} \cdot D^2 \cdot \bar{v}$$

Eq. (3.9), together with Eq. (3.5), presents a theoretical relationship between the flow rate $\dot{Q}$, the wall shear stress T_w and the pressure drop $\Delta P/\Delta L$. In this way we have available a set of equations which makes it possible, by means of measuring the flow rate and the pressure drop in the pipes, to determine the rheological data of a suspension and, vice versa, to calculate the pressure drop curves for pipes having determined the flow law in the laboratory. On this basis computer programs were developed which enabled not only the further processing of the pipes original measurement data, but also the evaluation of the experiments based upon the use of the laboratory viscometer /1/.

4. Investigation of empirical data

4.1 Rheological classification of highly concentrated coal-water slurries

Initial preliminary tests with highly concentrated coal water slurries were designed to enable rheological classification and thereby the determination of a flow law. For this purpose a slurry was produced using a non-ionic additive with a solid concentration c_w of 72 % and diluted in three stages to 70.7 %. The concentration of additive was 0.7 % by weight in relation to the slurry. The chemical analysis of the coal revealed the following composition:

Density	=	1 382 kg/m^3
Net Caloric Value	=	27 252 KJ/kg
Total Moisture	=	4.90 % by weight
Ash Content	=	13.89 % by weight
Volatiles	=	28.20 % by weight
Fe$_2$O$_3$	=	10.23 % by weight
SiO$_2$	=	51.83 % by weight
Al$_2$O$_3$	=	20.87 % by weight

Ultimate Analysis:

C	=	82.40 % by weight
H	=	5.20 % by weight
O	=	0.70 % by weight
N	=	1.50 % by weight
S	=	0.58 % by weight

Table (4.1): Composition of the coal used

The particle size distribution achieved from this coal in a one-stage wet grinding process is shown in Fig. (4.1): Here we have a unimodal distribution with the characteristic values x_{10} = 8 μm, x_{50} = 25 μm and x_{90} = 100 μm. In order to achieve constant test conditions after every dilution, a lengthy mixing and homogenization process of the suspension followed by intensive agitating in the storage tank with simultaneous pumping in the open circuit (see Fig. (2.1)). For each stage of concentration the pressure drop curves were recorded at a constant temperature in the DN 400 pipe. The results are shown in Fig. (4.2). The almost linear pattern and the point of intersection of the extended curves with the Y-axis above the origin of ordinates allow us to assume Bingham flow behaviour. This is confirmed by examining the rheological parameters such as yield stress T_0, fluid consistency index K and flow behaviour index n in accordance with Eq. (3.1), which are calculated by using Eq. (3.5) and (3.9) and shown in Table (4.2). The flow behaviour index n deviates only insignificantly in all four stages of concentration from the value of 1 which is valid for Bingham flow behaviour.

4.2 Comparison between measured pressure drop curves and pressure drop curves calculated based on the theoretical model

Following the preliminary tests, the large scale tests (Nos. 141507 and 141509), conducted in the

three pipelines in the order of DN 400, DN 300 and DN 200, confirmed the previously acquired knowledge regarding the linear flow behaviour. The pressure drop curves are plotted in Fig. (4.3) and (4.4), and the resulting rheological data obtained using the Buckingham Eq. (3.12) are listed in Table (4.3). In the present case, the fluid consistency index K is, as was demonstrated previously, equal to Bingham viscosity η_B and is a dimensioned magnitude. The high coefficient of regression (greater than 99 %) observed in all calculations indicates a very close agreement between the measured data and the arithmetic model used. The low variance indicates consistent behaviour of the sampled values.

For test No. 141507, the yield stress T_0 and the Bingham viscosity η_B agree closely with each other - for all three pipelines. The consistency of the suspension was constant during the entire test. In contrast, the drop in viscosity in the next test is noticable, ranging from 1.0859 Pas in the 400 mm line, 1.0002 Pas in the 300 mm line, to 0.9566 Pas in the 200 mm line. This variation of the slurry's flow behaviour was also observed in the analyses of samplings in the laboratory viscometer. The latter was carried out simultaneously with this series of measurements. This causes the test conditions to become inconsistent and originates from an improved dispersion of the coal-water mixture and thus an improved wetting of the coal surface which results from the transport process, the passage of the slurry through the pump and the mixing in the agitator tank.

In spite of the previously described changes in the consistency during the second test, a comparison between the measured pressure drops and the pressure drop curves (calculated for each test from the averaged rheological data of all three pipelines) reveals a very close agreement (Figs. (4.5) and (4.6)). A comparison between the measured and calculated pressure drops (Figs. (4.7) and (4.8)) emphasizes the statements made above, and justifies the application of the physico-mathematical model used for evaluation.

It was apparent that the highly concentrated fine-grained and homogeneous suspensions used for these tests did not differ in their flow behaviour from clear liquids; they can therefore be described by commonly used terms. In contrast to the two-phase, heterogeneous mixtures, this property enables one to describe the mathematical and physical characteristics of the suspension. The description is not dependent on the use of empirical approaches with their numerous limitations, rather one can apply a generally valid model. The so-called 'slip effects' which can be frequently observed in small diameter pipelines did not occur in the present case /2/, /3/. There was no need to modify the measured data.

The following demonstrates that, assuming the presence of Bingham flow behaviour, a simplified analytical method can be applied for evaluating the pressure drop measurements as well as possible extrapolation of the data for pipes of different diameters.

4.3 Simplified evaluation model for suspensions with Bingham flow behaviour

Extensive calculations are required if the Buckingham equation (Eq. (3.12), derived in Section 3) is to be used for determining rheological data on the basis of measured pressure drop curves, or vice versa, for obtaining pressure drop curves from rheological data:

$$(4.1) \quad \frac{\Delta P}{\Delta L} = \frac{32 \cdot K}{D^2} \cdot \bar{v} + T_0 \cdot \frac{4}{D} \cdot \left\{ \frac{4}{3} - \frac{1}{3} \left(\frac{T_0}{\frac{\Delta P}{\Delta L} \cdot \frac{D}{4}} \right)^3 \right\}$$

The above equation however, can be modified to such an extent that a simplified form is obtained for evaluating measurements made both on pipelines and in a laboratory, without incurring the risk of major errors. If the terms put in brackets are resolved, then the following relationship is obtained:

$$(4.2) \quad \frac{\Delta P}{\Delta L} = \frac{32 \cdot K}{D^2} \cdot \bar{v} + \frac{16}{3 \cdot D} \cdot T_0 - \frac{256}{3} \cdot \frac{\left(\frac{T_0}{D} \right)^4}{\left(\frac{\Delta P}{\Delta L} \right)^3}$$

The negative term in the above equation is responsible for a non-linearity of the pressure drop curve within the range of lower velocities ($\bar{v} \to 0$). For higher velocities, which are of practical interest and which result in higher pressure drops, this term approaches zero and can therefore be ignored. Thus Eq. (4.2) can be reduced to the following simplified equation

$$(4.3) \quad \frac{\Delta P}{\Delta L} = \frac{32 \cdot K}{D^2} \cdot \bar{v} + \frac{16}{3 \cdot D} \cdot T_0 ; \qquad K = \eta_B$$

which represents a linear relationship between pressure drops and velocities.

The results of the plotted values in Figs. (4.2), (4.3) and (4.5) can be expressed by means of a linear regression in the following form:

$$(4.4) \quad \frac{\Delta P}{\Delta L} = B \cdot \bar{v} + A$$

By comparing the coefficients in Eq. (4.3) with those in Eq. (4.4), the rheological data can be quite simply calculated from pressure drop curves, and conversely pressure drop curves from the rheological data for varying pipe diameters D:

$$(4.5) \quad \eta_B = \frac{D^2}{32} \cdot B \qquad T_0 = \frac{3 \cdot D}{16} \cdot A$$

$$(4.6) \quad B = \frac{32}{D^2} \cdot \eta_B \qquad A = \frac{16}{3 \cdot D} \cdot T_0$$

When comparing the rheological data obtained from the two tests with the data calculated according to the initial Eq. (4.1), one can see how well this calculation agrees with the numerical one (see Table (4.4)).

It has therefore been demonstrated that only a few mathematical manipulations are recquired (in the case of suspensions with a Bingham-type flow behaviour) to enable the effective comparison of sampled data and to apply these results to pipelines of different diameters.

4.4 Comparison between tube and laboratory viscometer

As mentioned earlier, samples of the suspensions were taken from the various pipelines during the second test series. These samples were subsequently analyzed in the laboratory in a rotational viscometer. The pressure drops extrapolated from the flow curves were then compared with those that were measured in the pipeline; they agreed closely, as can be seen from Fig. (4.8). This diagramm shows the readings for the three pipelines during the second test (No. 141509) and for the purpose of comparison the pressure loss curves which have been extrapolated from the analysis of individual samples. Moreover, Table (4.5) lists the associated rheological parameters, giving the respective confidence intervals for the various parameters; the confidence interval refers to the 95 % confidence limits and was calculated with the percentile value $t_{0.975}$ of the Student distribution.

It is clear from the table that the values for yield stress T_0 and Bingham viscosity η_B agree closely; also in the case of the laboratory measurements, Bingham's flow law, which is used to determine the regression calculation, leads to coefficients of determination greater than 99 %. Differences between tube and rotational viscometer readings can only be detected by examining the confidence intervals. These were considerably smaller for the actual pipeline tests than the laboratory tests, especially the yield stress T_0 values which can be equated with a lower variance of the pipeline readings. It is therefore quite clear that the tube viscometer yields data with a higher statistical reliability than the rotational viscometer.

5. Summary

It can be said that the mathematical relationship (worked out in Section 3) relating the flow rate $\dot{Q}$, the pressure loss $\Delta P/\Delta L$ and the wall shear stress T_w, can be readily applied to the highly concentrated fine-grained coal-water suspensions that were the subject for the present study. This is due to the fact that such homogeneous mixtures behave like single-phase, clear liquids in rheological terms. This facilitates the comparison of measurements in pipelines of differing diameters with each other as well as enables one to carry out additional extra-polations.

In addition close agreement between tube and rotational viscometer readings can be demonstrated. If in the future the investigations of suspensions, produced from different types of coal and additives, yield similar results, this technique can be used to determine the pressure drop in large-diameter pipelines by measuring the rheological data under laboratory conditions.

6. References

/ 1 /   Round, G. F.; El-Sayed, Essam
        A correlation for steady flows of Non-Newtonian fluids in pipes
        McMaster University, Canada

/ 2 /   Kenchington, J. M.
        The prediction of pressure drop in slurry pipelines
        Proc. Multi-Phase Systems Symp., Joint I. Ch. E. and I. Mech. E.
        University of Strathclyde, April 1974

/ 3 /   Antonini, G.; François, O.; Gislais, P.; Touret, A.
        A direct rheological characterisation of highly loaded coal-water slurries flowing through pipes
        Coal Slurry Combustion and Technology, June 25 - 27, 1984
        Orlando, Florida

Solid Concentration c_w	Yield Stress T_0 / N/m2	Fluid Consistency Index K	Flow Behaviour Index n
0.720	5.2346	1.6924	1.0267
0.717	4.5273	1.4692	1.0114
0.712	3.6367	1.3457	0.9560
0.707	4.2540	0.7095	1.0718

Table (4.2): Rheological parameters for various solid concentrations measured in a pipeline (DN 400 type)

Test No.:	Nominal Diameter D / mm	Yield Stress T_0 / (N/m^2)	Bingham Viscosity η_B / Pas	Coefficient of Regression	Residual Variance
141507	200	4.2413	0.8364	0.9982	0.1690
	300	4.3739	0.8745	0.9995	0.0485
	400	3.9457	0.8709	0.9939	0.0784
	all	4.3154	0.8468	0.9974	0.1653
141509	200	4.2104	0.9586	0.9970	0.2301
	300	4.8133	1.0002	0.9952	0.1529
	400	4.6318	1.0859	0.9963	0.0623
	all	5.4213	0.9354	0.9950	0.2503

Table (4.3): Rheological data calculated from pressure drop measurements (n = 1)

Test No.:	Nominal Diameter D / mm	Yield Stress T_0 / (N/m^2)		Bingham Viscosity η_B / Pas	
		Numerical	Analytical	Numerical	Analytical
141507	200	4.241	4.244	0.836	0.836
	300	4.374	4.281	0.875	0.879
	400	3.946	3.819	0.871	0.882
141509	200	4.210	4.262	0.959	0.956
	300	4.813	4.753	1.002	1.002
	400	4.632	4.509	1.086	1.097

Table (4.4): Comparison of rheological data calculated according to different equations

Nominal Diameter D / mm	Visco-meter	Yield Stress T_0 / N/m^2	Bingham Viscosity η_B / Pas	Confidence Interv. T_0 / N/m^2	η_B / Pas	Coeff. of Regression
200	Pipe	4.2104	0.9586	0.5300	0.0194	0.9970
	Rotat./11	4.5230	0.9743	1.4291	0.0448	0.9900
	Rotat./12	4.6520	0.9448	1.2433	0.0387	0.9900
300	Pipe	4.8133	1.0002	0.1733	0.0251	0.9952
	Rotat./07	4.2137	1.0091	1.1740	0.0368	0.9900
	Rotat./08	4.2523	1.0204	1.1468	0.0357	0.9900
400	Pipe	4.6318	1.0859	0.1733	0.0227	0.9963
	Rotat./01	5.0296	1.0411	1.2880	0.0403	0.9900
	Rotat./02	4.5968	1.0674	1.2839	0.0402	0.9900

Table (4.5): Rheological data from tube and rotational viscometer (n = 1)

36

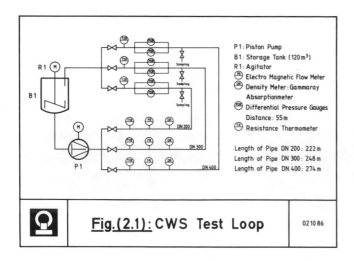

Fig.(2.1): CWS Test Loop   021086

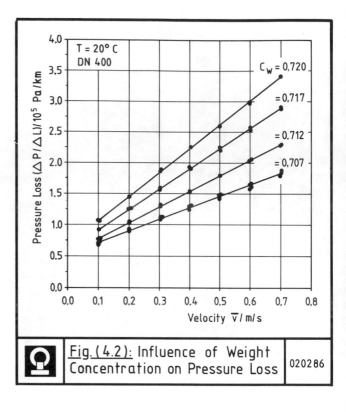

Fig.(4.2): Influence of Weight Concentration on Pressure Loss   020286

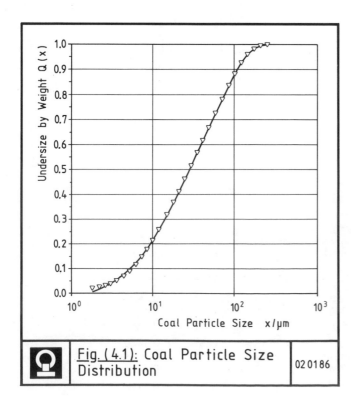

Fig.(4.1): Coal Particle Size Distribution   020186

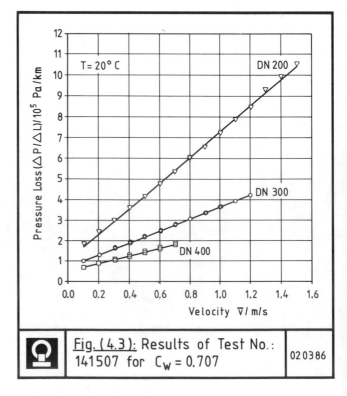

Fig.(4.3): Results of Test No.: 141507 for $C_W = 0.707$ 020386

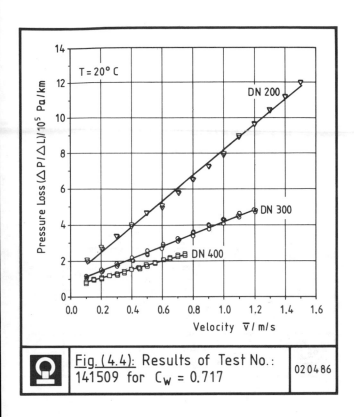

Fig.(4.4): Results of Test No.: 141509 for $C_W = 0.717$

020486

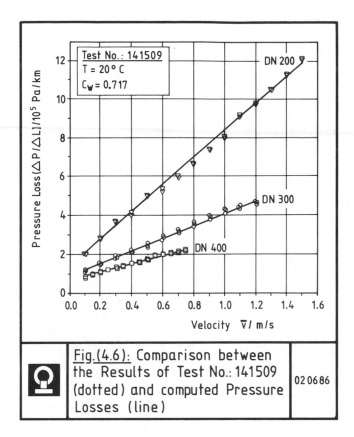

Fig.(4.6): Comparison between the Results of Test No.: 141509 (dotted) and computed Pressure Losses (line)

020686

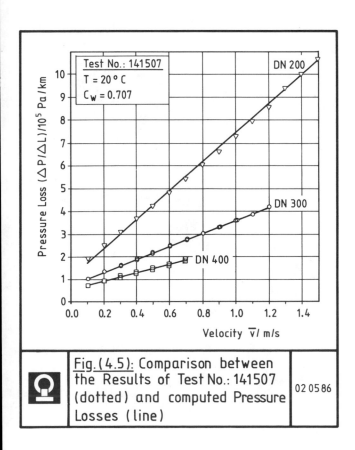

Fig.(4.5): Comparison between the Results of Test No.: 141507 (dotted) and computed Pressure Losses (line)

020586

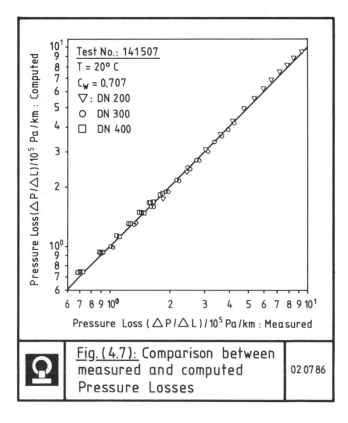

Fig.(4.7): Comparison between measured and computed Pressure Losses

020786

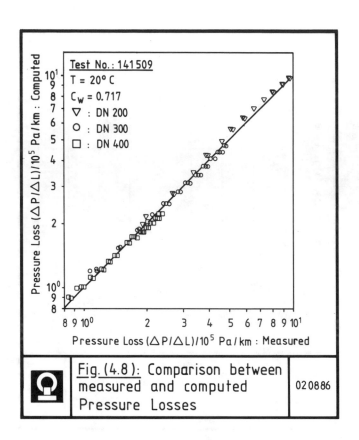

Fig. (4.8): Comparison between measured and computed Pressure Losses

020886

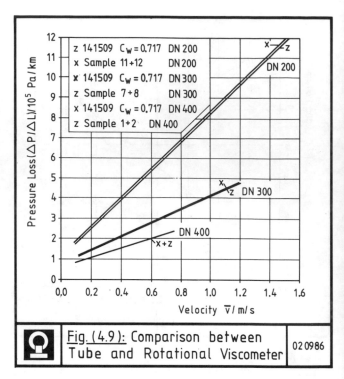

Fig. (4.9): Comparison between Tube and Rotational Viscometer

020986

10th International Conference on the
Hydraulic Transport of Solids in Pipes

HYDROTRANSPORT 10

Innsbruck, Austria: 29-31 October, 1986

PAPER B1

DENSE PHASE FLOW OF SOLIDS-WATER MIXTURES IN
PIPELINES: A STATE-OF-THE-ART REVIEW

M.STREAT

Department of Chemical Engineering and Chemical
Technology at Imperial College - London SW7 2BY

SYNOPSIS

The status of dense-phase transportation of
solids-water mixtures in pipelines is discussed
and the theoretical interpretation of energy loss
is given in detail. The technique is compared
with the concept of 'stabilised' flow and this is
illustrated by several case studies involving the
dense-phase flow of coarse gravel-water mixtures,
high concentration transportation of mine tailings and backfill and the transportation of high
concentration coal-water mixtures. It is concluded that the specific energy consumption for a
'stabilised' coal slurry is considerably less
than for dense-phase flow at the same mean velocity. Also, the hydraulic gradient is proportional to the inverse of pipe diameter for 'stabilised' flow and thus scale-up is likely to be advantageous.

NOMENCLATURE

A	Cross-sectional area of pipe	(m^2)
a	Fraction of pipe area above the interface defined by angle ϕ	$(-)$
A_b	Cross-section of pipe occupied by moving or stationary bed of solids	(m^2)
A_L	Pipe cross-section available to fluid flow	(m^2)
A_S	Pipe cross-section available to solids flow	(m^2)
C_b	Volume fraction of solids in moving or stationary bed	$(-)$
C_t	In-situ volume fraction of solids	$(-)$
C_V	Volume fraction of solids in delivered mixture	$(-)$
D	Pipe diameter	(m)
E_S	Specific energy consumption	$(kWh\ t^{-1}\ km^{-1})$

F	Friction force	(N)
F_D	Driving force	(N)
F_R	Resisting force	(N)
f	Fanning friction factor	$(-)$
f_B, f_T	Friction factors defined in Theoretical Section	$(-)$
f_S	Coefficient of sliding friction between solids and pipe wall (Dynamic)	
g	Acceleration due to gravity	
i	Pressure gradient, expressed as headloss in m of water m^{-1}. Also known as hydraulic gradient	$(-)$
i_B	Hydraulic gradient attributable to sliding bed of solids	$(-)$
i_f	Hydraulic gradient attributable to liquid-wall friction losses	$(-)$
i_L	Hydraulic gradient for water flowing at the mean slurry velocity in the same pipe	$(-)$
i_S	Hydraulic gradient attributable to the solids	$(-)$
i_T	Total hydraulic gradient of the solid-liquid mixture	$(-)$
i_R	Reversible component of hydraulic gradient (vertical flow)	$(-)$
i_{IR}	Irreversible component of hydraulic gradient (vertical flow)	$(-)$
K'	Consistency index in generalised power law (eq.30)	$(kg\ m^{-1}\ s^{-1})$
n'	Generalised power law parameter in eq.30	$(-)$
P	Pressure	$(N\ m^{-2})$
ΔP_f	Pressure drop due to friction	$(N\ m^{-2})$
ΔP_h	Hydrostatic head component of pressure drop	$(N\ m^{-2})$
ΔP_T	Total pressure drop	$(N\ m^{-2})$
R'_r	Hydraulic radius divided by radius of pipe	$(-)$
s	relative density (ρ_S/ρ_L), equivalent to specific gravity when fluid is water	$(-)$
V_B	Velocity of moving bed	$(m\ s^{-1})$
V_L	Mean liquid velocity	$(m\ s^{-1})$
V_M	Cross-sectional mean slurry velocity	$(m\ s^{-1})$

V_S Mean solids velocity (m s^{-1})

Z Height or depth (m)

θ Angle of bed deposit (rad)

ρ Density (kg m^{-3})

ρ_L Density of fluid (kg m^{-3})

ρ_M Mixture density (kg m^{-3})

ρ_S Density of solids (kg m^{-3})

τ Shear stress (N m^{-2})

τ_b Shear stress at bed surface (N m^{-2})

τ_W Shear stress at the pipe walls (N m^{-2})

τ_{WD} Wall shear stress for the bed associated section of the pipe invert (N m^{-2})

ϕ Angle of repose of flooded solids (rad)

SUBSCRIPTS

b   Refers to bed of solids

f   Refers to friction component

L   Refers to liquid

m   Refers to mean value

M   refers to mixture

S   refers to solids

T   refers to total

W   refers to water

INTRODUCTION

Slurry transportation is well established technology and many large-scale installations are in commercial operation. Applications are to be found in the mining and minerals industries and long distance pipelines have been installed to convey coal, iron ore concentrate, copper ore concentrate, limestone, phosphate rock and mine tailings (see Table 1).

Most slurry pumping applications involve the transportation of relatively fine material (<100µm) in homogeneous or heterogeneous suspension using custom designed centrifugal or positive displacement pumps in direct contact with the slurry. These suspensions are usually classified as dilute since the volumetric solids concentration is less than about 35%. Pipeline flow velocities are sufficiently large to avoid deposition of solids and pipeline blockage. This inevitably leads to the subsequent problem of slurry dewatering coupled with other environmental considerations.

In most conveying systems, the solids have to be prepared in an agitated tank prior to pumping so that the desired solids concentration can be matched to the pump characteristics. The suspension is normally passed directly though a special slurry pump of robust construction designed to withstand excessive wear. These pumps are therefore more expensive than conventional pumps used for the handling of single phase liquids or gases. The size of the solid particles to be handled is also limited by the clearances in the pump, otherwise both pump wear and particle attrition would become a major operating expense.

The pumping of high solids concentration slurries has generally been avoided due to excessive wear and attrition to the moving parts, e.g. impellors, valves, etc. in addition to the intuitive fear of pipeline blockage. Also the likely high pressure drops could offset the potential economic advantages. Nevertheless, several studies have shown interesting results for the transport of high concentration slurries without incurring these difficulties (1-9). Some workers have shown marked advantages in dense-phase flow with the added incentive of transporting coarse material in either a simple or in a non-Newtonian carrier fluid (10-17). The term 'dense-phase flow' was first used in gas-solids systems and relates to the transportation of a gas-solids mixture at high solids concentration into a gas fluidised bed.

The status of dense-phase flow in gas-solids systems has been comprehensively reviewed by Zenz and Othmer (18) in so far as it relates to fluidised bed reactors, though the technique has been developed for pneumatic conveying in the field of catalytic cracking, separation of minerals (19), production of vinyl acetate (20) and heat treating of granular materials. Solids have been transported within reactors by flowing across perforated plates (21,22) or bubble trays (23) and transported through heat-treating furnaces by flowing across perforated hearths.

The analysis of dense-phase flow in gas-solids systems assumes independent flow behaviour of the solid particles and interstitial gas flow so that co-current flow produces a net driving force on the bed of particles which will exceed gravitational and wall effects enabling the bed of particles to move in the direction of gas flow. Vertical or horizontal flow is possible at high solids concentration, about 60% by volume, and in unrestricted vertical flow the bed will move cocurrently upward at a velocity equal to the difference between the superficial fluid velocity and the incipient fluidisation velocity or the velocity corresponding to the observed pressure loss for a static fixed bed.

Dense-phase flow in liquid-solid systems is a more recent development and can be traced to early work on the concept of a mobile paste fuel in nuclear reactors, the development of continuous countercurrent ion exchange reactors and for the hydraulic transportation of minerals, especially coal slurries, as high concentration non-settling suspensions. The novel concepts of a mobile nuclear fuel and the idea of a continuous countercurrent moving bed reactor involve dense particle slurries in effectively creeping flow whereas hydraulic conveying over relatively long distances will involve operation over a range of throughputs and therefore the flow mechanism is rather more complex.

The case of creeping flow has been described by both Calvert and Miller (10) and Hancher and Jury

(24) for the case of particle suspensions in either upward or downward movement in vertical tubes. It is not possible to generalise this work since flow behaviour is dependent on the properties of the particles, i.e. size, size range distribution, shape, density and fluid properties, particularly density and viscosity. In creeping flow, a dense paste or slurry can be treated as two continua each capable of transmitting force and therefore independently responsible for energy dissipation.

Hancher and Jury (24) have investigated the intermittent movement of a packed bed of ion exchange particles (median size range 0.5mm) in an upwards direction by the application of an hydraulic pulse of process solution (an aqueous salt solution or water). These workers found little difficulty in moving a packed bed (solids concentration was about 60% by volume) over relatively short incremental distances in creeping flow. They presented a relatively simple force balance relationship for plug flow movement of the bed which included the fluid-wall and particle-wall frictional resistances. Calvert and Miller (10) also transported solids-liquid slurries at solids volumetric fractions of 60% for spherical particles and 50% for angular particles. Flow velocity for particles in the size range 40-300μm was in the range 0.0002-0.0003 m/s. In this work, sustained flow was performed in a hairpin tube flow configuration with measurements recorded in vertical upward and downward flow. A critical finding of this work relates to the design of the tube exit, since adverse design could cause excess pressure drop and ultimately

block the tube. The flow mechanism was interpreted by a simple modification of the standard correlations for laminar flow of fluids through porous media. The overriding effect of the tube exit was modelled by a force balance relationship which explained the restricting forces set up under adverse conditions. These two projects, though seemingly peripheral to the hydraulic conveying of high concentration slurries, nevertheless provide some credibility to the concept and represented the earliest account of both the practical and theoretical problems which were to be addressed vigorously in future years.

The mechanistic approach to the movement of dense pastes or slurries differs from that of comparable two-phase systems. Though it is appropriate to use the well developed theories of non-Newtonian fluid mechanics to dilute suspensions, this procedure is less rigorous in case studies where the particles can behave independently of the fluid and can exert intraparticle stresses and interact at the pipe wall. Early work by Delaplaine (25) has given a comprehensive treatment of forces acting on a downward flowing bed of solid particles. The momentum equations are presented to calculate the stress relationships for the special case of a bed of solids moving downwards under the influence of gravity. The stresses are calculated provided static friction at the pipe-wall can be determined. However, the author recognises that particle shear effects close to the pipe-wall may well invalidate the generalised stress theory. Brandt and Johnson (26) have also tried to predict the stress relationships in a moving bed of solid particles through which a fluid is flowing in a countercurrent or cocurrent direction. The work is confined to packed beds in which the particles remain in contact and

cannot of course be extrapolated to the case of an incipiently fluidised bed since here the particle stress components will tend to zero in all directions. Assuming a moving packed bed, then this work gives point values of the particle-wall friction and the ratio of the radial-to-boundary stresses.

This early theoretical work is important since it establishes ground rules for the analysis of two-phase conveying at high concentration in those case studies where particles are likely to behave independently of the motive fluid, i.e. sliding bed flow at low velocity or dense-phase flow with interstitial flow of the motive fluid. These case studies will now be discussed in some detail.

CLASSIFICATION OF DENSE PHASE FLOW

Dense phase slurries can be prepared in a number of ways. It is customary to prepare a high concentration mixture either by introducing a particulate solid into a carrier fluid so that the volumetric concentration of solids attains a value of about 60% or alternatively to pre-mix coarse particles into a fine particle non-Newtonian suspension of the same particles or of a compatible material. The former procedure is possible with discrete particles, usually of intermediate size (range 100-10,000μm) that will retain their identity during transportation and are not cohesive or liable to form a more complex non-Newtonian suspension. An assembly of particles in the configuration of a packed bed will possess a prescribed porosity (or solid concentration) which will lie close to the value obtained for a freely settled bed. Sustained flow of a high concentration suspension is possible because of interstitial flow of the motive fluid through the interstices of the bed. If the bed of particles occupies the entire cross-sectional area of the pipeline, then the motion of the slurry is opposed by particle-wall and fluid-wall frictional resistance. The concept is similar to a sliding bed of solids that in the limiting condition occupies the entire cross-sectional area of the pipe (see Fig.1). The idealised case study of a dense moving bed of solids in horizontal flow has been studied theoretically and it is possible to predict energy loss as a function of mean flow velocity if the coefficient of sliding friction of particles at the pipe-wall, the in-situ volumetric solids concentration, relative density of particles and fluid wall friction are known.

The alternative concept of achieving dense phase flow is to adjust the size distribution of the solid to give the greatest possible packing density. This procedure has been effectively demonstrated with coal-water mixtures whereby coarse coal is transported in a concentrated suspension of fine particles in a pre-mixed slurry. Coal-water mixtures of this type behave as Bingham plastics with laminar flow properties over a very wide range of velocities. The dense slurries settle without segregation i.e. they are stable and can be allowed to remain stationary in large vessels for long periods. Coal-water mixtures up to about 70% by weight have been transported in this manner. Recent pilot-plant trials by British Petroleum and Bechtel on a 2.5km pipe-rig have shown that it is possible to operate sustained flow of stabilised coal-water mixtures at relatively low pressure drop and at low linear velocities. This project opens up the possibility of both long distance land based transportation of coarse coal particles (~25mm) and the shore-to-ship transportation of even higher coal concentration slurries (about 85% by weight).

A similar procedure has been adopted in the mining industry for the hydraulic transportation of backfill (27). The fill material is transported as a dense paste at concentrations varying from 60-87% by weight. Coarse particles (about 25mm) are carried in a non-Newtonian slurry of finer particles (less than 30µm) of the same material. The technique is feasible, although difficulties have been encountered with wear in pumps, valves and associated plant and equipment. Fly ash slurries have also been transported as high concentration pastes (28). These pastes exhibit distinct non-Newtonian properties as the solids concentration is increased above about 70% by weight. Coarse fly ash (between 1-700µm) was difficult to pump due to scaling of the pipes and this problem was overcome by adding fine fly ash (1-200µm) to the paste. Also, the pressure drop was significantly reduced with an addition of undersize material.

THEORETICAL CONSIDERATIONS

Dense-Phase Moving Bed

In dense-phase flow, most, if not all of the solids travelling in horizontal flow are conveyed as contact load, with the submerged weight being transmitted to the pipe invert. It is therefore reasonable to assume that the main contribution to energy loss, especially at low velocities, is solids-wall friction. Indeed, many workers have used this fact to obtain semi-empirical correlations for predicting the energy loss in two-phase systems.

Newitt et al. (29) were pioneers in this field and they derived the following expression for the hydraulic gradient

$$i_T = 0.8 (s - 1) C_V + i_L \qquad (1)$$

using a 1-inch pipe. Though there are many such empirical correlations, only Wilson (30) has attempted to analytically relate the pressure gradient to the solids-wall friction coefficient (f_S) for cases of flow with a moving bed. He developed a bed-slip model to predict the point at which a stationary bed of solids in a pipe begins to move. His analysis is based on a force balance for the stationary bed at the point of incipient movement. For a steady flow, the driving force must be equal and opposite to the resisting force. Wilson assumed a hydrostatic pressure distribution of the solid particles on the pipe invert.

For a bed of particles occupying a portion of the pipe area defined by the deposit angle θ (see Figure 2) he found that at the slip point

$$i_S \left\{ \frac{\theta - \sin\theta\cos\theta}{4} + R_r' \left(\sin\theta - \frac{\theta f_S}{\tan\phi}\right) \right\}$$
$$= \frac{f_S (s-1) C_b (\sin\theta - \theta\cos\theta)}{2} \qquad (2)$$

where R_r' is the ratio of the hydraulic radius of the bed-associated portion of the waterway to the

pipe diameter i.e. $R_r' = \frac{\theta}{4\sin\theta}$

Consider the case of a sliding bed as shown schematically in Fig.1(a). There need not be a true suspension flowing above the bed, but saltating particles will always be present. The moving bed occupies a portion of the pipe defined by the deposit angle θ and is flowing at a mean velocity V_B. Figure 2 defines the symbols and geometry of the system.

For the moving bed the driving force is a sum of the shear stress acting at the bed surface and the pressure gradient due to water flow through the bed interstices. The velocity of this seepage flow is very nearly equal but slightly greater than the bed velocity (V_B). Otherwise high slip velocities would mean excessive pressure gradients (3) which are not encountered in practice. The forces exerted by this interstitial flow are of great significance. Almost all the pressure force causing the bed to move is transmitted to the particles by drag. Thus, the driving force for the bed section per unit length is

$$F_{DB} = \rho_L g i\left(\frac{\pi D^2}{4}\right)(1 - a) + \tau_b D \sin\theta \qquad (3)$$

where (1-a) is the ratio of the bed associated area (A_1) to the total area (A), and i is the total hydraulic gradient in metres of water per metre of pipe and τ_b the shear stress at the bed surface. The resisting force is the sum of the friction forces attributable to solids-wall and water-wall friction. To find the solids-wall effect, it is assumed that the granular mass exerts a hydrostatic type pressure on the pipe invert proportional to its submerged weight.

$$\frac{dP}{dZ} = \rho g (s - 1) C_b \qquad (4)$$

where Z is the distance measured downwards from the surface of the bed. Experimental verification of this hydrostatic type of normal stress distribution has been given by Wilson (30) from the observed slip point of deposits and from tests made with an inclined tube containing a flooded bed of particles. The granular pressure at the top of the bed need not be zero and depends on the definition of bed height. If bed height is defined as the level at which no particles are exchanged with the fluid, then Bagnold (3) has shown that the intergranular stress normal to the surface is equal to $\tau/\tan\phi$, where ϕ is the angle of repose.

Equation (4) can now be integrated to give the hydrostatic pressure of the particles at an arbitrary point on the pipe wall defined by the angle α.

$$P_\alpha = \frac{\tau_b}{\tan\phi} + \rho g (s - 1) C_b \frac{D}{2}(\cos\alpha - \cos\theta) \qquad (5)$$

For unit length of pipe the total sliding friction against the wall will equal

$$F_{sliding} = 2 f_S \int_0^\theta P_\alpha \left(\frac{D}{2}\right) d\alpha$$
$$= f_S \left\{\frac{D \theta \tau_b}{\tan\phi} + \rho g (s - 1) C_b \frac{D^2}{2}(\sin\theta - \theta\cos\theta)\right\} \qquad (6)$$

The friction losses due to fluid friction can be accounted for by defining a friction factor for the bed-associated section of the pipe, viz:

$$\tau_{WD} = \frac{1}{2} f_D \rho_{MB} V_B^2 \qquad (7)$$

where τ_{WD} is the shear stress at the bed associated section of the pipe invert,

f_D is the Fanning friction factor, modified for the bed section

$\rho_{MB} = \left[(s-1)\ C_b + 1\right]\ \rho_L$ is the mixture density of the bed.

The assumption is made here that the slip velocity in the bed section is negligible. The friction factor, so defined, takes account of the kinetic energy of the bed as a whole, recognising the fact that only liquid can induce shear stress at the wall; f_D should therefore be similar to the value for an equal discharge of pure water. This shear stress acts on the pipe periphery section θD and so a balance of forces for the bed section gives:

$$\rho_L\ g\ i(\frac{\pi D^2}{4})\ (1-a) + \tau_b \sin\theta\ = \theta\tau_{WD}D$$

$$+ f_S\left\{\frac{D\ \theta\ \tau_b}{\tan\phi} + \rho\ g\ (s-1)\ C_b \frac{D^2}{2}(\sin\theta - \theta\cos\theta)\right\} \qquad (8)$$

where a is the fraction of the pipe area above the interface defined by angle θ. Similarly, we can equate forces for the section of the pipe above the bed, of area ratio a.

$$\rho_L\ g\ i\ (\frac{\pi D^2}{4})\ a = \tau_{WT}D\ (\pi-\theta) + \tau_b \sin\theta \qquad (9)$$

τ_{WT} being the shear stress at the invert of the pipe above the bed section:

$$\tau_{WT} = \frac{1}{2}\ f_T\ \rho_{MT}\ V_T^2 \qquad (10)$$

where f_T is the modified friction factor for this section

ρ_{MT} is the mixture density and
V_T is the mean velocity of flow in the same section

Adding (8) and (9) gives an overall force balance for steady flow of the system

$$\rho_L\ g\ i\ (\frac{\pi D^2}{4}) = \tau_{WD}\ D\ \theta + \tau_{WT}\ D\ (\pi-\theta)$$

$$+ f_S\left\{\frac{D\ \theta\ \tau_b}{\tan\phi} + \rho\ g\ (s-1)\ C_b \frac{D^2}{2}(\sin\theta - \theta\cos\theta)\right\}$$

Substituting for τ_{WD} and τ_{WT} and simplifying gives:

$$i_T = 2\ f_S\ (s-1)\ C_b\left\{\frac{\sin\theta - \theta\cos\theta}{\pi}\right\} +$$

$$2f_D\ (\frac{\rho_{MB}}{\rho_L})\ (\frac{\theta}{\pi})\ \frac{V_B^2}{D\ g} + 2\ f_T\ (\frac{\rho_{MT}}{\rho_L})\ (\frac{\pi-\theta}{\pi})\ \frac{V_T^2}{D\ g} +$$

$$\frac{2f_S f_b}{\tan\phi}\ (\frac{\theta}{\pi})\ \frac{(V_T-V_B)^2}{D\ g} \qquad (11)$$

and τ_b has been defined as

$$\tau_b = \frac{1}{2}\ f_b\ \rho_{TB}\ (V_T - V_B)^2 \qquad (12)$$

where f_b is the friction factor for the bed surface shear defined by (12).

Equation (11) is not a simple expression to use for the general case which it describes. Many of the quantities required for the calculation of the hydraulic gradient are usually unknown, e.g. individual velocities and mixture densities.

The equation can, however, be simplified for special cases. The most interesting configuration is the horizontal flow of a dense moving bed of particles occupying the entire cross-section of the pipe. The following simplifications can therefore be made:
(1) $\tau_b = 0$ since $\theta = \pi$
(2) The liquid wall friction terms can be combined to give the headloss due to shear stress at the pipe invert:

$$i_f = 2\ f_W\ (\frac{\rho_M}{\rho_L})\ \frac{V_M^2}{D\ g}$$

(3) The overall hydraulic gradient for $\theta = \pi$ is thus given by:

$$i_T = 2\ f_S(s-1)\ C_t + 2\ f_W\left[(s-1)C_t+1\right]\ \frac{V_M^2}{D\ g} \qquad (13)$$

Since the hydraulic gradient for pure water flowing at a velocity V_M is given by

$$i_L = 2\ f_W\ \frac{V_M^2}{D\ g}$$

We can write:

$$i_T = 2\ f_S\ (s-1)\ C_t + (\frac{\rho_M}{\rho_L})\ i_L \qquad (14)$$

For suspension flow without a deposit the first and last terms of equation (11) cancel and we can combine the middle terms to get:

$$i_T = (\frac{\rho_M}{\rho_L})\ i_L \qquad (15)$$

This equation is obeyed reasonably well by homogeneous Newtonian suspensions.

Equation (14) is not rigorous due to the assumptions made in defining the friction factors, however, it is adequate for design purposes, though it will probably slightly over-estimate the hydraulic gradient.

For vertical transport the bed-slip model is not directly applicable. The submerged weight of the solids, acting vertically, does not transmit any stress to the pipe walls. At low velocities, however, when the bed is not fully fluidised, or in cases where there are downstream constraints to the flow, the slurry moves as a packed bed and solid-wall stresses are significant.

The energy losses are the sum of the potential energy changes which are reversible and the irreversible losses due to a drag and liquid-wall friction:

$$\Delta P_T = \Delta P_h + \Delta P_f \qquad (16)$$

The first term can be derived as follows:
If the energy dissipated in raising the slurry by unit height per second is E_h then

$$E_h = A_S V_S \rho_S g + A_L\ V_L\ \rho_L\ g \qquad (17)$$

where A_S, A_L are the pipe cross-sections occupied by solids and liquid respectively.

This energy loss is also equal to the delivered volume per second multiplied by the pressure loss due to the potential energy increase of the slurry.

$$E_h = A\ V_M\ (\Delta P_h) = A\ V_M\ i_R\ g\ \rho_L \qquad (18)$$

Equating (17) and (18) we get:

$$i_R = \frac{A_S V_S}{A \, V_M} \frac{\rho_S}{\rho_L} + \frac{A_L}{A} \frac{V_L}{V_M} \qquad (19)$$

Now the delivered volumetric concentration of solids C_V is:

$$C_V = \frac{A_S \, V_S}{A \, V_M} \qquad (20)$$

Hence, by substituting for C_V and $(1-C_V)$ into (19)

$$i_R = C_V (s-1) + 1 \qquad (21)$$

From a force balance over unit height of pipe, the driving force is provided by the pressure gradient.

$$F_D = i_T \, \rho_L \, g \cdot A \qquad (22)$$

The resisting forces are:

$$F_R = (1-C_t)\rho_L g \, A + C_t \, \rho_S \, g \, A + \tau_W \, \pi \, D \qquad (23)$$

where τ_W is the shear stress at the pipe-walls. For steady flow conditions we can equate the two forces to get:

$$i_T = C_t (s-1) + 1 + \tau_W \frac{\pi D}{(\frac{\pi D^2}{4}) \, \rho_L \, g} \qquad (24)$$

On defining τ_W as for horizontal flow

$$\tau_W = \frac{1}{2} \, \rho_M \, V_M^2 \, f \qquad (25)$$

where f is the modified Fanning friction factor

$$i_T = C_t(s-1) + 1 + (\frac{\rho_M}{\rho_L}) \cdot 2f \, \frac{V_M^2}{Dg} \qquad (26)$$

$$\text{where } \rho_M = (C_t(s-1) + 1) \, \rho_L \qquad (27)$$

If we assume that f is equal to that for flow of pure liquid at the same mean velocity V_M,

$$i_T = C_t (s-1) + 1 + (\frac{\rho_M}{\rho_L}) \, i_L \qquad (28)$$

where i_L is the hydraulic gradient for pure water at velocity V_M in the same pipe (flowing horizontally). The reversible hydraulic gradient i_R is given by equation (21) and we can write:

$$i_T = \underbrace{C_V(s-1)+1}_{} + \underbrace{(C_t-C_V)(s-1)}_{} + \underbrace{(\frac{\rho_M}{\rho_L}) \, i_L}_{}$$

| Total= | Reversible + | Friction Loss | + | Wall fric- |
| Head | Head | due to slip | | tion losses |

The friction losses due to slip arise from the drag forces on individual particles caused by the relative velocity of liquid and solids. This slip velocity is of the same order as the minimum fluidising velocity which in turn approximates to the free fall velocity of the particles.

Stabilised Flow

Several theoretical models have been proposed to relate shear stress, τ to shear rate γ, for viscous non-Newtonian fluids in the laminar flow regime. Stabilised suspensions of coal-in-water, containing in excess of 60% by weight of solids in suspension appear to behave as time-independent pseudo-homogeneous non-Newtonian fluids and can therefore be analysed by the more common theoretical treatments, e.g. the power law model and the Bingham plastic model (12). For fluids whose laminar flow behaviour does not fit well to any of the simple models, Metzner and Reed (22) have suggested that pipeline flow data can be treated using the following general expression:

$$\tau_w = K' \left(\frac{8V}{D}\right)^{n'} \qquad (30)$$

$$\text{where} \quad n' = \frac{d \, \ln \tau_w}{d \, \ln \, (8V/D)} \qquad (31)$$

In this model, the parameter n' is allowed to vary with shear rate. The shear stress at the wall is given by the expression

$$\tau_w = \left(\frac{\Delta P}{L}\right) \cdot \left(\frac{D}{4}\right) \qquad (32)$$

n' is determined from the slope of a log-log plot of $D\Delta P/4L$ against $8V/D$, where $\Delta P/L$ is the pressure gradient at velocity V in a pipe of diameter D under laminar flow conditions. The friction factor is given by the relationship

$$f = \frac{\tau_w}{\frac{1}{2}\rho V^2} = \frac{\Delta P D}{2L\rho V^2} \qquad (33)$$

If we define the generalised Reynolds number as:

$$f = \frac{16}{Re'} \qquad (34)$$

$$\text{then} \quad Re' = \frac{D^{n'} \, V^{2-n'} \rho}{K' \, 8^{n'-1}} \qquad (35)$$

The Metzner and Reed correlation is of great value in cases where the rheological behaviour of the fluid is not adequately described by one of the simpler constitutive equations or where scale-up is intended from data taken in small diameter pipe.

Friction factor-modified Reynolds number data has been obtained by Elliott and Gliddon (12) for coal-water mixtures and a typical set of data are reproduced here in Fig.3. The values of the flow indices for the slurry are given in Table 2 (reproduced from Elliott and Gliddon). The points on the graph are very close to the theoretical line for laminar flow and furthermore the apparent shear rates did not exceed 100 s^{-1} in an actual large scale installation. Thus n' is virtually constant in this case study:

$$\frac{D}{4} \cdot \frac{\Delta P}{L} = K' \left(\frac{8V}{D}\right)^{n'} \qquad (36)$$

and this equation can be rearranged in terms of hydraulic gradient as follows:

$$i_T = \frac{\Delta P}{\rho g L} = K' \left(\frac{8V}{D}\right)^{n'} \left(\frac{4}{D\rho g}\right) \qquad (37)$$

This theoretical treatment over-simplifies the complex flow behaviour of non-Newtonian suspensions but appears to be an adequate representation of the very few case studies that have been examined to date. Although there has been considerable work since Elliott and Gliddon, most studies have been confined to feasibility assessments and therefore theoretical understanding has not been advanced. This lack of fundamental knowledge must be speedily redressed in the light of the current interest in high concentration transportation of stabilised suspensions.

CASE STUDIES

Dense phase flow of coarse gravel-water mixtures

The most carefully researched area of dense phase flow involves the hydraulic transportation of sand and coarse gravel in horizontal and vertical pipelines. Theoretical and experimental work at Imperial College has validated the application of dense phase flow to particle suspensions at a volumetric solid concentration lying between the free and tap settled porosity of the particle assembly. The status of this work can best be described in relation to the hydraulic transportation of a coarse gravel (See Fig.5) in a 3.1 inch (78mm) diameter horizontal pipeline. In this case, the flowing solids do not necessarily occupy the entire cross-section of the pipe. For mean velocities in the range ~0.1 - 0.5ms^{-1} the sliding bed occupies the whole of the pipe cross-section but in the range 0.5-2.0ms^{-1} the bed only occupies a fraction of the pipe accompanied by a faster moving layer of pure water with saltating particles moving above it. Above 2.0ms^{-1} the distinction between these layers disappears and the particles are turbulently supported in the carrier fluid. The analysis of energy dissipation in this system is therefore interesting and can be facilitated by the mathematical treatment given earlier.

The coefficient of sliding friction, f_s, has been determined for gravel-water mixtures on several counterfaces and an appropriate value, suitable for predictive purposes is 0.32±0.02. Methodology to determine the coefficient of sliding friction is given extensively elsewhere (33).

The in-situ volumetric concentration of the flowing gravel-water mixture is close to the freely settled porosity of the system at all velocities in the range 0.1-2.0ms^{-1}. For gravel in the size range 5/16" - 4 mesh the relevant values are as follows:

free settled volumetric concentration = 0.472
pack settled volumetric concentration = 0.593

Knowledge of the coefficient of sliding friction and the in-situ volumetric concentration enables the hydraulic gradient to be predicted provided that all the physical properties of the solids and liquid are known. Let us attempt to predict the hydraulic gradient with the aid of equ.11. Four representative velocities have been chosen which cover the entire experimentally determined spectrum of measurements. In view of the foregoing comments, it is necessary to evaluate the deposit angle, θ, in each case. This is rather difficult and can only be done in an arbitrary manner. If one estimates the value of the solids concentration in the bed deposit, C_b, then it is possible to derive θ provided that the in-situ concentration, C , is known. In this simplistic way, it follows that C_t, C_b and θ are related by the following expressions:

$$C_t \cdot A = C_b \left(\frac{\theta - \sin \theta \cos \theta}{\pi}\right)A$$

$$\theta - \sin \theta \cos \theta = \frac{C_t}{C_b} \pi$$

C_b is not known from first principles and was found experimentally by careful observation of the two-phase mixture at start-up. In this case study, the value C_b= 0.55 is close to the solids fraction in the bed deposit. Using this information, the deposit angles for the four relevant velocities have been found and are given in Table 3.

Prediction of the hydraulic gradient in horizontal flow is given in Table 4. Equation 11 comprises four terms for the evaluation of the total hydraulic gradient. The contribution of each term is given in Table 4 and it is obvious that the major energy loss can be attributed to the first term which takes into account solid-solid sliding friction. Thus, discrepancies between the predicted and experimental results depends critically on the accuracy of the parameters C_b and f_s. The second term accounts for liquid-wall friction in the bed associated section of the pipe whereas the third term accounts for the pressure drop due to liquid-wall friction in the pipe region above the moving bed. The fourth term accounts for pressure drop due to liquid and particle shear at the bed surface.

The agreement between theory and experiment is satisfactory though by no means exact. Theory gives a minimum in the hydraulic gradient, i_T, in the range 0.1-2.6ms^{-1} and this is found in practice. The minimum occurs at about 1.3ms^{-1} and this point has been discussed in detail by Televantos et al (7). The qualitative agreement between theory and these results is encouraging and tends to validate the flow mechanism.

It is obvious that equ.11 is difficult to use and therefore it is worth examining the validity of equ.14. This equation can be used under the constraint that $\theta=\pi$, i.e. the solids occupy the entire cross-section of the pipe. This is a good approximation for this case study if the mean velocity exceeds about 1.5ms^{-1}. Equation 14 has been used to predict hydraulic gradient for five velocities in the range 1.5 - 3.5ms^{-1} and the data are given in Table 5. The comparison between predicted and experimental hydraulic gradient is good since the error is not larger than about 3%.

The experimental investigation of dense phase flow with sand-water systems has been carefully performed over many years and the results have been subjected to exacting theoretical analysis and it is therefore possible to use this system as a basis for comparison with other related case studies.

High Concentration transportation of mine tailings and backfill

The hydraulic transportation of mine tailings and backfill material has emerged strongly in recent years (27,34). This work is less meticulous in the sense that little if any theoretical work has been presented to support the growing amount of experimental data. It has been shown that very high concentrations of finely divided particles (<50μm) can be transported as pastes at relatively low velocities using extremely large operating pressures. This has encouraged several workers to examine means of lowering the hydraulic gradient. Of the more interesting suggestions are Horsley (35) who found that the addition of a relatively small amount of sodium tripolyphosphate to a slimes slurry containing about 75% by weight solids (53% by volume) approximately halved the pressure drop. Separately, Verkerk (27) has shown that the addition of coarse particles to a fine particle suspension lowers the total pressure drop. This observation is quite intriguing and will be examined further. Verkerk et al (27) have carried out tests with mixtures of waste rock and slimes slurries containing up to 85% by weight (68% by volume) solids in suspension. The tests were conducted in a 120mm diameter pipeline with top size material of −27mm. The experimental results are sparsely given in the original paper but all the data lie below a line drawn between the following points:

Pipeline Pressure Loss (kPa)	Paste Flow-Rate (tonne/hr)
200	10
1600	80

Converting this information into comparative data, then it follows that for a pipeline length of 120m the hydraulic transportation of an 85% by weight paste comprising waste rock and slimes is bounded by the following results:

Pipeline Pressure Loss (kPa)	Paste Flow-Rate (tonne/hr)
200	10
1600	80

Converting this information into comparative data, then it follows that for a pipeline length of 120m the hydraulic transportation of an 85% by weight paste comprising waste rock and slimes is bounded by the following results:

Velocity V_M (ms^{-1})	Hydraulic gradient i_T (m water/m pipeline)
0.12	0.34
1.00	1.36

Assume that a slurry of similar material is pumped at creeping flow (0.1–0.5ms^{-1}) at a concentration of 68% by volume (85% by weight) and let us calculate energy loss using the simplified equation for dense phase flow,

$$i_T = 2f_s(s-1)C_t + \frac{\rho_M}{\rho_L} i_L$$

Neglecting the second term and substituting $f_s = 0.32$; $s = 2.65$; $C_t = 0.68$, gives $i_T = 0.72$ m water/m pipeline

Note that this result lies between the limits given by Verkerk, and therefore supports his measurements. However, the effect of particle size is not anticipated directly in the theory and we therefore need an explanation for the surprising decrease in pressure drop in the presence of coarse particles.

Transportation of high concentration coal-water mixtures

The hydraulic transportation of coal-water mixtures has been extensively researched, e.g. the Saskatchewan Research Council has performed detailed rheological and pipeline investigations over a period of many years (36). However, this and similar work elsewhere is largely confined to the handling of finely crushed coal (<100μm) for direct pumping with conventional pumps in lean phase flow. Coarse particle work has also been performed with run-of-the-mine coal, in order to expand the scope of coal transportation. The most promising development involves high concentration slurries stabilised by careful coal preparation and currently under development in a joint venture by British Petroleum and Bechtel. It is too early to comment on the outcome of this work, suffice it to say that the indications are good as far as the fluid mechanics is concerned, though the peripheral mechanical engineering, especially the performance of pumps etc. warrants further work.

The most impressive work published to date is the CEGB study of high concentration coal-water mixtures presented sixteen years ago by Elliott and Gliddon (12). This work is unsurpassed in scope and clarity of presentation. The concept of stabilised suspensions differs from dense-phase flow as given in the first case study. Here the particulate phase does not behave independently from the carrier or motive fluid and it has been shown that the mixture is well described by non-Newtonian fluid mechanics. High concentration coal-water mixtures produced by premixing a carefully sized coal suspension appear to behave as a Bingham plastic in laminar flow or at least in a shear rate domain ranging up to a value of about 100s^{-1}. This discovery by Elliott and Gliddon, though unconfirmed elsewhere, allows the designer to use well established non-Newtonian theory for the prediction of energy loss and for predictive scale-up.

Sufficient experimental work has been given in pilot-plant test rigs to enable the practitoner to obtain reassurance that high concentration transportation is possible at modest pressure drop with the obvious associated advantages of improved dewatering properties and a suitable product for power station use.

Let us analyse a typical case study involving the transportation of a stabilised coal slurry containing 60% by weight (52% by volume) coal particles in water at the rate of 64 Te/hr coal over a distance of 2.5km. Assuming laminar flow, i.e. 8V/D is less than 100s^{-1}, then a typical pipeline diameter of 0.25m is realistic using a slurry velocity of 0.5ms^{-1}. The steady-state hydraulic gradient can be calculated using the equation and constants given by Elliott and Gliddon (see equ.37).

$$i_T = K' \left(\frac{8V}{D}\right)^{n'} \left(\frac{4}{D\rho g}\right)$$

where $K' = 8.323$kg m^{-1} s^{-1}
$n' = 0.167$

The slurry density is given by the expression

$$\rho_{MB} = \left\{(s-1)C_V + 1\right\} \rho_L$$
$$= \left\{(0.4)(0.52)+1\right\} 1000$$
$$= 1207\text{kg m}^{-3}$$

Given that D = 0.25m and V = 0.5ms^{-1}

$$i_T = 8.323 \left\{\frac{(8)(0.5)}{(0.25)}\right\}^{0.167} \left\{\frac{4}{(0.25)(1207)(9.81)}\right\}$$

$$= \underline{1.79 \times 10^{-2}} \quad \text{m of slurry/m pipeline}$$

The total pressure drop (ΔP) along a 2.5km pipeline is

$$\Delta P = (1.79 \times 10^{-2}) \, (2500) \, \underline{(1207)} \text{m } \underline{\text{water}}{1000}$$

$$= 53.92 \text{ m water}$$
$$= \underline{5.29 \text{ bar}}$$

It is difficult to generalise these calculations and compare them with actual results, though the derived hydraulic gradient given above is approximately equal to a pressure gradient of 10 psi/1000ft in the units given by Elliott and Gliddon. A typical set of results obtained in a 0.25m (10") diameter pipeline are given in Fig.6 taken directly from Elliott and Gliddon's work. It is not surprising that this data point would fit comfortably on the curves at V = 0.5ms^{-1} (= 1.64 ft s^{-1}).

Scale-up, i.e. the effect of throughput and pipe diameter can be anticipated using the simplified form of the Metzner-Reed equation. If we assume that the shear rate (8V/D) must remain below 100s^{-1} to maintain laminar flow, then this value can be substituted into equ.37 to give a limiting value of i_T as a function of D. Obviously, if shear rate is held constant (at 100s^{-1}) then,

$$i_T = K' \left(\frac{8V}{D}\right)^{n'} \left(\frac{4}{D\rho g}\right)$$

$$i_T \propto \left(\frac{4}{D\rho g}\right)$$

The variables are ρ (or ρ_{MB}) the mixture density and D. For a given slurry, say 60% by weight (52% by volume) then $\rho = \rho_{MB} = 1207$ kg m^3 and hence

$$i_T = \frac{6.07 \times 10^{-3}}{D}$$

The hydraulic gradient is therefore inversely proportional to the pipeline diameter. This observation is borne out by Elliott and Gliddon in work carried out in 1½", 2½", 4" and 10" diameter pipelines.

This result is particularly important, since the case study involving dense-phase flow of a sliding bed whereby energy dissipation is largely attributed to solid-solid sliding friction does not show this strong dependence on pipeline diameter, especially at relatively low velocities. It is obviously advantageous to prepare a homogeneous suspension which will behave as a non-Newtonian fluid, as opposed to the case study of rigid particles in contact with and sliding at the pipe wall. The latter frictional mechanism greatly enhances the dissipation of energy.

Let us calculate the specific energy consumption (E_s) for the transportation of a stabilised coal slurry containing 60% by weight coal particles in water as discussed above. The determination of E_s has been given by Streat (37) and is obtained directly from the expression

$$E_s = \frac{i_T g}{s \, C_v} \quad \text{(J kg}^{-1} \text{ m}^{-1})$$

or in more useful units

$$E_s = 0.2778 \quad \frac{i_T g}{s \, C_v} \quad \text{(kWh Te}^{-1} \text{ km}^{-1})$$

In this case study, we have calculated that

$$i_T = 1.79 \times 10^{-2} \text{ m of slurry/m pipeline}$$

$$= 2.16 \times 10^{-2} \text{ m of water/m pipeline}$$

$$E_s = \frac{(0.2778) \, (2.16 \times 10^{-2}) \, (9.81)}{(1.4) \, (0.52)}$$

$$= \underline{8.09 \times 10^{-2}} \quad \text{kWh Te}^{-1} \text{ km}^{-1}$$

The specific energy consumption for a stabilised pre-mixed coal suspension (0.08 kWh Te^{-1} km^{-1}) is extremely low and therefore this technique shows great promise for long distance hydraulic transporation.

A comparable calculation is possible for dense phase flow of a coal-water slurry containing coarse material at a weight fraction of 0.6. Using the simplified form of equation 14 and using only the solid-solid frictional terms (a reasonable assumption at low velocity) we can obtain an approximate value of i_T, i.e.

$$i_T = 2 \, f_s \, (s-1) \, C_t$$

Assuming a value of f_s = 0.32 (based on some independent measurements with coal) and s = 1.4, we obtain:

$$i_T = 2 \, (0.32) \, (0.4) \, (0.52)$$
$$= 0.13 \text{ m water/m pipeline}$$

$$E_s = \frac{0.2778 \, (0.13) \, (9.81)}{(0.52) \, (1.4)} = \underline{0.49 \text{ kWh Te}^{-1} \text{ km}^{-1}}$$

This value is approximately six times greater than the value obtained for a pre-mixed stabilised coal suspension assuming that pipeline diameter, flow velocity, etc. are the same. It follows, therefore, that dense-phase flow is only feasible over relatively short distances (less than 1 km in horizontal flow).

A comparison of stabilised flow and dense-phase flow leads to the following conclusions:

Stabilised Flow

Advantages

1. very high solids concentrations are possible e.g. coal-water mixtures can be transported at concentrations suitable for direct combustion. Backfill can be transported at a concentration suitable for direct placement.

2. a very low specific energy consumption is obtained in the laminar flow regime. Long distance transportation is therefore theoretically feasible.

3. stabilised flow techniques will minimise the problems of dewatering solids at the pipeline outlet.

Disadvantages

1. coal preparation is tedious and energy intensive requiring careful grinding, milling and pre-mixing. The associated capital and operating costs are not inconsiderable on a large throughput plant.

2. the selection of slurry pumps is still the subject of major project development. Operational difficulties and associated capital and operating costs are not yet fully optimised.

Dense-phase Flow

Advantages

1. high solids concentrations are possible with coarse particles. Run-of-the-mill material, requiring no special preparation can be handled. The method is suitable for materials such as aggregate, gravel, etc., which cannot normally be comminuted.

2. the specific energy consumption is acceptable for relatively short distance transportation. Energy loss is not strongly dependent on pipeline diameter.

3. indirect methods of feeding can be used and the solids-water mixture does not normally pass through a pump. Wear, attrition, etc. in pumps is greatly reduced.

Disadvantages

1. indirect feeding requires cyclic operation - this requires intermittent operating valves, timers, etc. Reliability of this equipment has yet to be evaluated.

2. pressure surges can be encountered in cyclic operation.

Acknowledgements

The author wishes to acknowledge the painstaking work of several former research students without whose valuable contribution this paper could not have been written.

References

1. Cloete, F.L.D., Miller, A.I. and Streat, M., 'Dense phase flow of solids-water mixtures through vertical pipes', Trans.Instn.Chem.-Engrs., 45, T392, 1967.

2. Bantin, R.A., and Streat, M., 'Dense phase flow of solids-water mixtures in pipelines', BHRA, Hydrotransport 1, G1-1, 1970.

3. Wilson, K.C., Streat M., and Bantin, R.A., 'Slip-model correlation of dense two-phase flow', BHRA, Hydrotransport 2, B1-1, 1972.

4. Bantin, R.A., and Streat, M., 'Mechanism of hydraulic conveying at high concentrations in vertical and horizontal pipes', BHRA, Hydrotransport 2, B2-11, 1972.

5. Streat, M., Televantos, Y., Carleton, A.J., 'Pilot-plant studies of hydraulic conveying of coarse materials at high concentration in pipelines', BHRA, Hydrotransport 4, F2-21, 1976.

6. Carleton, A.J., French, R.J., James, J.G., Broad, B.A. and Streat, M., 'Hydraulic transport of large particles using conventional and high concentration conveying', BHRA, Hydrotransport 5, D2-15, 1978.

7. Televantos, Y., Shook, C.A., Carleton, A.J., and M.Streat, 'Flow of slurries of coarse particles at high solids concentration' Can.J.Chem.Eng. 57, 225, 1979.

8. Wilson, K.C., Brown, N.P., and Streat M., 'Hydraulic hoisting at high concentrations: A new study of friction mechanisms', BHRA, Hydrotransport 6, 269, 1979.

9. Streat, M., 'A comparison of energy consumption in dilute and dense phase conveying', BHRA, Hydrotransport 8, 111, 1982.

10. Calvert, S., and Miller, R.A. 'Flow of dense pastes in vertical tubes', Ind.Eng.Chem. 50, (12), 1793, 1958.

11. Gay, E.C., Nelson, P.A., and Armstrong, W.P., 'Flow properties of suspensions with high solids concentration', A.I.Ch.E.J. 15, (6), 815, 1969.

12. Elliott, D.E., and Gliddon, B.J., 'Hydraulic transport of coal at high concentrations', BHRA, Hydrotransport 1, G2-25, 1970.

13. Petuit, P., 'Application of stabilized slurry concepts of pipeline transportation of large particle coal', Proceedings of Slurry Transport Association 3, Las Vegas, March 1978.

14. Duckworth, R.A., 'Hydraulic transport of coal', Proceedings international conference on bulk materials storage, handling and transportation, Newcastle, Australia, August 1983.

15. Brookes, D.A., and Dodwell, C.H., 'The economic and technical evaluation of slurry pipeline transport techniques in the international coal trade', BHRA, Hydrotransport 9, 1, 1984.

16. Navrade, D.H., and Klose, R., 'Densecoal pilot plant for production and transportation of highly concentrated coal fuels', BHRA, Hydrotransport 9, 47, 1984.

17. Lagana, V., Ercolani, D., Prassone, M., and Vercellotti, C., Snamprogetti's high coal concentration slurry preparation plant', BHRA, Hydrotransport 9, 77, 1984.

18. Zenz, F.A., and Othmer, D.F, 'Fluidization and fluid-particle systems', Reinhold, New York, 1960.

19. Sweet, C.T., British Patent 706, 257, 1951.

20. Kawamichi, K., Japan Patent 1863, 1953.

21. McCausland, J.W., US Patent 2,723,949, 1955.

22. Wilcox, M.J., US Patent 2,520,983, 1950.

23. Evans, J.E., US Patent 2,661,322, 1953.

24. Hancher, C.W., and Jury, S.H., 'Semicontinuous countercurrent apparatus for contacting granular solids and solution', Chem.Eng. Prog.Symp.Series, 55, (24), 87 1959.

25. Delaplaine, J.W., 'Forces acting in flowing beds of solids', A.I.Ch.E.J., 2, (1), 127, 1956.

26. Brandt, H.L., and Johnson, B.M., "Forces in a moving bed of particulate solids with interstital fluid flow', A.I.Ch.E.J., 9, (6), 771, 1963.

27. Verkerk, C.G., 'Hydraulic transportation as applied to backfilling in the South African mining industry', BHRA, Hydrotransport 9, 281, 1984.

28. Verkerk, C.G., 'Transport of fly ash slurries', BHRA, Hydrotransport 8, 307, 1982.

29. Newitt, D.M., Richardson, J.F., Abbot, M. and Turtle, R.B. 'Hydraulic conveying of solids in horizontal pipes', Trans.Insn. Chem.Eng. 33, 93, 1955.

30. Wilson, K.C., 'Slip point of beds in solid-liquid pipeline flow', Proc.A.S.C.E. 96, No.HY-1, Jan 1970.

31. Bagnold, R.A., 'Anto-suspension of transported sediment; turbity currents', Proc.Roy. Soc.Land.A., 265, 1322, 1962.

32. Metzner, A.B. and Reed J.C., A.I.Ch.E.J. 1, 434, 1955.

33. Televantos, Y., "The flow mechanism of hydraulic conveying at high solids concentration", Ph.D. Thesis, Imperial College, 1977.

34. Sauerman, H. "The influence of particle diameter on the pressure gradients of gold slurries pumping", BHRA Hydrotransport 8, 241, 1982.

35. Horsley, R.R., "Viscometer and pipe loop tests on gold slime slurry at very high concentrations by weight, with and without additives", BHRA Hydrotransport 8, 367, 1982

36. Schriek, W., Smith, L.G., Haas, D.B., and Husbund, W.H.W, "Experimental studies on the hydraulic transport of coal", Sashatchewan Research Council, Report E73-17, 1973.

37. Streat, M., 'A comparison of specific energy consumption in dilute and dense phase conveying of solids-water mixtures' BHRA Hydrotransport, 8, 111, 1982.

Table 1: Slurry Pipeline Installations (Courtesy of Bechtel)

System	LOCATION	LENGTH		OUTSIDE DIAMETER		CAPACITY (10^6 Metric Tons/Year)	YEAR IN OPERATION
		(mile)	(km)	(inch)	(mm)		
COAL							
1. Black Mesa	United States	273	439	18	457	4.5	1970
IRON CONCENTRATE							
2. Savage River	Australia	53	86	9 5/8	244	2.3	1967
3. Waipipi	New Zealand	1.5	2.4	12 3/4	324	1.0	1971
4. Pena Colorado	Mexico	30	48	8 5/8	219	1.8	1974
5. Las Truchas	Mexico	17	27	10 3/4	273	1.5	1976
6. Samarco	Brazil	247	398	20	508	12.0	1977
7. Sierra Grande	Argentina	20	32	8 5/8	219	2.1	1979
8. Kudremukh	India	41	66	18	457	7.5	1980
COPPER CONCENTRATE							
9. Bougainville	Papua, New Guinea	17	27	6 5/8	168	0.9	1972
10. West Iran	Indonesia	70	113	4 1/2	114	0.3	1973
11. Pinto Valley	United States	11	18	4	102	0.4	1974
12. Kenn.Chino	United States	7	11	5 9/16	141	0.7	1982
LIMESTONE							
13. Calaveras	United States	17	27	7 5/8	194	1.4	1971
PHOSPHATE							
14. Valep	Brazil	74	119	9 5/8	244	2.0	1979
15. Chevron	United States	96	154	10 3/4	273	2.6	1985*
TAILINGS							
16. Western Mining	Australia	4	6	4 1/2	114	0.1	1971
17. Kenn.Chino	United States	9	14	3x16	3x416	12.0	1982

*In design.

TABLE 2

Values of Flow Indices for a 60% by Wt.
Concentration
Slurry of Mixture C (Fig.4C).

Apparent Shear Rate (sec^{-1})	Shear Stress lbs/ft sec^2	n'	K' lbs/ ft sec.
25	9.5	.167	5.56
50	10.7	.167	5.56
75	11.5	.167	5.60
100	12.0	.223	4.29
150	14.5	.334	2.35
200	14.5	.344	2.35
300	17.0	.445	1.34
400	19.5	.448	1.06
500	22.0	.488	1.06
600	24.3	.545	0.747
800	34.0	.625	0.525
1000	41.0	1.058	0.0275

*(Note: $1 \text{ lb ft}^{-1} \text{ s}^{-1} = 1.486 \text{ kg m}^{-1} \text{ s}^{-1}$)

TABLE 3

Bed Deposit Angles, θ

Velocity ms^{-1}	C_t	C_b	Angle of deposit, θ, rads.
0.18	0.59	0.59	π
0.9	0.505	0.55	2.39
1.3	0.485	0.55	2.28
2.6	0.470	0.55	2.21

TABLE 4

Hydraulic Gradient Prediction Using Eqn.11

Velocity ms^{-1}	1st Term	2nd Term	3rd Term	4th Term	i_T Predicted	i_T Experimental
0.18	0.597	0.002	--	--	0.6	~2.0
0.9	0.430	0.022	0.005	0.003	0.46	0.59
1.3	0.397	0.028	0.013	0.011	0.449	0.46
2.6	0.375	0.072	0.042	0.032	0.521	0.66

TABLE 5

Hydraulic Gradient Prediction Using Eqn.14

Velocity ms^{-1}	Eqn.14	i_T Experimental result	% Error
1.5	0.543	0.55	1.3
2	0.59	0.59	--
2.5	0.635	0.65	2.3
3.0	0.714	0.72	0.9
3.5	0.76	0.74	2.8

(a)   sliding bed regime

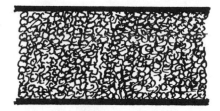

(b)   dense-phase moving bed

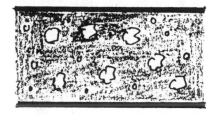

(c)   stabilised flow

Fig. 1.   Schematic representation of dense-phase flow.

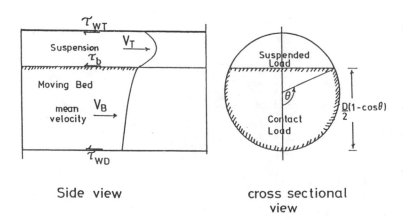

Side view   cross sectional
                                  view

Fig. 2.   The bed-slip model.

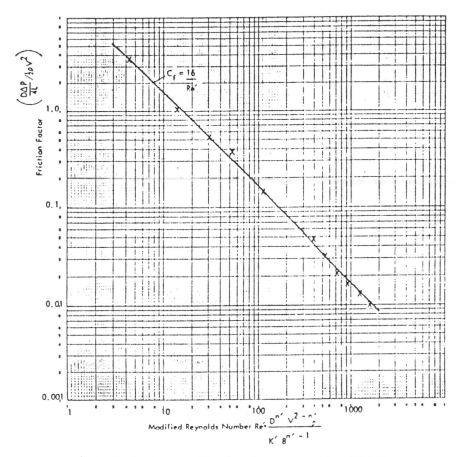

Fig. 3. Friction factor-modified Reynolds number plot (see Table 2).

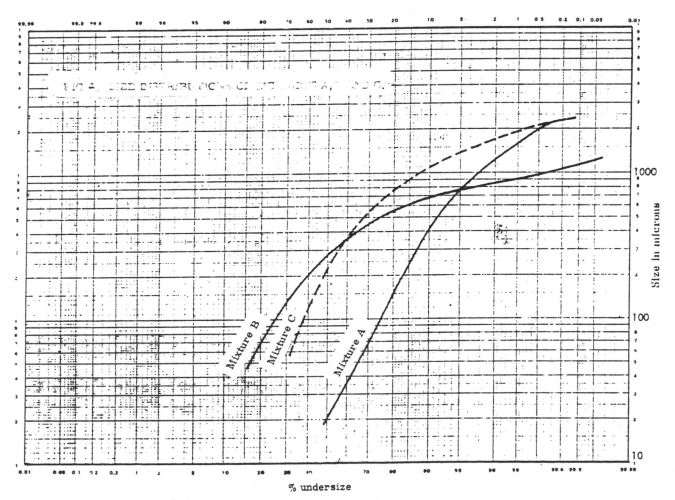

Fig. 4. Partial size distribution for $\frac{5}{16}''$ 4 mesh gravel.

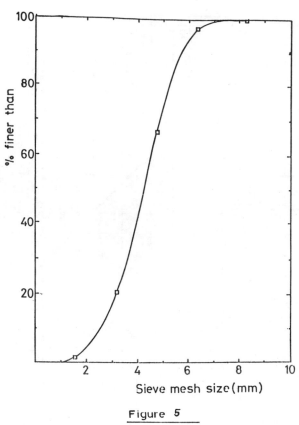

Figure 5

Particle Size Distribution
for ⁵/₁₆″–4 mesh Gravel

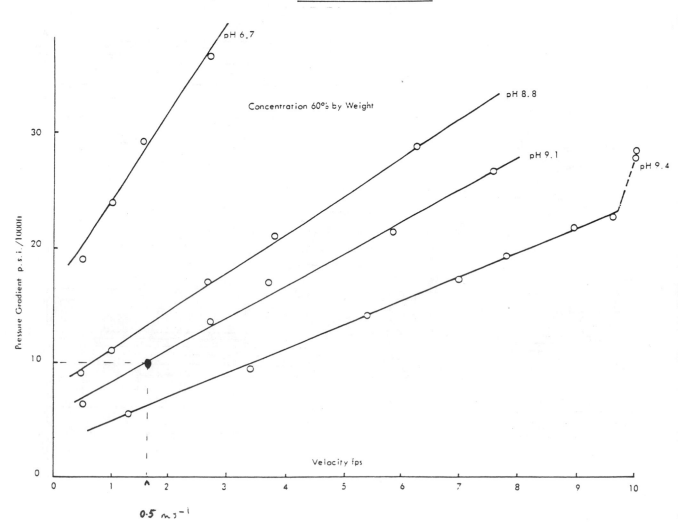

Fig. 6.   Pressure drop velocity curves in 10″ diameter pipeline loop.

10th International Conference on the
Hydraulic Transport of Solids in Pipes

HYDROTRANSPORT 10

Innsbruck, Austria: 29-31 October, 1986

PAPER B2

COARSE COAL HIGH CONCENTRATION SLURRY TRANSPORTATION

C G VERKERK

UNIVERSITY OF THE WITWATERSRAND TECHNOLOGY CENTRE, JOHANNESBURG, SOUTH AFRICA

SYNOPSIS

Hydraulic conveying of dense phase slurries is finding new applications in the mining industry in South Africa. One of the latest applications is the implementation of a 160 tonnes/hr high concentration coarse coal slurry pipeline.

The paper outlines some of the experimental results obtained from test work conducted in a 120 mm diameter pipeline on slurries with concentration C_w in the range of 73 to 85% and a coal top size of 30 mm.

1. INTRODUCTION

Conventional fine coal slurries have been used since the late 1950's with the most well known pipeline using this technology being the Black Mesa pipeline system. This system transports coal particles less than 1mm at concentrations of C_w = 50% at relatively high velocities.

Coarse coal, up to 150 mm top size, has been pipelined at concentrations of about C_w = 35%. Concentrations of C_w = 60% have been pumped with a smaller top size of 50 mm (Ref.1).

The problems of transporting coarse coal slurries, ie. high velocities, risk of blockages etc. have been overcome by the advent of stabilized slurry systems which use a fine coal as a carrier to prevent the coarse coal from settling. Research by Duckworth (1983) at the CSIRO demonstrated that C_w = 70% slurries could be successfully pumped with low energy consumptions. Subsequent work by BP with BHRA has indicated that stabilized slurries could be transported at C_w's in the range of 67% - 83% without significant settlements.

The stacking of coal at a South African mine posed several problems. To provide a possible solution, the use of a stabilized coal slurry for the transportation stage was considered. This report outlines the research conducted to determine the pumpability, stability and performance characteristics of a high concentration slurry formed from the coarse coal and dilute fine coal slurry available at the mine (Ref 4,5).

2. TEST FACILITY LAYOUT AND DESCRIPTION

The layout of the test facility utilised in this programme is illustrated in Fig. 1. As is shown a closed loop system was utilised to ensure consistency of the material in the system and to minimise the material requirement.

The test facility encompasses a series of pipe racks and horizontal sections of 120 mm I.D pipe, to facilitate the generation of sufficient back pressure to simulate longer length pipelines and to accurately measure the pressure losses over a straight length of pipe.

Two pumps were used in this test series, a Schwing KSP 80 and a Putzmeister KOD 2180. Both pumps are twin cylinder single acting positive displacement pumps and both pumps utilise a kidney shaped transfer valve between the pumping cylinders. The basic operation of the valve arrangement is illustrated in Fig. 2. These large pumps were tested to determine their suitability for high concentration coal transportation in the commercial applications that they were earmarked for.

The instrumentation in the facility was selected so that pressures, concentration temperature and flowrates could be continuously monitored. Pressures in the system were monitored with pressure transducers while a magnetic flow meter measured the average flowrates through the system.

3. MATERIAL DESCRIPTION

The coal utilized in the test programme consisted of a coarse coal and a fine coal slurry. Size distributions of the two components are illustrated in Fig. 3. The two components were mixed in various ratios so that the effect, of the ratio of fines to coarse material and the overall solids concentration, on the pressure gradients could be established. The range of size distributions tested were based on the supply ratios of the slurry and coarse coal fraction and curves developed for transport of aggregate systems for the manufacture of mortar and concrete. Furthermore maximum packing in the matrix to ensure a stable mixture was considered prior to the definition of the boundaries of the size distribution envelope tested. The total range of size distributions tested is illustrated in Fig. 3.

4. TEST RESULTS

This transportation method is essentially based on a two phase concept with the first phase being the carrier composed of the fines and water and the second the coarse material or heterogeneous component. For the purposes of this test all material below 212 um was considered to be the fines component. This was based more on convenience from a sieving point of view rather than a rigorous approach to determine the fines cut off particle size and from a size point of view was found to be consistent with the approach adopted by BP Coal (Ref. 3).

At high concentrations the viscosity of the mixture is controlled by the particle grading and the concentration of the fines component. As the particle gradings were fixed, variations in viscosity and pumping pressures were achieved by changing the ratio of fines to coarse material on a mass basis and by varying the concentration of the total mixture.

The total percentage of fines on a mass basis was varied from 16% to 41% and the mixture concentrations were varied from $C_w = 72\%$ to 85%. The minimum fines value was dictated by the minimum packing requirements. Below this value a solid coarse material matrix resulted that often "hung up" in the pump hopper and resulted in blockages in the pipeline. The upper boundary was established to determine a correlation between the pressure gradient and the percentage of fines in the system.

The minimum and maximum concentration tested were defined by the range of quantities of fine slurry envisaged by the mine. This range resulted in a predicted variation of the total mixture $C_w = 78,3\%$ to 87%.

The test work soon established that the temperature of the slurry contributed to the system pressure gradient. To determine the effect, the slurry was circulated around a short pipeline, approximately 130 m in length at a constant flowrate and concentration while the temperature and pressure gradient were continuously monitored. A pump outlet pressure versus temperature curve of a test is illustrated in Fig. 4 ($C_w = 81,5\%$ and flowrate of 40 m^3/hr). Maximum slurry temperature for these tests was limited to 40°C. As is illustrated by the relationship between the pressure and the temperature in Fig. 4 there is an increase in pressure loss in the system with an increase in temperature. All the other tests conducted displayed a similar characteristic.

A variation in the percentage fines in the system also resulted in a change in the pressure losses in the system. An indication of the relationship is illustrated in Fig. 5. A general relationship of increasing pressure loss with increasing fines contribution is illustrated. A partial explanation for this phenomenon results from the increasing concentration and viscosity of the carrier as the quantity of coarse material is reduced for mixtures having similar overall concentrations. By decreasing the quantities of larger size material the wetting surface area of the total mixture increases resulting in less water being available to fulfill a lubricating function.

The relationship between the pressure gradient and the flowrate appears to be linear, (Fig. 6), indicating a Bingham type fluid. The actual slope of the curve for a slurry flowrate flowing at constant temperature and concentrations would be below the curves plotted in Fig.6. These pressure gradient figures were obtained from the short pipeline tests (130m) in which a large temperature variation was experienced when varying the flowrates from minimum to maximum. The temperature profile of each curve is therefore also included in Fig. 6.

To improve the pressure gradient responses of these high concentration slurries an additive was added to the mixture in one of the tests.

The polymer VERSA TL4 was added (0,32% on a wet mass basis to a mixture with $C_w = 72,4\%$). The visual changes of the paste as it leaves the conveying pipe before re-entering the storage hopper are illustrated in Fig. 7. The paste with the additive appeared to flow more like a fluid.

One anomaly from the additive test run is an apparent increase in solids concentration after the addition of the additive. Prior to the addition of the additive $C_w = 72,4\%$ and after addition $C_w = 79,9\%$. A subsequent test without further additions revealed a similarly higher concentration and also that there appeared to be little decaying of the viscosity modifying characteristics of the additive. It would appear that this results from a changing input ratio of fine to coarse material to the pipeline from the storage hopper. Size distribution tests of the delivered slurry prior to and after the addition of the additive revealed a decrease in the fines component from 35% to 19%.

The addition of the additive resulted in a decrease in the system pressure loss (Fig. 8), care should however be exercised when using viscosity modifiers because if the yield stresses and viscosities are reduced to such an extent that sedimentation and segregation occur then blockage problems can occur.

The mixture stability was determined from tests in which the material was left stationary in the pumping circuit for periods of up to 10 days. The term stability in this case refers to resistance of the mixture to sediment out in the pipeline and the pump hopper and hence the ease of starting up the pumping system after a shutdown period.

The extended shut down tests indicated virtually no sedimentation and no problems in start-up of the system. Sedimentation was determined from the extent of the layer of water forming on the surface of samples taken from each test run. In most case no visible layer would form and in the extreme case only a fine layer of water would appear on the surface over the periods of shut down.

A further aspect of high concentration coal transportation revealed in the tests is the importance of wetting the internal surface of the pipeline before the mixture is transported. If the coal mixture is pumped without wetting, the changing characteristics of the annulus flow of the initial plug as the pipe is wetted results in the formation of a dry plug at the end of the slurry column. This is illustrated in exaggerated form in Fig.9. These dry plugs result in higher pressures being required for start-up then for transportation once the pipeline has been filled. Wetting can be achieved by initially pumping water or a dilute slurry through the pipeline. Care must however be taken with the interface between the dilute wetting slurry and the high concentration paste to prevent segregation. If the two faces are not separated the dilute slurry tends to dilute the leading plug of coarse and fine material resulting in the coarser material dropping out of suspension and possible blockages.

5. CONCLUSIONS

The testwork indicated that coarse coal mixtures up to $C_w = 85\%$ could be successfully transported. High stability of the slurries were recorded and provided that the minimum fines (16%) were present in the mixture no blockages

occurred in the system. Temperature influences the pressure gradient and for longer pipelines the paste temperature could increase along the pipeline and this effect will need to be accounted for at the design stage.

The successful conclusion of the testwork has resulted in the installation of a pumping system at a mine which is currently undergoing commissioning trials.

ACKNOWLEDGEMENTS

The author wishes to acknowledge the assistance provided by the Councils of the University of the Witwatersrand, Goldfields South Africa (Pty) Ltd, Putzmeister S A (Pty) Ltd and Westinghouse Bellambie (Schwing) and for their permission to publish this paper.

REFERENCES

1. R Gillies et al. Coarse Coal in Water Slurries - Full Scale Pipeline Tests, SRG Publication E-725-10-C-82/1982.

2. Duckworth R A et al. Hydraulic Transport of Coal. Proc International Conference on Bulk Storage Handling and Transportation, Newcastle, Australia, August 1983.

3. Brookes D A et al. The Economic and Technical Evaluation of Slurry Pipeline Transport Techniques in the International Coal Trade. 9th International Conference on Hydraulic Transport of Solids in Pipe, BHRA, Cranfield, Bedford (1984).

4. Verkerk C G, Putzmeister KOD 2180 Pump Evaluation. Project Report. University of the Witwatersrand Technology Centre, Johannesburg, 1986.

5. Verkerk C G. Civils Discard Pump Evaluation. Project Report, University of the Witwatersrand Technology Centre, Johannesburg, 1986.

58

FIG.1: TEST FACILITY LAYOUT

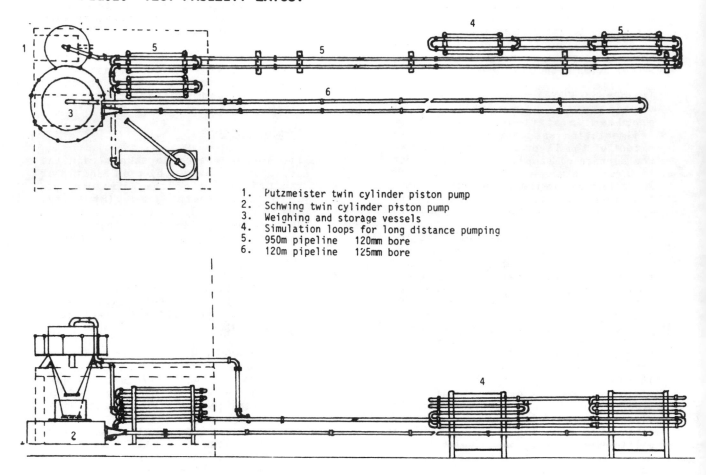

1. Putzmeister twin cylinder piston pump
2. Schwing twin cylinder piston pump
3. Weighing and storage vessels
4. Simulation loops for long distance pumping
5. 950m pipeline   120mm bore
6. 120m pipeline   125mm bore

FIG.2: VALVE OPERATION

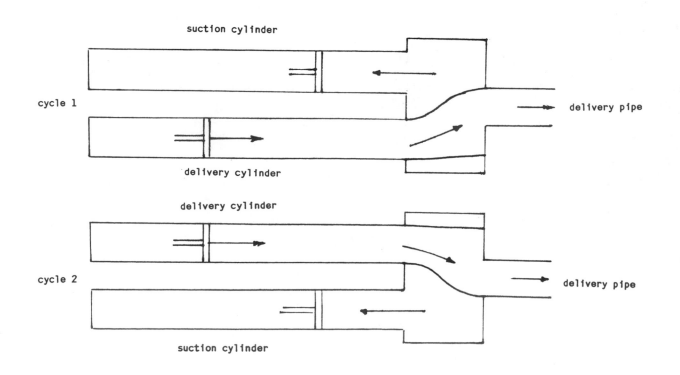

suction cylinder

delivery pipe

cycle 1

delivery cylinder

delivery cylinder

delivery pipe

cycle 2

suction cylinder

FIG.3: COMPONENT AND MIXTURE SIZE DISTRIBUTIONS

1   mixture envelope

2   coarse coal

3   fine coal

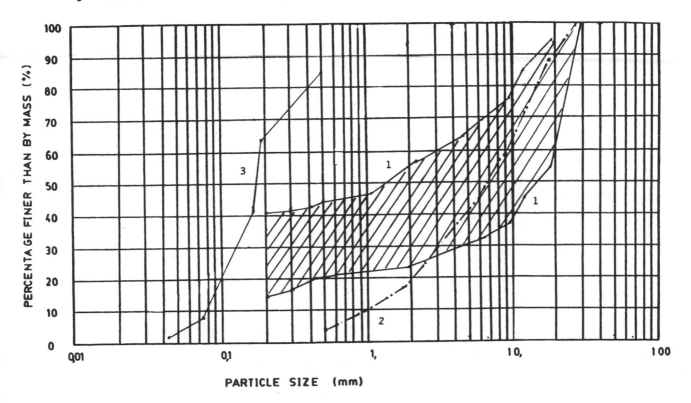

FIG.4: TEMPERATURE INFLUENCE ON SYSTEM PRESSURE LOSS

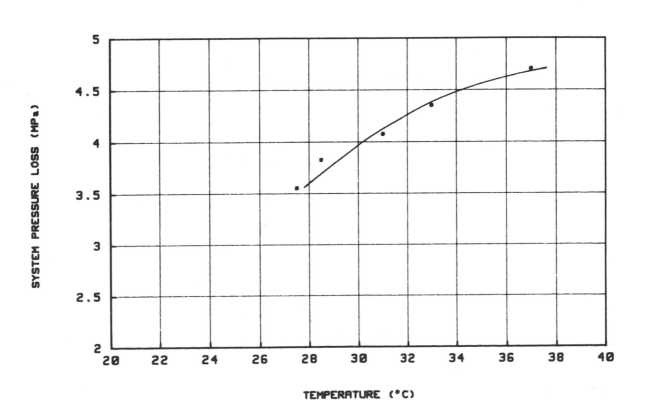

FIG.5:   INFLUENCE OF FINES CONTRIBUTION ON SYSTEM PRESSURE LOSS

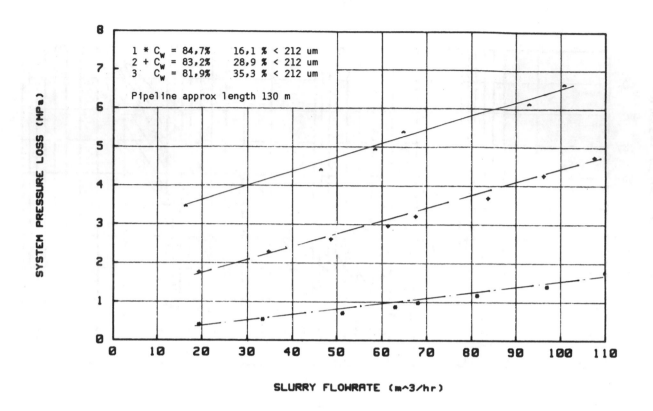

FIG.6:   TYPICAL PRESSURE GRADIENTS

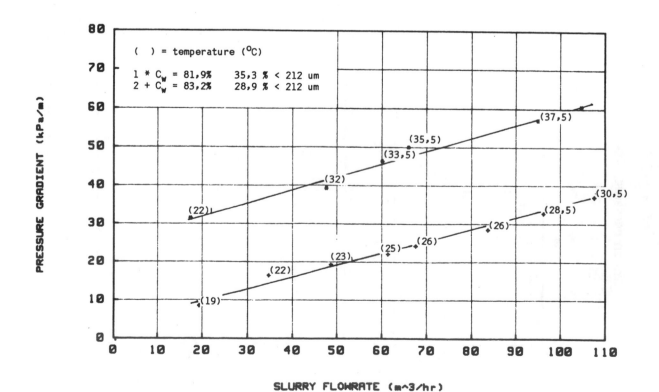

FIG.7: VISIBLE FLOW CHARACTERISTIC CHANGES

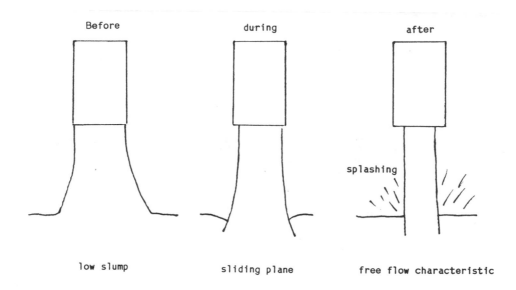

FIG.8: SYSTEM PRESSURE LOSSES WITH ADDITIVE

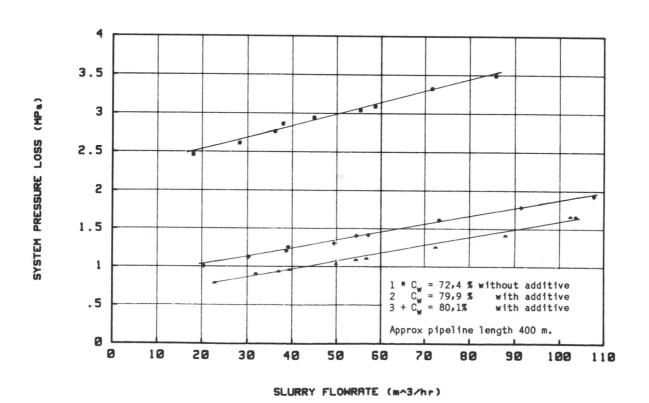

FIG.9: CHANGING PLUG FLOW CHARACTERISTICS ON PIPELINE FILLING

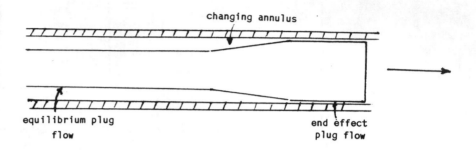

10th International Conference on the
Hydraulic Transport of Solids in Pipes

HYDROTRANSPORT 10

Innsbruck, Austria: 29-31 October, 1986

PAPER C1

DEVELOPMENT OF THE ASEA MINERAL SLURRY TRANSPORT SYSTEM FOR COARSE COAL

The authors are Ajay Bhattacharyya, ASEA-ATOM, Sweden and Ian Imrie, ASEA Pty Limited, Australia

1.0 INTRODUCTION

The ASEA Mineral Slurry Transport (MST) system is a pipeline transportation concept, initially being developed for coking and steaming coals. Topsize of the material to be transported is limited only by pipe diameter. For example, a 250 mm diameter pipe can be used to transport coal with a topsize of 50 mm.

The ASEA system is based on a novel pump station design and features

- low cost coal preparation and dewatering based on conventional equipment

- high coal concentration, typically 70 % by weight

- variable flow rate

- large particle size

- wide range of economically and technically viable applications.

- minor impact on the environment

2.0 BACKGROUND

The ASEA Group with headquarters in Västerås, Sweden, and active in many countries of the world is continuously involved in developing new technologies for effective production, transmission and utilization of energy.

ASEA-ATOM, a subsidiary of ASEA has been the prime supplier of nuclear power stations to Sweden and Finland. During the last fifteen years ASEA-ATOM has delivered eleven nuclear power plants with a combined electrical rating of 8500 MW. Approximately half of the electrical energy needs of Sweden are supplied by nuclear power plants.

The ASEA Group in general and ASEA-ATOM in particular has a strong tradition in developing system oriented products. As an example, it may be mentioned that ASEA-ATOM is the only supplier outside the USA of light water reactor power plants who has never been dependent on licensing arrangements with US suppliers. ASEA-ATOM has developed its nuclear technology on its own.

Late in 1970's when the need for new technologies for effective and safe utilization of coal became apparent, the ASEA Group embarked on various ambitious development projects. ASEA-ATOM over and above its heavy involvement in manufacturing nuclear fuel for and providing maintenance services to existing nuclear plants has looked for ways to utilize its inherent technical strength to develop new technologies for coal utilization.

Together with other subsidiaries of the parent ASEA Group, ASEA-ATOM has participated in pioneering work with Pressurized Fluidized Bed Combustion. Related to this development an effective feeding system was sought for loading coal into the pressurized fluidized bed combustor. It was at this stage the ASEA became aware of an Australian concept of a Rotary Ram Pump. The concept was demonstrated by the inventor in a rudimentary prototype with 25 mm cylinders. Recognizing the merits of the design ASEA-ATOM embarked on an extensive programme to develop a coarse coal transportation system.

The ASEA Group believes in developing new products in close cooperation with the end-users. Australia has a very dynamic coal industry. It was therefore very natural that the Australian subsidiary of ASEA became an active partner in this effort. Today, ASEA Pty based on its intimate knowledge of the coal mining industry is making important contributions towards developing a complete coarse coal transportation system.

A fullscale pump station with a capacity of 250 ton/hour is in operation in a test site in Australia. This size of the pump is well suited to handle the flow required for most applications.

3.0 MST SYSTEM

The purpose of the MST system is transportation of coal from one location to another location. It could be a situation where a totally new transportation system is needed on a virgin site. It could also be a retrofit case where the MST system is incorporated in an existing infrastructure.

Irrespective of the application, a primary design objective is that the system can be utilized without affecting the end-user's way of utilizing the delivered coal.

Main functional blocks of the MST system are shown below.

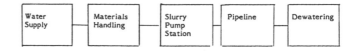

Important back-up functions are electrical power and control.

3.1 Slurry Pump Station

The heart of the MST system is the slurry pump station and the principal component in the pump station is the ASEA Rotary Ram Pump (RRP). Figure 1 shows the main features of the pump station and Figure 2 shows the basic action of the RRP. Dimensions and ratings given below refer to the pump station in operation in the ASEA test site in Australia.

The RRP consists of a cylinder barrel rotated by a 17,5 kW electrical variable speed drive and contained in a steel casing. The barrel contains four parallel 250 mm cylinders each containing a piston. When one of the four cylinders passes the slurry inlet port its piston is forced back by hydraulic action and the cylinder is charged with slurry. The barrel rotates and when opposite the discharge port high pressure water forces the piston to discharge the slurry from the cylinder into the pipeline. The hydraulic action for piston return is separate from the two main flow circuits i.e slurry flow and high pressure flow.

A large multi-stage centrifugal pump is driven by a 450 kW variable speed motor. This pump produces the high pressure water necessary for pumping the slurry. At one end of the barrel there are inlet and discharge ports for high pressure water. Inlet and discharge ports for the slurry are at the other end of the barrel.

Inlet and discharge ports on both the water and slurry sides overlap the cylinders in a way that gives a constant cross-sectional flow area. The cylinders work in pairs. When one cylinder is charging the opposite cylinder is discharging. As a result of the overlapping and the cylinder interaction the flow is continuous and smooth.

To reduce wear on the ports and to protect against clogging the RRP is lubricated and sealed hydrostatically with clean water. The water pressure is always greater than the slurry pressure. This ensures that the fine coal particles from the slurry are not forced into the pump.

The 250 mm pump station in operation in Australia has demonstrated that the concept works in full scale. At present development efforts are devoted to

- maintainability aspects
- higher capacity versions
- wear characteristics of RRP components

A Failure Mode and Effect Analysis (FMEA) has been performed on the slurry pump station. Purpose of the analysis has been to identify possible failure modes, both critical and non-critical. The effects on the pump station are decided for each of the possible failure modes.

The availability assumptions have been based principally on the failure modes identified in the FMEA. Estimated failure rates have been used for non-standard parts of the pump.

Availability figures have been calculated for various typical operational modes. The results show that the pump station can be expected to fulfill very adequately the expected requirements in a coal mining complex.

In Västerås, Sweden, a 50 mm pump installation and test loop have been established together with special test rigs to conduct detailed testing on

- design alternatives
- pump materials

Some of these tests are continuous in rigs designed to accelerate the interaction between wearing surfaces.

3.2 Material Handling

As mentioned earlier, a design objective of the MST system is that the system can be utilized without affecting the end-user's way of utilizing the delivered coal. The functional block "Materials Handling" is typically built of the following components

- Coal receiving
- Coal preparation
- Slurry Mixer
- Slurry Agitator

The ASEA test facility in Thornton is a case in point. A schematic diagram of the 250 mm plant is shown in Figure 3. The materials handling system is designed for 250 ton per hour of coal.

Coal with a top size of 50 mm is received by road. The supply is usually in 150 ton samples and held in stockpiles until testing commences. From the stockpile the coal is moved by a 3 m³ capacity front end loader and placed in a ground hopper.

Coal discharge from the ground hopper is controlled by a chain feeder with the feed rate set by the desired pipeline flow rate. The chain feeder is driven by a 55 kW variable speed hydraulic drive. Coal passes from the chain feeder onto a belt conveyor and is elevated to the mixer tank. Flow rate of coal on the conveyor is measured by a conveyor belt scale.

Water is added to the coal in the mixer at a rate set by the desired concentration and the flow signal from the belt scale. From the mixer the slurry passes into the feeder tank which is mounted on precision load cells. The slurry in the tank is agitated by a paddle driven by a variable speed hydraulic motor. The feeder tank can be isolated from the RRP by a gate valve which is normally closed when slurry pumping is not in progress. When pumping, slurry passes through the RRP and is discharged into the pipeline.

3.3 Water Supply

It is envisaged that the functional block Water Supply will typically consist of following components

- Raw water supply
- Raw water treatment
- Hydraulic water treatment
- Slurry water system

This part of the MST System is very application specific. In the Thornton test site, water for the slurry is obtained from the main water supply pond of the adjacent mine.

3.4 Pipeline

In the Thornton test site two test loops are available

- 130 meter for set up test
- 550 meter main loop

The pipeline with a wall thickness of 6.3 mm is laid above the ground on concrete supports. Pipe segments which are approximately 10 meters long are joined with Victaulic couplings. The profile of the main loop includes two eight meter rises and an eight meter fall. The final rise to the mixer tank is at a 15 degree slope. Bends have a seven meter radius.

Using a recirculating pipe loop system of 50 mm diameter, experimental determination of pipe wall erosion - corrosion rates of a number of different candidate materials has been made in Västerås, Sweden. The wear has been evaluated as an average wear rate and as a wear distribution around the circumference.

The initial tests were conducted in thousand hour segments and preliminary results confirm that wear can be kept within acceptable limits without going to costly pipe materials.

Further investigations are planned to establish the importance of slurry flow conditions to the wear.

3.5 Dewatering

Once again, this part of a MST plant is application specific. Basically, there is no difference between the slurry leaving the pipeline and wet coal normally handled in a conventional coal washery. At this time, no need is foreseen for developing special equipments for dewatering.

In Thornton site the pipeline discharge is arranged so that slurry can either be recirculated or handled straight through. In the "straight through" mode the slurry is discharged over static screens onto a concrete pod. Various combinations of operation can be accommodated such that coal concentration can be varied throughout a test program.

3.6 Electrical power and control

The MST system does not impose any special requirements on the electrical equipment. Standard equipment which fulfil local requirements such as mine site regulations, external power grid demands etc can be utilized.

Slurry flow rate can be typically turned down to 50 % design operation value.

At Thornton where ASEA MASTER programmable process control equipment is installed control algorithms are being tested for different coal types.

4.0 INTEGRAL PERFORMANCE

Efforts in Australia are devoted primarily to gain an understanding of the integral performance of a full scale MST system. ASEA Pty, the Australian subsidiary, has recruited specialists form the coal industry and other sources to complement their "in house" experience.

Also, ASEA Pty has a collaborative arrangement with the Commmonwealth Scientific Industrial Research Organisation of Australia (CSIRO) to study the behaviour of the slurry under disturbance conditions and to develop models to predict slurry behaviour.

Tests at the Thornton site have been conducted and are continuing to establish

- component suitability
- control algorithms
- slurry stability
- energy consumption and optimisation
- coal degradation
- mass dynamics

Steaming and coking coals from various sources in Queensland and New South Wales, both washed and unwashed, have been tested. Concentrations have ranged from 65 % by weight for a high ash steaming coal to 82 % by weight for a low ash coking coal.

All tests have so far been conducted without any special preparation of the coal i.e normal size distribution as a result of crushing to the 50 mm top size. Size distributions for some of the tested coals are shown in Figure 4.

The various control algorithms have ensured a slurry stability such that the slurry can be left in the pipeline for up to 5 days before restarting. No attempt has yet been made to establish the upper time limit on stability.

5.0 CONCLUSION

A total concept of coarse coal transportation is being developed and tested in full scale by ASEA. The ASEA Rotary Ram Pump has enabled coarse coal slurries to be pumped with high concentration. Both steaming and coking coals have been tested. No special equipment are needed for either coal preparation or dewatering. Conventional equipment used for coal preparation and dewatering are adequate.

These features mean that slurry transportation of coal is both economically and technically available to a greater market segment than previously considered feasible.

REFERENCES

1.   Imrie, I

      ASEA Mineral Slurry Transport System for Coarse Coal Transportation.
      Presented in 11th International Conference on Slurry Technology, Hilton Head, South Carolina (March 1986).

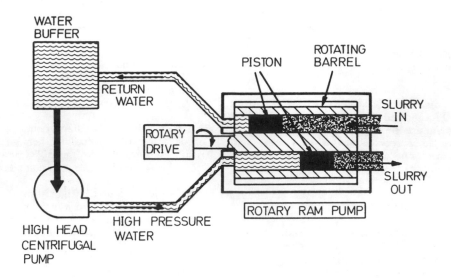

FIGURE 1   MST PUMP UNIT – MAIN FEATURES

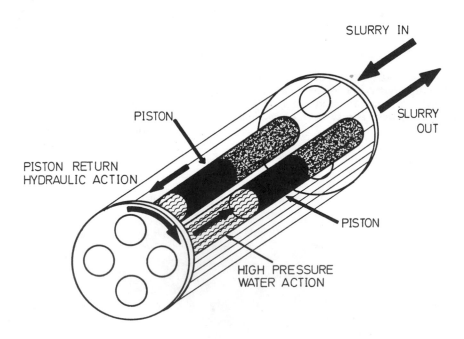

FIGURE 2   BASIC ACTION OF THE ROTARY RAM PUMP
(only two cylinders shown for clarity)

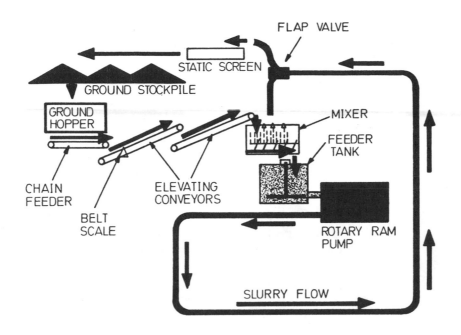

FIGURE 3   SCHEMATIC DIAGRAM — 250 MM PLANT THORNTON

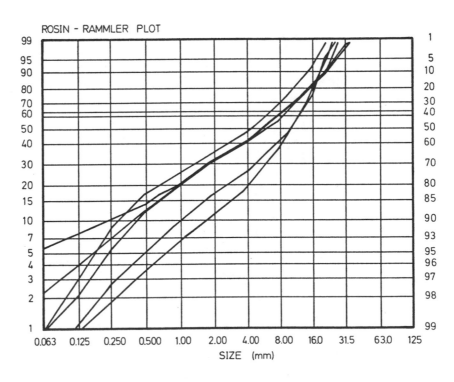

FIGURE 4   ROSIN - RAMMLER SIZE DISRIBUTION PLOT
EXAMPLES OF COALS PUMPED AT THORNTON

10th International Conference on the
Hydraulic Transport of Solids in Pipes

HYDROTRANSPORT 10

Innsbruck, Austria: 29-31 October, 1986

PAPER C2

THE PIPELINE TRANSPORT OF COARSE MATERIALS
IN A NON-NEWTONIAN CARRIER FLUID

By

R. A Duckworth, Consultant
11B Great Austins
Farnham, Surrey GU9 8JY
ENGLAND

L. Pullum
Research Scientist
Division of Mineral Eng.
C.S.I.R.O
P. O. Box 312
Clayton, Victoria
AUSTRALIA

Graeme R. Addie
GIW Industries, Inc
5000 Wrightsboro Road
Grovetown, Georgia 30813
U.S.A.

* C. F. Lockyear   ⎛ * Now with:
  Division of Mineral Eng. ⎜ B. P. Research Centre
  C.S.I.R.O.   ⎜ Sunbury on Thames
  P.O. Box 312   ⎜ U.K.
  Clayton, Victoria   ⎝
  AUSTRALIA

ABSTRACT

The transport of coarse materials by pipeline has, until recently, been restricted to relatively short distances, this being due largely to the high pressure gradients and specific energies required. Work by Duckworth et al in Australia has demonstrated that, by using a non-Newtonian carrier fluid, coarse coal having a top-size of 20-25 mm may be transported at acceptable pressure gradients and specific energies under laminar flow conditions at velocities of 1 to 2 m/s. This work has now been extended by the authors to the field of mine-waste disposal in which the material specific gravity may be as high as 2.65 (cf., S.G. = 1.4 for coal).

In this paper, the underlying physical concepts which make possible this mode of pipeline transport, the experimental studies and the analysis of the experimental data are discussed for both coal and mine waste slurries. Design equations for the prediction of velocity, pressure gradient and the specific energy associated with the transport of coarse materials in a non-Newtonian carrier fluid are given.

NOTATION

C_w	-	Concentration by mass
C_v	-	Concentration by volume
$\hat{d}$	-	Diameter of top sized particle
D_{50}	-	Geometric mean diameter
d_y	-	Yield diameter
D	-	Pipe diameter
E_s	-	Specific Energy
g	-	Gravitational acceleration
H_{ea}	-	Hedstrom number - $(\rho_m \tau_{ya} D^2 / n_a^2)$
i	-	Hydraulic gradient
L	-	Pipe length
M	-	Mass flow rate, Mass
M_c	-	Mass flow rate-coarse material
M_F	-	Mass flow rate-fine material
M^*	-	Mass flow ratio-M_c/M_F
N_{ya}	-	Yield number - $(\tau_{ya} D / n_a V_m)$
Q	-	Volumetric flow rate
S	-	Specific gravity
S^*	-	Specific gravity ratio-$((S_m-1)/(S_{mca}-1))$
V	-	Velocity
α_a	-	Shear stress ratio - (τ_{ya}/τ_o)
α_{ca}	-	Critical shear stress ratio-(τ_{ya}/τ_{oc})
η	-	Plastic viscosity of carrier
η_a	-	Apparent plastic viscosity of coarse slurry
η^*	-	Plastic viscosity ratio-(η_a/η)
γ	-	Friction factor
μs	-	Coefficient of sliding friction

ρ - Density

τ_o - Wall shear stress

τ_y - Yield stress

τ_{ya} - Apparent yield stress of coarse slurry

τ - Yield stress ratio - (τ_{ya}/τ_y)

Suffixes

ca - Carrier fluid

m - Slurry

s - Solid

w - Water

INTRODUCTION

The transport of coarse material by pipeline has been practiced for several decades, one of the earliest examples being in 1914 when coal from barges in the Thames was transported to Hammersmith power station. More recently there has been interest in the prospect of transporting coarse material such as coal over much longer distances, notably in cases where there is a need to transport coal from a mine to a port for export, Duckworth (1). However, for such transport to be practically feasible the pressure gradients and specific energies need to be considerably lower than those normally associated with the pipeline transport of coarse material in water. These latter considerations led Duckworth (2) to recommend that the work of Elliott & Gliddon (3) should be extended to provide the basis for the development of a pipeline technology for the transport of coarse material, notably coal.

In this paper, the authors discuss their work which shifts the emphasis from the need for a maximum packing density distribution, as recommended by Elliott & Gliddon (3), to the need for the carrier fluid to be a non-Newtonian fluid of the Bingham type. The coarse material pipeline technology, based on the use of such a carrier fluid, and currently referred to as stabilized flow, (stab-flow), following the use of the term by Lawler et al (4), is shown to be more generally applicable. This is illustrated by its application to both coarse coal and coarse mine waste.

PREVIOUS WORK

The restriction of coarse material pipeline transport to relatively short distances in current practice is quite understandable when it is realized that in order to transport such material reliably (i.e. without the danger of blockage), low concentrations, velocities of 4-5 m/s and pressure gradients of the order of 1.0 MPa/km (10 atmospheres/km) are required, these conditions leading to high wear rates and specific energies of 2 kWhr/Tonne km. These values are to be compared with velocities of 1.5 m/s, pressure gradients of 0.1 → 0.2 MPa/km (1 → 2 atmospheres/km) and specific energies of 0.05 kWhr/Tonne km in the case of fine coal slurry technology.

Fortunately, however, during the late 1960's, Elliott & Gliddon (3), during studies of the pipeline transport of 13 mm coal in water, discovered that it was possible to transport coal at low velocities and at relatively low pressure gradients with the concomitants of low wear rates and low specific energies. For this to be possible they concluded that:-
(i) the concentration should be of the order of 60% w/w,
(ii) fine coal should be present and be about 25 - 30% of the coal,
(iii) the size distribution should conform to the condition of maximum packing density. However, very marked differences in flow behavior between coals from different locations was observed, though not explained. Lawler et al (4) carried out some small scale experiments to confirm the findings of Elliott & Gliddon and concluded that there were practical advantages to be gained from this mode of transport for distances of up to 80 km.

In view of the conclusion of Elliott & Gliddon (3), concerning the importance of the packing density in relation to the static settling and dynamic behavior of coal/water slurries at high concentration, Duckworth & Pullum (5) carried out some qualitative experiments. In these experiments, two coal samples, one being so sized to accord with the condition of maximum packing density and the other being a standard sample from a coal preparation plant, the size distributions being as shown in Table I. The samples were mixed with water in a helical mixer to give similar concentrations by weight and as expected sample A settled quickly to yield supernatant water, whereas sample B showed no tendency to settle over several months. In a subsequent experiment the coal of less than 20 microns,

Table I

Size (μm)	Sample A % less than	Sample B % less than
10,000	79	84
1,000	8	40
500	3	28
20	-	7
% Voids	34	28

Sample A -   as received from a coal preparation plant - top size 20 mm

B -   after adjustment to maximum packing density with a top size of 20 mm and a lower size of 20 μm.

(i.e. 7%) was replaced by cement and the mixture allowed to solidify in a plastic cylinder.

Sections of solidified cylinder were cut horizontally and vertically and showed a very uniform distribution of the coarse material in the form of a matrix in the fine coal. The differing behavior of the two samples was considered to be due not to the size distributions per se, but due also to the rheological properties of the fine slurry formed from the particles of d < 500 μm and the water.

It thus seemed reasonable to suppose that the rheological properties arising from the mixture of fines and water were likely to play a major role in both the static and dynamic characteristics of highly concentrated coal/water mixtures and subsequent research work carried out by the authors placed great emphasis on the rheological properties of the carrier formed from fines and their interaction with the coarse material. In their paper, Duckworth & Pullum (5), proposed also that as in the case of concrete, in which a coarse aggregate is mixed with a fine sand and cement water slurry, the configurations adopted, because in their view it could not be otherwise, consists of a core of coarse material, the interstices of which are filled with fine slurry,

surrounded by a lubricating film of fine slurry. Subsequent, studies by Pullum, Lockyear et al (6), in which they used the geometry of the collimated beam of a nuclear density gauge to assess the density distribution at a pipe cross-section, showed that there was a tendency, at low velocities, for the coarse material to settle to form a loosely packed sliding bed.

A recognition of the extreme importance of the rheology of the carrier fluid on coarse material pipeline transport led Duckworth et al to carry out a detailed study of the rheology of fine coal slurries. This study was reported by Lockyear et al (7), and this showed in both quantitative and qualitative terms the influence that the origin of the coal, the size distribution, the pH value, the ionic strength of the water used to form the slurry, the zeta potential and the volumetric concentration had on the rheology of a number of fine coal slurries. In parallel with these rheological studies, extensive studies have been made in an attempt to understand the flow behavior of suspensions of coarse materials in a non-Newtonian Bingham type fluid. Such studies, carried out for coal by Duckworth et al (6), in Australia and more recent studies of mine waste by Duckworth and Addie will now be considered.

THEORETICAL CONSIDERATIONS

In their work, Duckworth et al, have shown that for coarse coal to be transported under laminar flow conditions the carrier fluid should have a yield stress, τ_y; that is the fluid should be of the Bingham or yield pseudoplastic type. Moreover, they showed that suspensions of coarse coal in such a fluid flowed as stable suspensions provided that the condition for the static equilibrium of the suspension was fulfilled. The latter condition, discussed by Duckworth, is defined by the equation:

$$\tau_y \geq 0.1 \, \rho_w g \hat{d}(S_s - S_{mca}) \tag{1}$$

where $\hat{d}$ is the size of the largest particle, S_s and S_{mca} are the specific gravities of the particles and the slurry respectively, and $\rho_w g$ is the specific weight of water.

However, the authors now consider that the use of

$\hat{d}$ in equation 1 is somewhat conservative and in their later work with mine waste $\hat{d}$ is replaced by the geometric mean diameter, d_{50}. This approach is based on the assumption that the density and the yield stress of the coarse material suspension formed from the fine slurry carrier and the coarse particles of $d < d_{50}$ provide additional support for the particles having $d > d_{50}$.

Adding coarse material to a carrier fluid having Bingham plastic properties has the effect of increasing the yield stress and the plastic viscosity and it is convenient to refer to these pseudoproperties as the apparent yield stress, τ_{ya}, and the apparent plastic viscosity, η_a. The phenomenological behavior of such coarse slurries which produces the increase in the properties is not wholly understood but is considered to arise both from the tendency, in horizontal pipes, for the coarser material to occupy the lower portion of the pipe cross-section and from the interaction of the coarse particles with the fine slurry. We might expect the tendency to settle to dominate the behavior at low carrier concentrations when the yield stress tends to disappear and the plastic viscosity tends to the viscosity of water. Under the latter conditions the flow would tend to the familiar sliding bed regime. Nevertheless, whatever phenomena influence the values of τ_{ya} and η_a, the experience of work to date has shown that we may predict these values from equations of the following generalized forms:

$$(\tau^* - 1) = \phi_1[(S^* - 1), M^*, d_y/\hat{d}, d_{50}/\hat{d}, \hat{d}/D, \mu_s, S_s] \quad (2)$$
$$(\eta^* - 1) = \phi_2[(S^* - 1), M^*, d_y/\hat{d}, d_{50}/\hat{d}, \hat{d}/D, \mu_s, S_s] \quad (3)$$

where

$$\tau^* = (\tau_{ya}/\tau_y), \quad \eta^* = (\eta_a/\eta)$$

and

$$S^* = (S_m - 1)/(S_{mca} - 1) \quad (4)$$

$$d_y = 10\tau_y/\rho_w g (S_s - S_{mca}) \quad (5)$$

and the functional relationships ϕ_1 and ϕ_2 are determined from an analysis of the experimental data.

The wall shear stress, τ_o, may now be determined from the Buckingham equation,

$$8/N_{ya} = [1 - 4 \alpha_a/3 + \alpha_a^4/3]/\alpha_a \quad (6)$$

where the yield number N_{ya} is given by,

$$N_{ya} = \tau_{ya}D/\eta_a V_m \quad (7)$$

and

$$\alpha_a = \tau_{ya}/\tau_o$$

It has been found that the onset of the transition from laminar to turbulent flow may be predicted by the following equation, due to Hanks (4),

$$\rho_\omega S_m \tau_{ya}D^2/\eta_a^2 = 16800 \, \alpha_{ac}/(1 - \alpha_{ac})^3 \quad (8)$$

where α_{ac}, is the critical shear stress ratio. The Buckingham equation only applies to laminar flow and may therefore only be used for shear ratio values greater than the critical value. In the turbulent region the wall shear stress is found to follow the Colebrook-White equation and its simplified explicit form, the Jain equation (5), given by

$$1/\lambda_m^{\frac{1}{2}} = 1.14 - 2 \log_{10} [e/D + 21.25 \, R_e^{-0.9}] \quad (9)$$

where the friction factor, λ_m, and the Reynolds number, R_e, are given by,

$$\lambda_m = \tau_o/\rho_m V_m^2/8 \quad (10)$$

and

$$R_e = \rho_m V_m D/\eta_a \quad (11)$$

and e/D is the roughness ratio. We now consider the experiments, the experimental results, and the analysis of these in the light of the above considerations.

COARSE COAL STUDIES

Although the transport of suspensions of coarse coal in a carrier fluid made from finely ground coal has been studied by the authors over an extended period, the final report on these studies, Duckworth et al (6), including those carried out at a full scale prototype demonstration facility, has only recently been declassified for publication. In the following, the recent work covered by the report is combined with the results of the previous studies to give as complete a picture as possible of our understanding of the pipeline transport of coarse coal.

EXPERIMENTAL TEST FACILITIES

Three test facilities, designated 'A', 'B', and 'C' have been used to provide the experimental data that is considered in this paper, diagram layouts of these facilities being given in figures 1(a), (b) and (c). The facilities 'A' and 'B' have been described previously, Duckworth et al (8, 9, 10, 11), and therefore only a brief description will be given. The facilities, 'A' and 'B', are located at the CSIRO Division of

Mineral Engineering Laboratory at Clayton Victoria, Australia, the facility 'B' being a development of the original facility, 'A', which provides for either a continuous circulating system or a once through system, and makes it possible to change the fine coal/coarse coal fraction during operation in this circulating mode.

In facility 'A', a centrifugal slurry pump driven by an electric motor via a hydrostatic variable speed drive serves to pump the coarse slurry through the pipeline, 300 m in length and 152 mm in diameter, in a circulating mode only. The delivered density and the velocity of the slurry were determined by measurements of the volume and mass of slurry collected in a measuring tank over known time intervals, following diversion of the flow into the tank. In the case of facility 'B', however, the centrifugal slurry pump was driven by a diesel engine, variations in flow being obtained by throttle control. A nuclear density meter and a magnetic flow meter were used for the continuous monitoring of in-line density and velocity, respectively. Additionally, facility 'B' was provided with 300 m in length of pipeline of 203 mm in diameter and a 400 m length of pipeline of 254 mm in diameter.

In the case of both facilities, 'A' and 'B' pressure tappings at six pairs of equally placed stations along the horizontal test length of some 200 m were connected through slurry isolators to differential pressure transducers, to provide a means of determining the spatial averaged pressure gradient. The transducers were connected to a data logging system which provided for the time and space averaging of the gradient for both facilities. Velocity and in-line density measurements were made continuously in the case of facility 'B'. The transducer scans along with auxiliary data such as time, etc. were stored in addition to being displayed on the monitor.

The demonstration facility, facility 'C' is located at the R. W. Miller coal mine at Minmi, near Newcastle in NSW Australia and consists of a pipeline some 2 km in length and 300 mm in diameter fitted with short-circuiting valves to change the pipe length to 500, 1000, 1800 or 2000 m, and a diagram of this facility is given in figure 1(c).

Two 8" x 6" centrifugal slurry pumps in series, driven by a 250 hp diesel engine draw a fine coal slurry or a coarse/fine slurry mixture from a flat bottomed fine slurry storage tank fitted with an agitator or from a small conical bottomed coarse/fine slurry mixing tank, respectively. In order to charge the systems with fine slurry the pump draws from the fines tank and discharges the fine slurry at the end of the line into the coarse/fine mixing tank, until the latter is partially filled. The coarse coal may then be added to the fines in the mixing tank by means of the conveyor belt which is loaded from the coal bin, and the mixture transported via the pumps through the pipeline. Clearly during the coarse coal charging process it was necessary initially to remove a volume of fine slurry to maintain a sensibly constant volumetric inventory of coarse/fine slurry. To achieve the latter the slurry was returned to both the fine slurry tank via the sieve bend and to the coarse/fine mixing tank, either directly or via the sieve bend. Clearly the inventory of coarse coal in the system may be altered by either removing coarse coal by means of the sieve bend and the reversible conveyor which may be used to transport coarse coal to the mixing tank or to a stockpile, or by adding coarse coal as discussed above.

The facility is fitted with instruments for the

measurement of the following:-

1) velocity - a magnetic flowmeter,

2) the in-line density at three stations-
   nuclear density meters,

3) the pressure gradients at six stations for
   each test length of 500, 1000, 1500 and 2000
   m - differential pressure transducers

4) the weight of the coarse coal input-load
   cells fitted to the coal bin,

5) the slurry temperature - temperature probe,

6) the engine speed - tachometer.

All signals from the above instruments were fed
to the computer data logger for continuous
monitoring, storage and computing.

THE MATERIAL USED

During the pilot plant test program, the carrier
fluid used was formed from dry ball milled coking
coal, having a size less than 90 μm, this being
mixed with Melbourne tap water at an appropriate
concentration, approximately 50% w/w, to provide
the required rheological properties. Coking coal
from the same coal mine (R. W. Miller Minmi: Coal
Mine), of 20 mm top size and having the typical
size distribution shown in figure 3, was used as
the coarse fraction of the slurries examined.
Clearly, some degradation took place in the size
distribution before and after pumping and this is
shown in figure 4. The coarse material was
replaced periodically to maintain a sensibly
constant size distribution similar to that shown
in figure 3.

Although it was intended that the fine and coarse
particles used in the pilot and prototype plants,
shown in figures 1(a) and 1(b) and 1(c) respectively, should be the same, this was not in the event
possible, the size distributions for the fine
being shown in figure 2 and the coarse coal being
shown in figures 3 and 5.

However, the rheological properties of the
carrier fluids formed from the fine coal and
water is in fact considered to be more important

than the size distribution per se and these
properties are summarized in Table II and III for
the pilot and prototype plants, respectively.

THE EXPERIMENTS
Rheological Studies

During both the pilot plant and the prototype
studies the rheology and the specific gravity of
the particle used in the carrier fluid were
measured initially before coarse coal addition
and by sampling throughout the experiments. A
Haake type rotational viscometer was used for the
former and in the case of the specific gravity,
screened samples were weighed before and after
the thermal drying of samples of known slurry
concentration. The values of the rheological
data thus obtained are given in Tables II and
III, and a comparison of the viscometer and
pipeline data is given in figure 6.

Table II
Rheological properties of carrier fluid used in pilot plant trials

Bingham Yield Stress-τ_{yB} Pa	Plastic Viscosity-η Pas x 10^3	C_w %	C_v %
0.010	2.9	30.0	23.7
0.13	3.2	35.0	28.1
0.14	7.6	35.0	32.6
0.21	15.0	45.0	37.2
1.16	40.6	50.0	42.0
5.30	90.0	53.2	45.2
15.28	121.0	55.0	46.9

pH value = 8.4

Table III

Bingham Yield Stress-τ_{yB} Pa	True Yield Stress-τ_y Pa	Plastic Viscosity-η Pas x 10^3	C_w %	C_v %
7.3	5.0	56	56.4	43.4
7.2	4.0	80	54.0	44.0
8.2	4.9	88	56.5	47.2
10.6	5.2	130	57.1	47.8
15.5	6.8	130	57.5	48.2
11.6	5.4	159	58.1	49.0
12.0	6.0	255	58.2	49.1

pH value = 8.4

PIPELINE STUDIES
Pilot Plant

In the first series of experiments the test

facility 'A' shown in figure 1(a), was used and the carrier fluid and each of five coal slurries of differing concentrations of coarse coal were pumped over a range of velocities. The various concentrations and the coarse coal fractions of the several slurries were obtained by the sequential addition of 800 kg of coarse coal to the slurry in the system. In the second set of experiments, the slurry corresponding to a coal concentration of 65.6% w/w was successively diluted to yield a sensibly constant coarse coal concentration in the range $47.6 < C_{wca} < 53.2\%$. The data logging system was used to record and process the data related to the velocity and the density of the delivered slurry and the corresponding pressure gradients. As an additional check on the delivered density of each of the slurries, a one litre sample was taken at the beginning and end of each run. These samples were also used for the determination of the size distribution of the coal and the rheological properties of the carrier fluid.

In the second set of experiments the test facility 'B', shown in figure 1(b), was used to transport the carrier fluid alone and slurries composed of the carrier fluid plus varying concentrations of coarse coal. The variation in the coarse coal concentration was achieved by passing the slurry over a sieve-bend to remove some of the carrier fluid, thus increasing the concentration of coarse material. The instrumentation provided by this facility allowed for the continuous monitoring of the in-line slurry density, the velocity and the pressure gradient, the outputs from each of the instruments being connected to a computer data logger for record and analysis. Some 35 sets of experiments, involving 1000 measurements per set, and covering pipe sizes of 152, 203 and 254 mm in diameter, were carried out with the equipment of facility 'B'.

The data when taken together with the data obtained from facility 'A' forms the basis of the analysis, and provides also the basis for the prediction of the pipeline transport of coarse coal in the prototype test facility.

PROTOTYPE PLANT

It will be clear from the diagrams of figures 1(b) and 1(c) that the facilities 'B' and 'C' operate in essentially the same way, though the increased pipe length and diameter require that longer periods are required for the system to stabilize. Fully established steady flow conditions could be observed on the computer monitor, which provided a continuous record of the velocity, pressure gradient distribution and slurry density distribution. Steady flow and homogeneous distribution of the slurry could be inferred from the pressure gradient and slurry density spatial and time distributions. As in the pilot plant studies, samples of the slurry were taken for a range of velocities to establish the delivered concentration, size distribution of the coarse coal and the rheology of the carrier fluid, throughout the range of experiments. Experiments covering six sets of operating conditions were conducted in the 500 m and 2000 m long test loops, to provide a basis for the comparison of the data with that predicted by the pilot studies.

THE EXPERIMENTAL RESULTS

The experimental data obtained during the pilot and prototype plant studies are conveniently expressed in the form of the pressure gradient velocity characteristics and wall shear stress-shear strain rate (8 V/D), typical examples of such data being given in figures 7, 8, and 9. It is evident from these figures, that in all cases the slurries whether fine slurry carriers a suspensions of coarse coal in a fine slurry carrier, exhibit non-Newtonian behavior of the Bingham plastic type, characterized by a yield stress, τ_{ya}, and a plastic viscosity, n_a.

We also see from these figures that both the yield stress and plastic viscosity for the suspensions of coarse coal in the carrier fluid, τ_{ya} and n_a, respectively, are greater than the corresponding values for the carrier only, τ_y and n, the values of τ_{ya} and n_a increasing as the concentration and the slurry specific gravity increases. This suggests that if a functional relationship between the slurry specific gravity and the yield stress τ_{ya}, and the plastic viscosity, n_a, can be obtained, and if the behavior of such coarse slurries is indeed of the Bingham type, then the pressure gradient may be predicted from the Buckingham equation, equation 6, and the critical velocity from equation 8. We may also expect to be able to scale the data to pipes of a larger size. However, consideration of the correlation of the data is deferred until the corresponding data for mine waste has been discussed.

COARSE MINE WASTE STUDIES

The relatively successful development of coarse coal pipeline technology by Duckworth et al led, Duckworth and Addie to carry out a research program to extend the technology to coarse mine waste slurries, for which the particle specific gravity could be as high as 2.6 (c.f. 1.4 for coal). This is a particularly attractive application in that the underflow from the flocculation tanks, hycrocyclones, etc., provides a ready made Bingham type carrier fluid have the appropriate rheology for the transport of coarse mine waste.

Since much of the work has already been published (14), and since the intention in this paper is to compare the experimental data for coal and mine waste slurries, each of which appear to behave similarly, only a brief description of the work will be given.

THE EXPERIMENTS

In this work, because of the friable nature of mine waste, the novel type of extrusion rheometer, shown in figure 10, was used. This rheometer consists essentially of two pressure vessels A and B flexibly interconnected and suspended from load cells, the weight of each vessel being balanced by dead weights to improve the sensitivity of the load cells. Provision is made for the vessels to be filled with slurry to any required depth and for the slurry to be transferred from vessel A to vessel B and vice versa by the application of an air pressure differential across the free surfaces in the vessels. Since a sensibly constant low pressure differential, 6 m of water, was required, a large air reservoir of some 17 cu.m. in capacity was provided. Homogenization of the slurries was achieved by slow moving agitators in each vessel.

A ball valve allows for interconnection or isolation of the two vessels and a gate valve enables the system to be drained down. The interconnecting pipe of 76 mm in diameter is used as the test pipe, pressure tappings fitted across a 4.57 m test length being connected to a pressure transducer via slurry isolators. The outputs from the pressure transducer and the load cells were fed to a computer data logger, the data being scanned continuously at intervals of some three seconds to provide the wall shear stress and the corresponding shear rate 8V/D. However, the pressure differential and the load cell readings were only recorded when the levels in vessels A and B were the same. A pre-set null reading was used to provide the trigger for the readings to be recorded.

In carrying out the experiments a sample of some 227 kg of the fine slurry was loaded into the rheometer and the levels allowed to equalize in order to set the load cell null reading at a value of some 113 kg. The slurry was then driven into vessel A until 80% of the slurry had been placed in this vessel. At this point the ball valve was closed, the air relief valves on vessel B opened to atmosphere and the air pressure in vessel A adjusted to provide the required flow rate of slurry. After the outputs from the transducers had stabilized the ball valve was opened to allow the slurry to flow from vessel A to vessel B, the valve being closed when the level in vessel A had fallen below the null point. This procedure was repeated for a range of flow rates to provide the data for the rheogram of the carrier. Coarse material was then added in equal increments of some 18 kg up to some 108 kg and the above procedure followed to provide a set of rheograms, a typical set being shown in figure 11. A similar procedure was followed for a number of coarse materials that are typical of mine waste and for a number of fine slurry carriers.

An examination of the several rheograms of figure 11 indicates quite clearly that the slurries formed from the addition of coarse material to a Bingham plastic carrier appear to behave in substantially the same way as the carrier fluid, that is they appear to behave as Bingham plastics. However, we see that as in the case of similar coarse coal slurries, the apparent yield stress, τ_{ya}, and the apparent plastic viscosity, η_a, vary as the amount of coarse material is added, that is, as the specific gravity of the slurry increases. By selecting appropriate values of τ_{ya} and η_a and using these in the Buckingham equation, curves have been fitted to the experimental data, these curves being given in figure 11. The curves are seen to exhibit a gradual rather than a sudden transition from laminar to turbulent flow and the locus for the onset of the transition obtained experimentally is seen to be in reasonable agreement with that predicted by equation 8, due to Hanks (12), from which the locus has been plotted.

CORRELATION OF THE DATA

We have seen that the experimental data for the pipeline transport of a coarse material in a Bingham plastic type of carrier fluid suggests that the coarse slurry so formed also behaves as a Bingham plastic of higher yield stress, τ_{ya}, ($\tau_{ya} > \tau_y$), and plastic viscosity, η_a, ($\eta_a > \eta$). Whilst it is unlikely that such a slurry behaves in a way phenomenologically similar to a Bingham plastic, nevertheless it is convenient to treat it as a Bingham plastic and accordingly to use the equations such as the Buckingham equation. However, we must first attempt to correlate the apparent rheological properties, τ_{ya} and η_a of the coarse slurry, from the experimental data, using for this purpose equations 2 and 3.

An analysis of the data of Table IV, for a slurry made from coarse coal suspended in a Bingham plastic carrier made from fine coal and water and for a slurry formed from the suspension of coarse mine waste in a Bingham plastic carrier made from the underflow from the flocculation tank at a coal preparation plant, will now be briefly discussed.

It is convenient to first of all simplify and recast equations 2 and 3 in the following forms,

$$\tau^*-1 = \phi_{1A}(S^*-1) \times \phi_{1B}(M^*) \times \phi_{1C}(d_y/\hat{d}) \times \phi_{1D}(S_s-1) \quad (2a)$$

$$\eta^*-1 = \phi_{2A}(S^*-1) \times \phi_{2B}(M^*) \times \phi_{2C}(d_y/\hat{d}) \times \phi_{2D}(S_s-1) \quad (3a)$$

This simplification is based on the assumption

that the dependent variables may be expressed in terms of the products of the functions of the several independent variables and on the assumption that $(d_{50}/\hat{d})$ and $\hat{d}/D)$ are sensibly constant. We now consider the analysis of the several functions such as ϕ_{1A}.

ANALYSIS OF $\phi_{1A}(S^*-1)$

In order to determine the function, $\phi_{1A}(S^*-1)$, from the variables listed in Table IV, we note from equation 2a that,

$$\phi_{1A}(S^*-1) = (\tau^*-1)/\phi_{1B}(M^*) \times \phi_{1C}(d_y/\hat{d}) \times \phi_{1D}(S_s-1) \quad (12a)$$

By a judicious selection of the data such that M^*, $(d_y/\hat{d})$ and S_s are sensibly constant, equation 12a reduces to,

$$\phi_{1A}(S^*-1) = (\tau^*-1) \quad (12b)$$

In fact, it transpires that the influence of M^* and $(d_y/\hat{d})$ is insignificant over the range of variables covered in the analysis, this being evident from the curves of figure 12. From a regression analysis, the function $\phi_{1A}(S^*-1)$ may be expressed in the form,

$$\phi_{1A}(S^*-1) = \text{Constant } (S^*-1)^{1.5} \quad (12c)$$

and thus by combining equations 12b and 12c we obtain,

$$(\tau^*-1) = \text{Constant } (S^*-1)^{1.5} \quad (12d)$$

where for coal the constant = 7.26, and for mine waste the constant = 20.4.

It is interesting to note that the ratio of the constants is equal to 2.8, this being equal to the ratio of the submerged specific gravities, (e.g. (S_s-1))

$$(S_s-1)_{Mw}/(S_s-1)_{Coal}{}^2 = 1.13/0.4 = 2.8$$

This suggests that a general expression for (τ^*-1) is likely to be of the form,

$$(\tau^*-1) = 18 (S_s-1)(S^*-1)^{1.5} \quad (13)$$

ANALYSIS OF $\phi_{2A}(S^*-1)$

As in the previous analysis, equation 3a may be reduced to the form,

$$\phi_{2A}(S^*-1) = (\eta^* - 1) \tag{14}$$

Values of the function, $\phi_{2A}(S^*-1)$, for a range of (S^*-1) values have been calculated by using selected data from Table IV, these values being plotted in figure 13. From a regression analysis the relationship may be expressed in the form,

$$\phi_{2A}(S^*-1) = \text{Constant } (S^*-1)^3 \tag{14b}$$

Table IV
Values of Dimensionless Groups Used in the Analysis
Material - Mine Waste S.G. = 2.13

τ^*	η^*	S^*	M^*	$dy/\hat{d}$	$d_{50}/\hat{d}$	d/D
3.40	1.10	1.21	0.31	0.25	0.20	0.25
3.70	1.10	1.27	0.47	0.25	0.20	0.25
5.00	1.20	1.33	0.61	0.25	0.20	0.25
6.67	1.20	1.45	0.76	0.25	0.20	0.25
10.84	2.0	1.69	0.89	0.25	0.20	0.25
6.27	2.0	1.44	0.39	0.38	0.32	0.25
11.76	2.0	1.56	0.57	0.38	0.32	0.25
14.63	2.0	1.67	0.78	0.38	0.32	0.25

Material-Coal S.G. = 1.40

τ^*	η^*	S^*	M^*	$dy/\hat{d}$	$d_{50}/\hat{d}$	d/D
1.19	1.03	1.07	0.14	1.27	0.27	0.125
1.30	1.25	1.14	0.30	1.45	0.27	0.125
1.99	1.56	1.25	0.60	1.64	0.27	0.125
1.71	2.54	1.20	0.52	1.76	0.27	0.125
1.52	1.59	1.16	0.37	2.10	0.27	0.125
1.85	3.02	1.24	0.57	1.35	0.185	0.09
1.54	1.82	1.19	0.45	2.24	0.185	0.09
1.75	2.23	1.23	0.57	2.30	0.185	0.09
2.46	3.67	1.34	0.93	2.30	0.185	0.09
1.30	1.60	1.16	0.37	3.93	0.185	0.09
1.51	1.72	1.19	0.46	4.04	0.185	0.09
1.83	1.32	1.25	0.59	1.50	0.185	0.09
2.65	2.39	1.37	0.89	0.43	0.185	0.09
3.55	1.79	1.48	1.35	0.73	0.185	0.09
3.27	2.00	1.48	1.42	0.94	0.185	0.09
3.55	2.60	1.51	1.68	1.22	0.185	0.09
4.15	2.60	1.50	1.35	0.37	0.185	0.09
2.97	2.84	1.43	1.17	0.76	0.185	0.075

ANALYSIS OF $\phi_{2B}(M^*)$

Now since the group $(dy/\hat{d})$, for the mine waste material data is constant, we may combine equations 3b and 14b to give,

$$\phi_{2B}(M^*) = (\eta^* - 1)/(S^*-1) \tag{15a}$$

This relationship obtained from equation 15a and the mine waste data has been plotted in figure 14 and this is given by,

$$\phi_{2B}(M^*) = \text{Constant } (M^*)^{-2} \tag{15b}$$

ANALYSIS OF $\phi_{2C}(dy/\hat{d})$

At this point, we may combine the relationships of equations 14b and 15b with equation 3b to give,

$$\phi_{2C}(dy/\hat{d}) = (\eta^*-1)M^{*2}/(S^*-1)^3 \tag{16a}$$

If we now use equation 16a and all of the data of Table IV, we find that the function, $\phi_{2C}(dy/\hat{d})$, is sensibly constant, having a mean value of 24.3 at a standard deviation of 4% for coal and a mean value of 1.7 at a standard deviation of 4% for mine waste. We thus see that (η^*-1) is independent of the yield diameter ratio.

ANALYSIS OF $\phi_{2D}(S_s - 1)$

Only one independent variable, the submerged specific gravity of the coarse material, (S_s-1), remains to be examined. Unfortunately, because this study has only examined two coasre materials, we cannot do other than speculate on the form of the functional relationship, $\phi_{2D}(S_s-1)$. For this purpose we use the two values of the function, $\phi_{2D}(S_s-1)$, which has the values of 24.3 for coal and 1.7 for mine waste material, and note that the ratio of these values is inversely proportional to $(S_s-1)^{2.56}$. This leads us to the speculative conclusion that the final form of the relationship for η^* is as follows:

$$\eta^* - 1 = 2.33 \, (S^*-1)^3/(S_s-1)^{2.56} \, M^{*2} \tag{17}$$

It is interesting to note that the ratio $(S^*-1)/S_s-1)$ is related to the volumetric concentration, C_v, by the equation,

$$(S^*-1)/S_s-1) = (C_v/(S_{mca}-1)) - (1/(S_s-1)) \tag{18}$$

and it thus follows from equations 17 and 18 that the plastic viscosity ratio, η^*, is related to C_v. Such relationships are characteristic of non-Newtonian fluids and the relationship given by equation 17 would appear to be reasonable.

PRESSURE GRADIENT-VELOCITY CHARACTERISTIC

The correlation of the experimental data has shown that the yield ratio, τ^*, and the plastic viscosity ratio, η^*, may be determined from equations 13 and 17, respectively. These equations, when combined with the yield stress and the plastic viscosity of the carrier fluid, the S.G. of the coarse material and assigned values of D and V, provide us with the means to determine the apparent yield stress and plastic viscosity, τ_{ya} and η_a, respectively.

If we now invoke the Buckingham equation, equation 6, and the Hanks equation, equation 8, and solve these equations using the apparent values, τ_{ya} and η_a, we may obtain the rheogram relating the wall shear stress, τ_o, and the pseudo shear rate, 8V/D, and the critical value, $8V_c/D$, which defines the boundary of the laminar flow zone. Furthermore, since the pressure gradient is related to the wall shear stress by the equation,

$$\Delta p/L = 4\tau_o/D \qquad (19)$$

the rheogram may be transformed into the practically important pressure gradient-velocity characteristic for the laminar flow carrier for a pipe of arbitrary diameter.

SCALE UP

The fact that it has been possible to correlate the data covering pipe sizes ranging from 76 to 254 mm in diameter, suggests that the scale up condition for true Bingham plastics, (i.e. the equivalence of 8V/D for model and prototype), applies equally to the pseudo Bingham coarse slurry discussed in this paper.

We now examine the application of the results of the foregoing analysis to the prediction of the pressure gradient-velocity characteristic for the flow of a coarse slurry in a pipe of 300 mm in diameter, this being compared with the experimentally determined characteristics shown in figure 15.

The agreement between the predicted and measured curves is considered to be reasonable and we may conclude that the equations developed provide a means of predicting the performance of slurries of coarse coal and coarse mine waste in Bingham plastic carrier fluids flowing in pipes of arbitrary size.

STATIONARY BED FORMATION AND START-UP

Although the foregoing analysis has provided a means of estimating the pressure gradient it is equally important that we should establish a criterion for the onset of the formation of a stationary bed. In previous work by Lockyear et al (11), it was reported that neither the formation of a stationary bed nor difficulties with the start-up of a pipeline after shut-down had been experienced provided that the condition,

$$dy/\hat{d} \geq 1.0 \qquad (20)$$

was fulfilled. We thus see that although the yield diameter ratio does not appear to influence either τ^* or η^* it does provide a criterion for ensuring that a stationary deposition condition does not occur.

However, it should be noted that whether or not stationary deposition occurred during the course of the studies, no difficulties have been experienced in restarting a pipeline after shutdown for extended periods. In the case of the pipeline test facility 'C', of 2km in length and 300 mm in diameter, the pipeline was restarted after a shutdown period of 46 days.

CONCLUSION

This paper has shown that slurries formed from coarse coal (S.G. - 1.4) and mine waste (S.G. = 2.13) of top size 25 mm, suspended in Bingham type carrier fluids, appear to behave in very much the same way as the carrier fluid. This has been demonstrated by the fact that the experimentally determined rheograms are in close agreement with those predicted by the Buckingham equation and the onset of transition from laminar flow has been shown to be predicted by the Hanks equation.

The yield stress, τ_{ya}, and the plastic viscosity,

η_a, are given by the equations,

$$\tau^* - 1 = 18 \, (S_s-1)(S^*-1)^{1.5}$$

$$\eta^* - 1 = 2.33(S^*-1)^3/(S_s-1)^{2.56} \, M^{*2}$$

These equations when used in conjunction with the equations due to Buckingham and Hanks provide a means of determining the pressure gradient-velocity characteristic for the laminar flow zone and for the boundary of this zone for such slurries. Furthermore, the criterion for the scale-up of data for Bingham fluids, based on the equivalence of the pseudo shear strain rate, 8V/D, has been shown to be applicable to coarse slurries.

The benign nature of slurries of coarse materials in a Bingham carrier fluid, their stability and pipeline restart characteristics, plus the fact that design can now be based on the equations of this paper, means that we now have an alternative method for the pipeline transport of coarse materials. This method is both practically feasible and economically attractive.

ACKNOWLEDGEMENTS

The authors wish to acknowledge the support given to the work by the Australian National Energy Research Development and Demonstration Program of the Department of Energy and Resources, the R. W. Miller Pty., Ltd., the C.S.I.R.O Executive, Mr. Thomas W. Hagler, Jr., President of GIW Industries, Inc., and the Peabody Coal Co., and for their permission to publish the paper. They also wish to thank the engineering, scientific and technical staff of the CSIRO Pipeline Engineering Group and the GIW Industries, Inc. Hydraulic Laboratory.

REFERENCES

1. Duckworth, R. A., ' The hydraulic transport of coal in Australia', Rheology in the Conversion and Conservation of Energy Symposium, British Society of Rheology (Australian Branch), R.M.I.T., March 1980, pp 17-29.

2. Duckworth, R. A. 'Mineral pipeline research study - Final report', Division of Mineral Engineering, CSIRO, Melbourne, August 1978.

3. Elliott, D. E. and Gliddon, B. J., Paper G2, Proc. 1st International Conference of Hydraulic Transport of Solids in Pipes, BHRA, Cranfield, Sept. 1970

4. Lawler, H. L., Pertuit, P., Tennant, J. D. and Cowper, N.T., 'Applications of stabilized slurry concepts of pipeline transportation of large particle coal', 3rd International Tech. Conf. on Slurry Transportation, Organized by STA, Washington, DC, March 1978.

5. Duckworth, R. A. and Pullum, L., 'The hydraulic transport of coarse coal', 2nd National Conference on Rheology, British Society of Rheology (Australia Branch), May 1981.

6. Duckworth, R. A., Lockyear, C. F., Pullum. L. and Littlejohn, M.H., 'Transport of coal by pipeline parts I and II' End of Grant Report, No 464, NERDDP, Dept. of Resources and Energy, Australia, August 1985.

7. Lockyear, C. F., Pullum, L., Lenard, J. and Duckworth, R. A., 'The rheology of fine coal suspensions', Third National Conference on Rheology, British Society of Rheology, (Australian Branch), Melbourne, May 1983.

8. Duckworth, R. A., Pullum, L., and Lockyear, C. F., 'The hydraulic transport of coarse coal at high concentraton', Presented at 4th International Symposium on Freight Pipelines, Atlantic City, N.J., October 1982, Journal of Pipelines, 3(1983) pp 251-265, Elsevier Science Publishers, B. V. Amsterdam.

9. Duckworth, R. A., Pullum, L., Lockyear, C.F., and Lenard, J., 'The long distance transport of coarse coal by pipeline', paper presented at Chemeca 1983, 11th Australian Conference on Chemical Engineering, September 1983.

10. Duckworth, R. A., Pullum, L., Lockyear, C. F. and Lenard, J., 'Hydraulic transport of coal', paper presented at International Conference on Bulk Materials Storage, Handling and Transportation, Newcastle, Australia, Inst. Eng. Aust., August 1983, Bulk Solids Handling, Vol. 3, No 4, November 1

11. Lockyear, C. F., Pullum, L., Duckworth, R. A., Littlejohn, M. H. and Lenard, J., 'Prediction of pressure gradients for the transport of coarse coal in a fine coal carrier', paper presented at the Transportation Conference of the Inst. of Engineers Australia 1984, Perth, W.A. October 1984

12. Hanks, P.W., 'A generalized criterion for laminar-turbulent transition in the flow of fluids: Union Carbide Company, November 1962.

13. Jain, A. K., 'Accurate explicit equation for friction factor', Technical Notes, Journal Hydraulic Div. Proc. A.S.C.E., Vol. 102, HY.5, May 1976.

14. Duckworth, R. A., Addie, G. R. and Maffett, J., 'Mine waste disposal by pipeline using a fine slurry carrier', 11th Int'l Conf. on Slurry Technology-The Second Decade organized by STA, Washington, DC, Hilton Head, SC, March 1986.

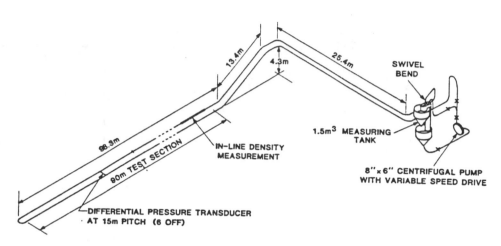

Fig. 1a. Diagrammatic layout of test facility 'A'.

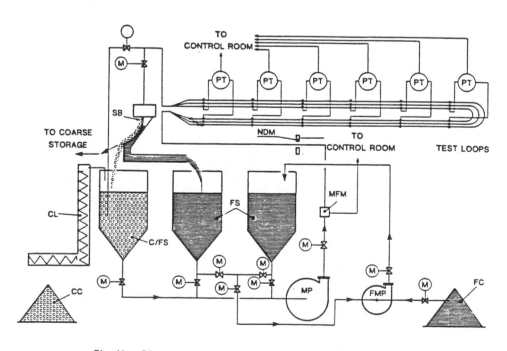

Fig. 1b. Diagrammatic layout of test facility 'B': CL, Coal loader; FMP, Fines Mixing Pump; MP, Main; M, Manually Operated Valve; SB, Sieve Bend; CC, Coarse Coal; C/FS, Coarse/Fine Slurry; FC, Fine Coal; FS, Fine Slurry; MFM, Magnetic Flow Meter; NDM, Nuclear Density Meter; PT, Pressure Transducer.

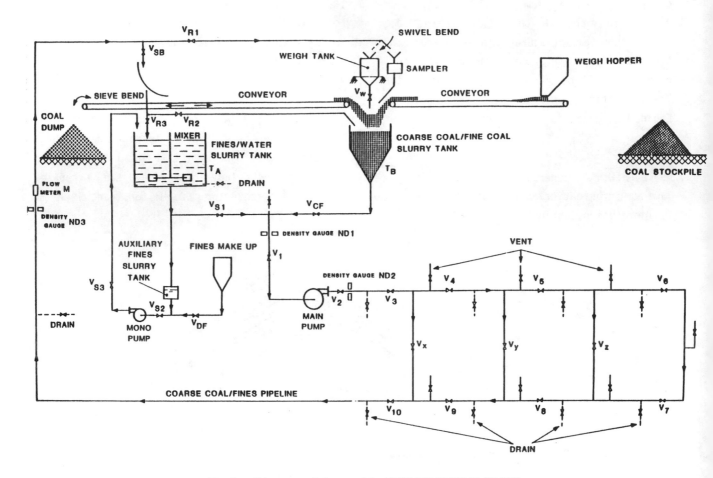

Fig. 1c. *Diagrammatic layout of the NERDDP/CSIRO/R. W. Miller test facility 'C'.*

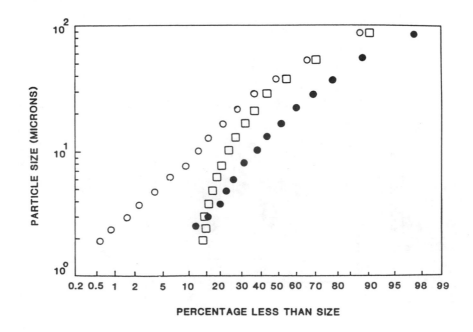

Fig. 2. *Particle size distributions for the fines used to form the fine coal slurries.* ● *Typical size distribution used in pilot plant.* ○ *Original size distribution used in prototype plant.* □ *Size distribution used in prototype plant after the addition of clay (Calculated).*

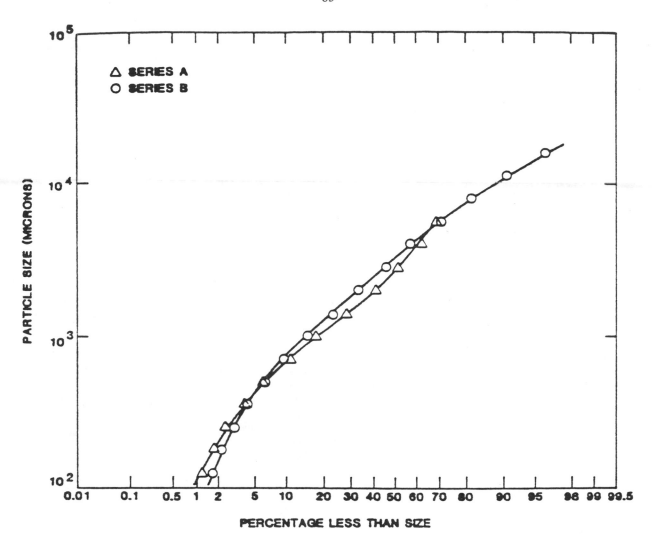

Fig. 3.   The initial size distribution of the coarse coals used in the
pilot plant experiments.

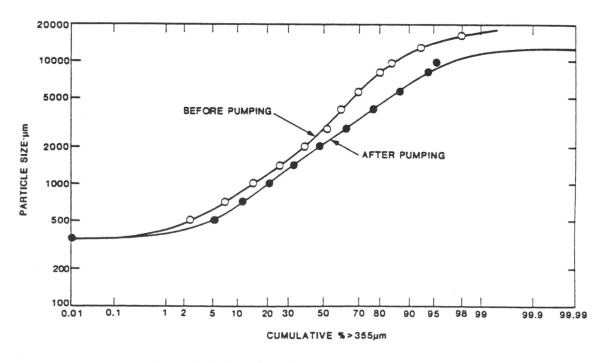

Fig. 4.   Typical size distribution showing the influence of pumping.

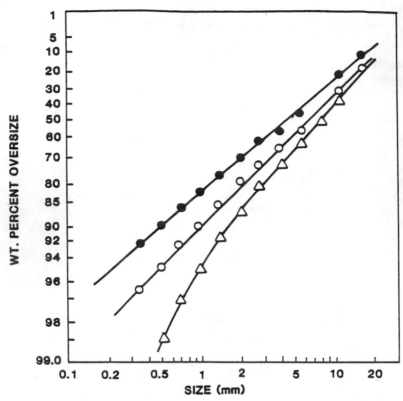

Fig. 5. Initial size distributions of the coals added to the prototype plant. ● Size distribution of coal added during initial trials. △ Size distribution of coal added after initial trials. ○ Size distribution after initial trials (Estimated).

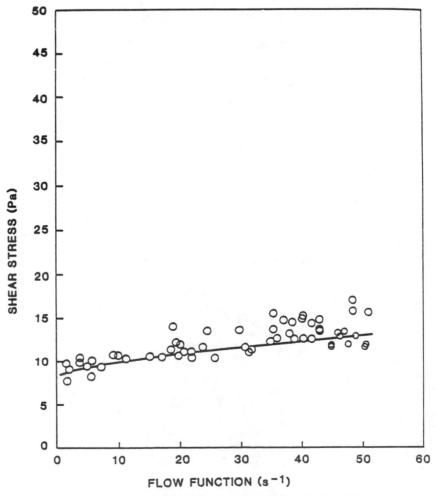

Fig. 6. Comparison of the rheograms for the carrier fluid, used in the prototype experiments, based on the pipeline tests and the Haake rotational viscometer.

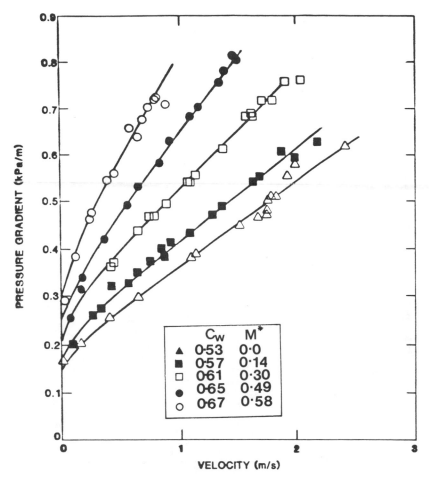

Fig. 7. *Pressure gradient-velocity characteristics for the transport of coarse coal in a pipe of 152 mm in diameter using test facility 'A'.*

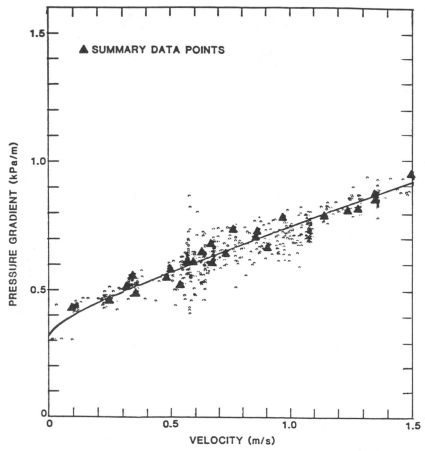

Fig. 8. *Typical pressure gradient-velocity characteristic for the transport of coarse coal at a concentration C = 60% in a pipe of 200 mm in diameter using test facility 'B'.*

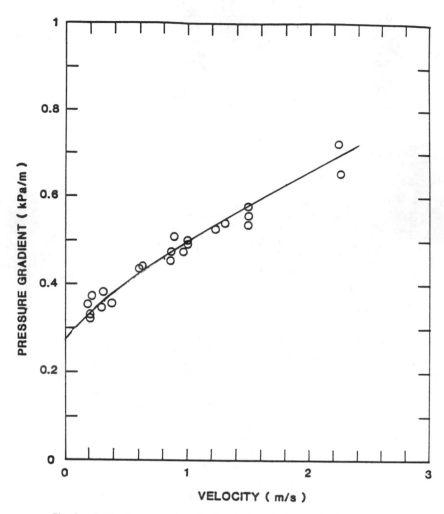

Fig. 9. Typical pressure gradient-velocity characteristic for the transport of coarse coal at a concentration C = 69% in a pipe of 300 mm in diameter using test facility 'C'.

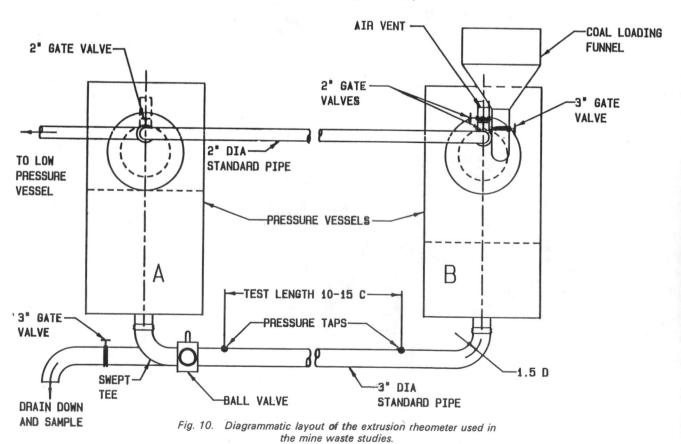

Fig. 10. Diagrammatic layout of the extrusion rheometer used in the mine waste studies.

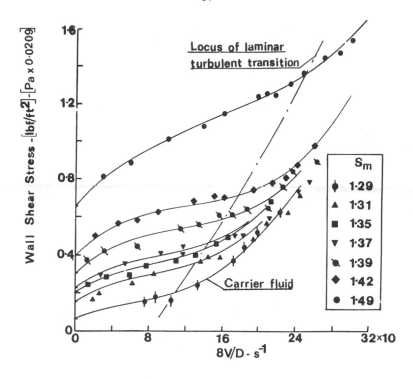

Fig. 11. Experimentally determined rheograms for coarse mine waste slurries.

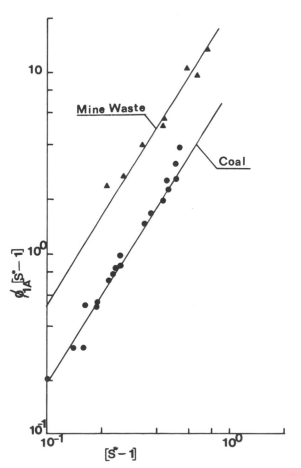

Fig. 12. Relationship between $\phi_{1A}(S^* - 1)$ and $(S^* - 1)$.

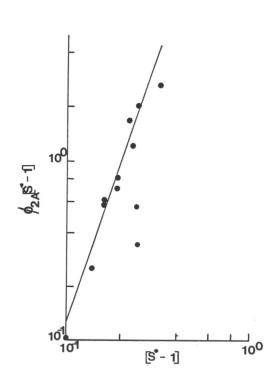

Fig. 13. Relationship between $\phi_{2A}(S^* - 1)$ and $(S^* - 1)$ for coarse coal $M = 0.42$.

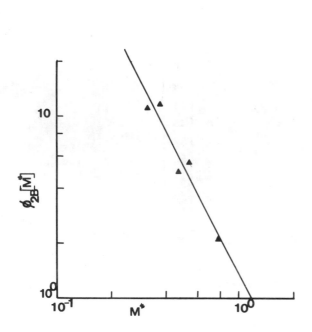

Fig. 14.   Relationship between $\phi_{2B}(M^*)$ and (M^*) for coarse mine waste slurries.

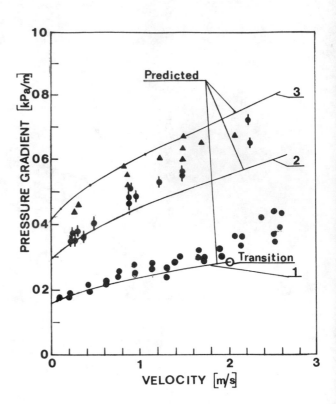

Fig. 15.   Comparison of the experimental and predicted pressure gradient-velocity curves for the transport of slurries of coarse coal in a fine coal carrier in a pipeline of 300 mm in diameter.

Test	C_w	M^*	τ_y	η
● 1	0.67	0.5	7.2	0.08
● 2	0.69	0.46	13.75	0.14
▲ 3	0.74	0.64	13.75	0.14

10th International Conference on the
Hydraulic Transport of Solids in Pipes

HYDROTRANSPORT 10

Innsbruck, Austria: 29-31 October, 1986

PAPER C3

STABFLOW SLURRY DEVELOPMENT

D.A. BROOKES
MANAGER, STABFLOW DEVELOPMENT,
BUSINESS TECHNICAL SUPPORT DEPARTMENT,
BP INTERNATIONAL LIMITED, U.K.

P.E. SNOEK
MANAGER, SLURRY TECHNOLOGY,
BECHTEL INC.,
U.S.A.

SYNOPSIS

Stabflow is a stable slurry mixture of coarse and
fine coal with water which is analogous to a
concrete mix, the principles and advantages of
which were reviewed at Hydrotransport 9.

BP and Bechtel have formed a joint venture to
develop this technology. A large scale pilot
plant has been constructed and commissioned at the
former refinery at the Isle of Grain, England.
The major principles of the plant and the eight
months operating experience are reviewed. The
results of pumping trials with thin and thick
mixtures are examined and the main broad
conclusions are dicussed. The sophisticated
instrumentation and control equipment is described
including the data acquisition and analysis
system. The operating shock pressure problems
with the main slurry pumps are described and the
results of successful modifications presented. A
subsequent economic study has confirmed the wide
range of applications.

The test work has demonstrated the advantages of
the Stabflow technology for coarse coal transport.

1. INTRODUCTION

   At the last Hydrotransport Conference
   (Ref. 1), BP described the principles of the
   Stabflow coarse coal pipelining technology and
   favourably compared this with other coal
   transport and coal slurry systems. This paper
   describes the large scale Stabflow test
   facility and the experience gained in the first
   phase of operations.

   Stabflow coal slurries are essentially coarse
   coal slurries produced by mixing coarse coal
   with a smaller amount of wet milled fine coal

in water. The slurry is intended for two
applications. For long distance overland
applications as a relatively dilute slurry at
approximately 70% Cw. A high concentration
+80% Cw slurry is proposed for direct offshore
shiploading via an SBM into conventional coal
carriers for grab unloading. These two types
of slurry have potentially the same particle
size distribution, but pressure drops and
orders of magnitude different.

The test facility was designed to handle both
types of slurry and obtain accurate pressure
drop measurements.

2. STABFLOW TEST FACILITY

BP and Bechtel formed a Joint Venture in 1984
to develop the Stabflow technology. A large
scale test facility was constructed on the Isle
of Grain, Kent, UK to demonstrate the
technology at near commercial scale and obtain
design data for a first commercial application.
Some conceptual design was done in 1982/83, but
detail design did not commence until May, 1984.
Civil construction was well underway in October
of that year and commissioning commenced in
May, 1985.

The test facility is located on a part of the
redundant BP Oil Kent Refinery at the Isle of
Grain, Medway, Kent, UK. This site had the
advantage of an open site in an industrialised
area with extensive redundant tanks, 12"
pipelines, electrical power and water supplies.
It was close to the CEGB Kingsnorth coal
burning power station, and an arrangement was
made to "borrow" large tonnages of coal for
testing and return it as slurry. Up to 1,000
tons of Westoe coal was stockpiled at the test
site. This coal came all from one seam, was
partly washed and was similar to that used for
prior small scale tests.

2.1 Previous designs for a test facility in
    Australia had considered a full scale once
    through system with a fines carrier return
    pipeline.

    There are some disadvantages with such a
    system, mainly the high cost of fines
    recovery and concentrating necessary to
    keep a mass balance of fines in the forward
    coarse coal pipeline. The system chosen
    for the Kent site used a batch production
    principle with the facility for
    recirculation and it was designed on a
    short life, optimum cost basis. It was
    able to monitor the following parameters:-

    2.1.1 Slurry friction losses as a function
          of velocity, slurry concentration,
          pipeline length and coarse/fines/
          water ratio.

    2.1.2 Settling and stability limits during
          pumping, shutdown and start-up.

    2.1.3 Production of fines during pipelining
          (attrition).

2.1.4 Pipe wear.

2.1.5 Pump and other equipment performance.

2.2 Detailed Plant Description

Figure 1 is a photograph of the main plant. Figure 3 is a schematic flow diagram. This is described in some detail below.

2.2.1 Fines Preparation

To minimise costs, a batch slurry preparation system was installed. Fine coal slurry is made first and then the same basic coal feeding system is used to prepare the final Stabflow mixture.

A front end loader was used to transfer coarse 50 mm x 0 coal from a stockpile to a 10 tonne weigh-feed hopper. The metered coal was fed to an inclined conveyor belt which delivered the coal to a chute situated above a shuttle conveyor. Coal was then directed to either a hammermill for fine coal slurry production or to a continuous paddle mixer for final Stabflow production.

In the fines production mode, 5–15 TPH of coal is fed to a primary hammermill, from here −8 mm product drops with water added into the feed spigot of a 5 m long x 2 m diameter rubber tyre driven ball mill. This operated on a once through basis with no recirculation; the discharge from a trommel being pumped away to two existing 400 m^3 storage tanks by a variable speed mono pump controlled on discharge sump level. The tanks, which were previously used for lube oil, were closed with no agitation and had a shallow cone bottom.

A larger, variable speed, mono pump was arranged to pump the fines slurry at a higher rate back to the mixing plant.

2.2.2 Slurry Production

The usual pattern of operation was to make sufficient fines slurry at a low flow rate on the first day which would then be stored. This fines slurry would be mixed at a high flow rate with coarse coal and some water on the second day in the continuous concrete type paddle mixer. The Stabflow slurry could then either be pumped into a large holding storage tank or directly into the pipeline.

2.2.3 Stabflow Feed Tanks and Pump Suction Pipework

Two open top, non-agitated slurry feed tanks were utilised. The smaller tank, positioned directly under the mixer discharge chute, had a capacity of about 11 m^3.

Immediately adjacent to this was a large 780 m^3 holding tank, both tanks had conical bottoms discharging into a common suction header to the main slurry pumps – see Figure 2.

2.2.4 Main Pumps

The two main slurry pumps were Putzmeister, reciprocating single acting duplex piston units each equipped with 300 Kw electro-hydraulic drivers. These pumps were essentially standard concrete type pumps adapted for coarse coal and were rated at 125 m^3/hr @ 50 bar with an override to 60 bar. They are the largest concrete pumping units commercially available and were capable of varying the flow rate from zero to the maximum rating. Each pump consists of two cylinders and a sequenced, hydraulically actuated changeover or "Delta" valve as illustrated in Figure 4. The main slurry pistons are driven by a rapid reversing swash plate hydraulic pump via hydraulic cylinders. The oil pumps operate on a closed circuit free-flow basis. A comprehensive control system was provided to control the two pumps in parallel. A diesel driven concrete pump from Schwing with a maximum pumping capacity of 90 m^3/hour was also used.

2.2.5 Test Loops

One or both of the main slurry pumps could discharge stabilised slurry to either a 50 m, 10 inch loop which recycled slurry back to the slurry holding tanks or to the main 12 inch pipeline. The lines were constructed of welded unlined steel pipe with occasional flanges for relocation and inspection purposes. The 12 inch line was constructed from the redundant refinery pipe so that lengths of 0.5 km, 2.5 km, 4 km or 7 km could be utilised. The lines are laid out in an L shaped configuration approximately 0.5 km x 0.6 km long with six lines being linked by 180° return loops. In addition a 20 metre high structure is fitted with sloping and vertical lines to check behaviour in these configurations. At the end of the line, slurry could be recycled to either of the two slurry feed tanks, returned to the mixer, or discharged to other slurry storage tanks for eventual disposal by truck back to the Kingsnorth power station. The lines were re-rated and modified for the increase to a 60 bar maximum operating pressure. A large pipeline damper of over 2 m^3 capacity was mounted downstream of the main pumps manifold. Considerable scraper pigging was carried out intially to clean the pipe loops.

2.2.6 Tankage

In addition to the two new conical bottom Stabflow tanks and the fines slurry storage tanks described above, other 12 m diameter x 6 m high former oil tanks were adapted. Two were used for water storage, three were used for Stabflow storage and two were cut down to 1 m high upstands for use as large settling ponds. Access doors and weirs were fitted into the tank walls to enable front end loaders to be used to remove the slurry.

2.2.7 Electrical System and Utilities

To suit the old refinery system all major drives were 3.3 kv with transformers down to 440 v for smaller drives. Refurbished refinery and new switchgear was housed in a prefabricated concrete MCC building; total installed power was approximatley 1.5 MW. A large 600 m^3/hr and small jockey water pump rated at 60 m^3/hr were arranged to provide a common 10 bar water supply for normal usage and flushing water. Compressed air was supplied from a packaged screw compressor, with a desicant drier and an ultra oil filter side stream for instrument air.

2.2.8 Instrumentation

While some short life economies were made on some aspects of the design, considerable efforts were made to ensure that good data was obtained from the extensive instrumentation.

The main coarse coal weigh-feeder system used a batch ratio system driven by a belt weigher with facilities for cross check on loss in weight of the feed hopper.

Stabflow and fine coal slurry flows were measured by magnetic flow meters. These were calibrated with both water and slurries.

Nearly all the pipeline pressure drop measurements were made with Honeywell "SMART" pressure transducers using 75 mm barrier diaphragm seals and silicone fluid pressure impulse lines. The design of the attachment was such that the diaphragm was essentially flush with the pipe inner wall. These transducers were either absolute, gauge or differential and can be re-ranged for span, range and response direct from the control room using a simple key pad. Approximately 30 transducers were positioned at intervals around the loop measuring a mixture of gauge and differential pressures, some

more sensitive transducers were fitted across the return bends. The robust nature of these transducers, their ability to withstand shock pressures up to 70 bar, and an accuracy of over two orders of magnitude proved invaluable. To monitor rapid transient pressures associated with the pumps, small "Kistler" piezo resistive transducers were mounted directly on the pipe, close to the pipe inner surface. These were linked to multi-channel storage oscilloscopes and to a BHRA transient capture unit.

Slurry densities were measured routinely with single beam nuclear density meters. Slurry concentrations could then be obtained from a laboratory measured coal density. In addition a single beam scanning densitometer was used to measure two dimensional density/concentration profiles. This was fitted into a flanged section of pipeline towards the end of the loop.

Slurry tank levels were measured using ultrasonic monitors, mostly fitted with plastic shielding tubes. Main pipeline temperatures were monitored using thermowell probes at the beginning and the end of the loop.

2.2.9 Scada System

Process control, monitoring, data logging and data analysis were achieved by the use of a Ferranti Argos process control minicomputer. A major consideration in selecting this system was its flexibility. Displays could be built and modified, controllers configured, etc. by engineers who were not software specialists. The Scada system was able to perform the following tasks:-

a. Scan at 1 second intervals and store data from the plant. Typically this was flow rate, pressure and density information. The majority of the 1 second data was averaged over 30 seconds and the average, maxima and minima were stored. The system was able to record and store up to 36 channels for 24 hours of data per tape received as 8 second and 30 second averages. It could also store some information as 1 second measurements. All the data was stored on a 25.6 megabyte Winchester disc unit with transfer to tape for permanent storage. The 'averaging' facilities were particularly useful in analysing pulsating flows.

b. Display information derived from

both analog and digital inputs to the operator on demand.

c. Provide control functions to the plant. These were predominantly flowrate, level and blending ratios.

d. Sound alarms to the operator.

e. Allow the operator to produce a coloured hard copy of recorded data.

f. Allow off-line historical data reduction and graphics to develop correlations with proposed flow models.

g. Display real and historical hydraulic gradient and batch tracking diagrams.

2.2.10 Onsite Laboratory

A comprehensive laboratory facility was installed for concentration, particle size analysis, coal quality, ph and rheological measurements.

3. Operating Experience

3.1 Equipment and Ancilliary Processes

Extensive prior small scale test work had been carried out on the fine slurry production plant. This was an advantage as the hammermill and wet ball milling system performed well other than simple material handling problems. The fines slurry concentration was kept in the range 55-58% Cw. Below this concentration the slurry would tend to settle out in the unagitated storage tanks after about five days, above 58% Cw it became too thick to pump or reclaim. These limits were specific to the batch process used at the test facility and would not be a limitation on a commercial application where the wet milling would be at a much more economic 50% Cw.

While the Stabflow mixing preparation plant generally worked well, the weigh-feeder was inaccurate. In practice the feed rate accuracy was ± 10% at best and this made accurate mix ratio control difficult. Because of this, the ratio controllers could not be used during batch production because they relied on the inaccurate weigh-feeder signal. For the future a more accurate weighfeeding system is recommended, together with online coarse coal moisture measurement to enable a more consistent control on the overall mix concentration. The continuous paddle mixers had a minimum effective flow of about 30% of rated capacity.

The majority of the instrumentation worked well. Only two of the SMART transducer diaphragms failed despite severe shock and solids loadings. They proved very accurate with little drift. The Kistler pressure transducers did block up after prolonged use, but were generally robust and accurate. The magnetic flowmeters consistently underpredicted the true Stabflow slurry throughputs after calibration with water. It is not known whether this is due to this type of meter or the flow characteristics of Stabflow slurries. The main loop density meter required several calibration checks. While density measurements were consistent subsequent concentration predictions are highly sensitive to the base coal s.g. which was variable. There is some evidence that the inverted 'U' piping configuration used in the main flow loop for the flow meter and density meter may induce density waves and consideration is being given to mounting the instrumentation elsewhere and removing that loop.

The Scada system proved of immense value for both plant control and data interpretation. An example of a typical VDU display is shown in Figure 5.

Problems with the main slurry pumps were the main limitation to the phase I operations. Extensive prior works performance tests on 'dead' concrete with both pumps in parallel on a test loop had not identified these problems. In operation at the Stabflow facility, transient shock pressures associated with the 'Delta' valve changeover and 'square wave' output were relatively severe; transient peak pressures approaching five times the average discharge pressure were recorded - see Figure 6. Extended operations with these shocks would damage both pumps and pipeline and considerable efforts were made to eliminate them. It is apparent that the transient shock pressures experienced with this type of pump and valve are a function of the larger pipe diameter used. They are of a much lower magnitude in concrete pumping applications which use pipes up to 125 mm bore.

By using the BHRA HYPSMOP transient analysis computer programs the fundamental cause of the shocks was identified. The large main pipeline damper and various ancilliary dampers effectively eliminated shocks propagating downstream, but did not improve matters upstream and shocks were transmitted throughout the pump hydaulics system and frame. At moderate pressures low volume air injection direct into the pump discharge was partially effective, but adversely affected the recirculating slurry characteristics and was ineffective at higher pressures. A solution was developed, incorporating a special valving arrangement, which effectively eliminated the shocks and a shock free trace for pumping at 58 bar is shown in Figure 7. Both pumps were adversely affected by the shocks and required extra maintenance before the shocks were eliminated. Once eliminated, the reliability was much improved. On completion of the first phase of the test work, both pumps have been refurbished at the manufacturers works and a production version of the special valving system has been fitted. Detailed examination of the pump slurry hydraulics showed the wear to be low. Development programmes have been identified for a larger pump for a commercial application.

3.2 Pipeline Process Experience

3.2.1 In all, some 50 different consistency Stabflow mixes were flow tested in the 0.5 km and 2.5 km loops; in addition ten high concentration batches were flow tested in a shorter 10"/12" once through system. Concentrations tested were in the range 65% Cw to approximately 83% Cw and flow velocities in a range from zero to 0.9 m/sec. To reduce the number of variables, the first phase of tests have been conducted with coal from the same mine, a part washed steam coal from NCB Hatton Colliery. In practice there was some variation in the quality and the amount of fines in the as delivered coal. Later batches had significant amounts of large heavy shale pieces and the range of size distributions for the coarse coals are shown in Figure 8. Efforts were made to hand pick large coal pieces (+50 mm) and shale from the main conveyor belt. Test work with this scale of plant requires large quantities of coal, the 2.5 km loop requires over 200 tons of coal to fill it, in all some 2,500 tonnes of coal were used during this first phase, all of which had to be returned to the CEGB in either wet or dry slurry form. During the summer months the slurries from the pipeline dried out naturally and were easily handled at 13% to 14% total moisture.

3.2.2 The majority of the tests were aimed at the lower concentration/lower pressure drop slurries for overland pipelines. Test work was directed towards optimising a low fines content, low pressure drop, stable slurry. Criteria for stability were developed as the tests progressed. For commercial pipelines, the slurry stability must be within limits for satisfactory operations under all conditions, this is a key factor in expertise gained by the JV. The criteria used included shutdown and restart after several days, absence of restart overpressure, stable pressure drops at very low flowrates, consistent concentration over the velocity range and other more fundamental lab scale tests using soil mechanics criteria. The scanning densitometer provided a measure of insitu concentration profile to identify settled beds.

3.2.3 Tests were conducted with a range of fines (-200 µm) to coarse ratios; the normal pattern of operations was to recirculate the slurry and progressively dilute it until there was some evidence of instability. There was clear evidence of some degradation with repetitive recycling, the increased fines

content tending to diminish the reduction in pressure drops with lower concentration. Future work will be aimed at reducing the number of passes through the loop during testing, long loops up to 7 km are esssential for this work and to distinguish between fines generated by pump systems and by pipelining.

3.2.4 The pressure drops measured were a strong function of concentration and a weak function of velocity as can be seen on the respective Figures 9 and 10. Pressure drop per unit length was essentially constant throughout the loops including the large 5D bends. Reducing fines content reduced pressure drops, in one test a 1.1% reduction in the fines content reduced the pressure drop by nearly 10% at a given concentration. However, with too few fines there is insufficient yield stress in the carrier fines and the slurry becomes unstable. Further work is required to establish the absolute lowest fines possible for a range of different concentrations. This is important for shorter overland pipelines where terminal costs comprise a higher proportion of the overall system.

3.2.5 Pipeline temperatures were monitored and the range of temperatures was 33°C to 8°C. Due to the large pipeline volume the temperatures remained essentially constant (+ 2°C) throughout a test, the flow characteristics were generally independent of temperature over the range covered.

3.2.6 The test work proved the inherent advantages of Stabflow slurries with ease of shutdown and restart and no minimum velocities. The vertical concentration profiles are relatively constant as can be seen in Figure 11 for a typical scanning densitometer profile at 0.18 m/sec. It is clear from Figure 11 that during flow in a horizontal pipe the density distribution in a vertical direction is sensibly constant.

3.2.7 Ultrasensitive measurements of pipeline wear were made using Thin Layer Activation techniques. First results indicate an even wear pattern, but further work with larger tonnages under more closely controlled conditions is required.

3.2.8 Sampling and Particle Size Analysis

Considerable efforts were made during the test programme to develop the sampling and testing arrangements. One of the major problems that had to be overcome was obtaining truly representative samples.

To get good statistical data with 50 mm x 0 coal, a sample of about

150 kg must be weighed and sized. To perform this task, two full line sample points were provided; the first at the start of the loop and the second at the end of the loop. Initially, samples were dumped into a 1 cubic meter hopper where both the weight and volume could be measured. From that hopper, the sample was fed to a set of 'Sweco' screens which divided the whole sample into five components: a portion of the wet finest fraction was then examined in a laser particle size analyser.

Experience soon showed that accurate weight percent determinations could not be achieved with this method due to entrapped air and variations in base coal sg.

For this reason, an oven drying technique was employed to determine slurry solids concentration. This allowed a direct determination to be made and gave more reliable results. To avoid temporary shutdowns for side stream sampling, later samples were taken from the loop return pipeline discharge. The integrated scanning densitometer gave an additional check on slurry density. Extensive use of computers also was of great help in speeding up the turn around of the laboratory analysis.

4. High Concentration Slurries for Shiploading

Pipeline test work with the main slurry pumps and associated suction system was limited to less than 79% Cw due to hang-up and blockages. A modification using a direct feed into the auxilliary Schwing pump hopper improved matters and once through tests were conducted with a 10"/12" system using concentrations of up to about 83% Cw at a range of fines contents. A consistent feed is required to avoid blockages and a screw auger feed is required for concentrations in excess of 83% Cw. The slurries are stable with much reduced fines content and the latter tend to have lower friction losses. Pressure drops in general rise steeply with concentration and for these tests they were an order of magnitude greater than the low concentration tests.

As the aim of the slurry shiploading is to have a slurry which is both pumpable and ship stable, extensive lab scale tests were carried out for the latter criteria. Tests derived from soil mechanics and ship cargo stability methods were developed in conjunction with Warren Springs Laboratories, Manchester University and Delft LGM. These have employed ship motion simulators, index tests, wall friction testers, fines and moisture migration measurements, Rowe consolidation cells and large triaxial tests to increase our understanding of fundamental cargo behaviour. Provisional analysis suggests good correlation with simple index tests and these have been used on samples from the pipeline test facility.

Further pipeline and lab scale test work over a wider range of concentrations and fines

contents are required to establish a wider band of stable, pumpable slurries.

5. Viscometry and Pipeline Flow Models

Stabflow slurry samples from the pipeline test facility and fines slurry samples have been routinely measured for rheological properties. A significant portion of the total 'fines' in the slurry come from the coarse coal feed, this made representative viscometry mesurements on the true carrier slurry alone difficult. Viscometry techniques have been developed for the specific characteristics of Stabflow. A number of different rheological/solids pipelining flow models have been developed and tested against pipe loop data. To date the simpler rheological models give the best correlations, but further work is required in different sized pipelines to confirm scale-up characteristics. For economic studies, a range of scale up predictions have been used.

6. Economics Study Update

In the previous paper (Ref. 1), detailed economic comparisons were given between Stabflow and competitive slurry pipeline and conventional systems. Using data from the test facility these site specific cases have been re-evaluated and an additional generic economic study has been completed. Specific transport energies of the order of 0.03 to 0.06 Kw hr/ton km are foreseen. Overall the studies have confirmed that Stabflow systems are potentially competitive for applications less than 100 km to those approaching 1,000 km. For the shorter distances the breakeven point is specific to the site and insensitive to the scale up predictions which impact the long distance comparisons.

7. Conclusions

The test work has proved that stable Stabflow slurries with economic pressure drops can be pipelined up to 2.5 km and probably for much longer distances. The ease of preparation, slurry handling, shutdown and restart have been demonstrated.

Stabflow pipeline systems offer an economically attractive alternative to other systems for medium and long distance coal haulage with a final product which is acceptable to the international coal trade. Further work is required to fully demonstrate the technology for commercial applications.

8. Acknowledgements

We would like to particularly thank Owen Davies, Mike Weston and John Carney of Bechtel, together with James Whitmore and Nigel Brown of BP and Tom Grove of BHRA for their efforts during the Project.

References

1. Brookes D.A.; Dodwell C.H.; 'The Economic and Technical Evaluation of Slurry Pipeline Transport Techniques in the International Coal Trade'; Paper A1 BHRA Hydrotransport Conference 9; Rome, Italy, October 1984.

Fig. 1.   View of Stabflow test facility.

Fig. 2.   Main pumps with tank discharge and suction manifold.

Figure 3
STABFLOW TEST FACILITY
SIMPLIFIED FLOW DIAGRAM

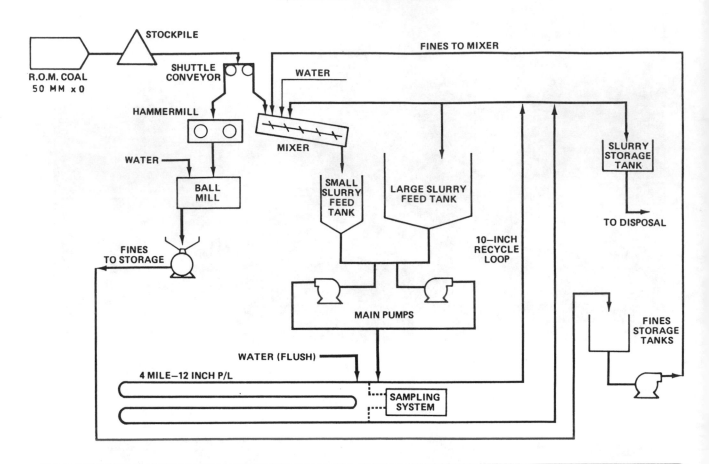

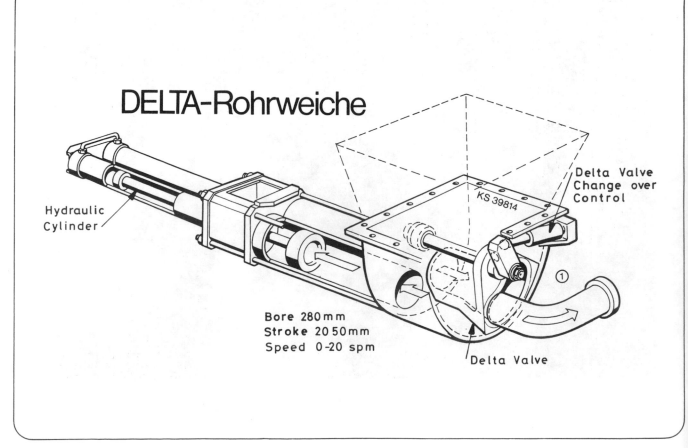

FIG.4 SCHEMATIC ARRANGEMENT OF MAIN SLURRY PUMPS

97

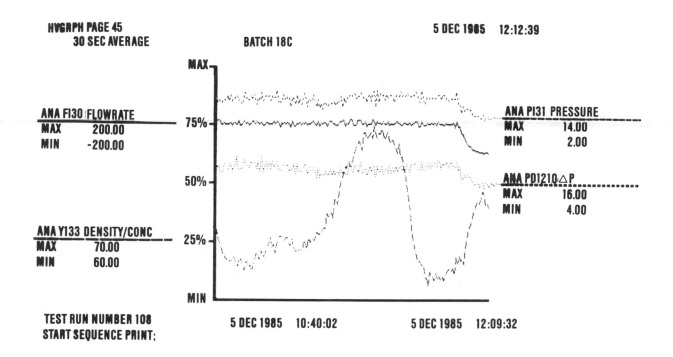

FIG. 5   TYPICAL SCADA HISTORICAL DATA DISPLAY

FIG 6. TYPICAL UNMODIFIED PUMP DISCHARGE AND HYDRAULIC OIL PRESSURE
TRANSIENTS FOR ONSITE & WORKS TESTS

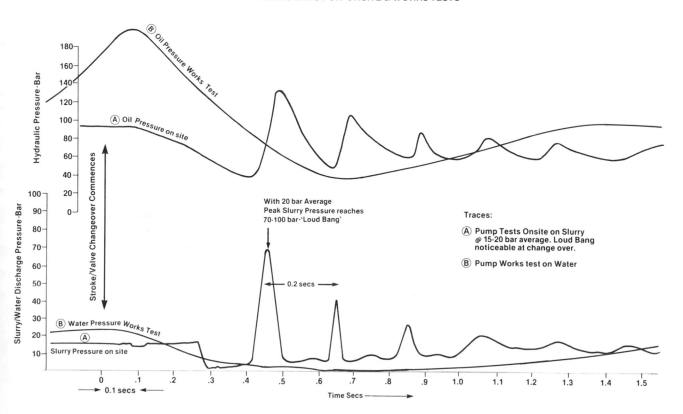

Fig 7 TYPICAL MODIFIED PUMP DISCHARGE PRESSURE TRANSIENT - PULSE FREE

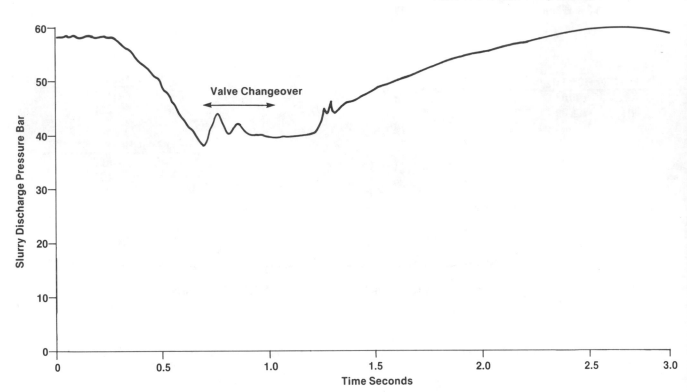

Tests Onsite with Slurry at 58 Bar

Valve Changeover

FIG 8 COARSE COAL P.S.A.

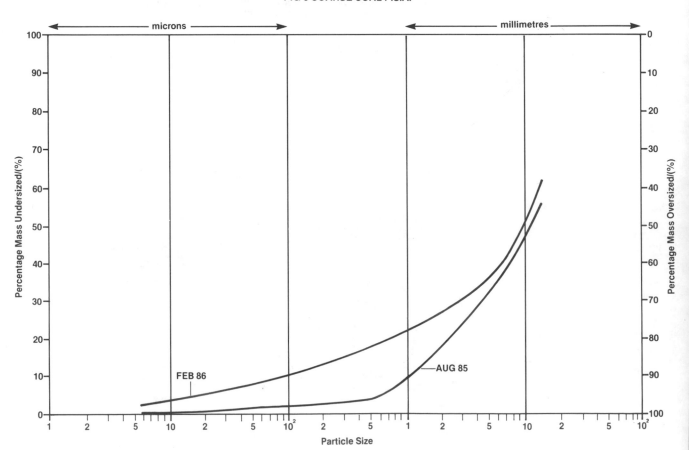

FIG.9 GRAPH OF COAL CONCENTRATION % Cw.v. PRESSURE GRADIENT RATIO

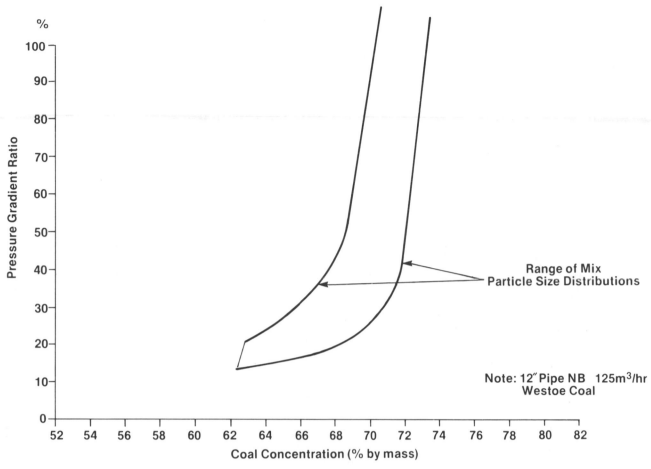

FIG.10 GRAPH OF FLOW RATE v. PRESSURE GRADIENT RATIO FOR VARIOUS CONCENTRATIONS (% Cw)

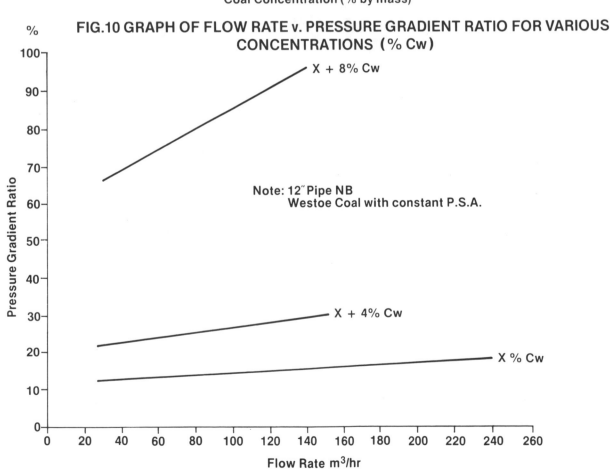

Figure 11

**VERTICAL SOLIDS CONCENTRATION
SCANNING DENSITOMETER RESULTS**

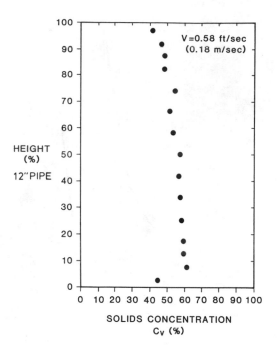

10th International Conference on the
Hydraulic Transport of Solids in Pipes

HYDROTRANSPORT 10

Innsbruck, Austria: 29-31 October, 1986

PAPER D1

VERTICAL HOISTING OF SETTLED SLUDGE SOLIDS FROM
THE 2500 LEVEL VIA A SINGLE PUMP LIFT

J. Hastings
M. Zoborowski

Mr. Hastings is Underground Project Construction
Supervisor at Westmin Resources Limited, Myra
Falls, British Columbia, Canada.

Mr. Zoborowski is the Industrial Sales Manager for
Mining and Slurry Transport at Zimpro Inc.,
Rothschild, Wisconsin, U.S.A.

SUMMARY

Described are design considerations, installation,
and operation of a tubular diaphragm positive
displacement high pressure pump for the vertical
hydrohoisting of a fine-grained complex sulphide
ore containing 20% to 75% pyrite, mineralized with
chalcopyrite (Cu), galena (Pb), and sphalerite
(Zn).

The pump is located at the 2500 level of the mine,
approximately 328 m (1077 feet) below sea level,
and pumps to the tailings deposition pond, a
vertical distance of 699 m (2292 feet) plus a
horizontal distance of 762 m (2500 feet).

Specific gravity of the solids is 4.3 maximum,
density is controlled to nominally 35% - 45% (C_w)
solids by weight which results in a slurry
specific gravity of approximately 1.4 to 1.5.

Startup conditions, modifications, and operating
experiences are detailed with special attention to
wear characteristics. Suggestions and ideas that
may be implemented within the design of a vertical
slurry pipeline are detailed. Approximately one
year of operating experiences are reviewed.

Theoretical design calculations are coupled with
an actual operating installation which confirms
and also supplies meaningful operating data for
the design of pumping systems for vertical slurry
pipelines.

NOMENCLATURE

C_w	Concentration by Weight	%
d	Particle Diameter	m
D	Internal Diameter of Pipe	m
L	Length of Pipe	m
H	Total Head	m
P	Pressure	bar
V	Velocity	m/s
p	Density of Slurry	kg/m^{-3}
s	Relative Density - Density Ratio of Solids to Carrier Fluid	
Q	Volumetric Flow Rate	m^3/s

INTRODUCTION

Westmin Resources completed, in September of 1985,
a 250 million (Canadian Dollars) capital program
to develop its new H-W orebody located on
Vancouver Island, British Columbia, Canada (Ref.
1) (Schematic B).

A showpiece in British Columbia's mining industry
- and probably nationally as well - this mine/mill
complex was officially opened about five years
after discovery of the orebody.

The mine is located in the middle of a park so
addressing environmental concerns was a key
element. Almost $20 million was spent on engi-
neering and environmental studies implementing
systems for the protection of the area.

Westmin decided on a sub-aerial technique for
tailings disposal, a land based system which will
be sealed and revegetated at the completion of
mining. All water flows from the mining and
milling complex (including waste stockpiles) pass
through two treatment plants and it is confirmed
that government performance expectations have been
more than satisfied.

Water is pumped from the underground workings to
surface in one lift. The settled sludge from two
parallel sumps is passed to a mix tank (Fig. 4).
The density is controlled to between 35% to 45% C_w
and the sludge is pumped at .0035 to .0038 m^3/s
(55 to 60 U.S. GPM) vertically 699 m (2292 feet)
and then horizontally 762 m (2500 feet) via a
single pump lift with a tubular diaphragm positive
displacement high pressure slurry pump.

OPERATION

The theory of operation of the tubular diaphragm
slurry pump is fairly simple. (Ref. 2) The slurry
side of the pump consists of the inlet piping
connection and inlet check valves, the outlet
check valves and outlet piping connection, and the
pressure vessels containing the tubular diaphragms
and their holders. The pumping power is transmitted from an electric motor or motors driving
one or more hydraulic pumps (depending on required
capacity - photo "A" has three motors and three
hydraulic pumps) through the hydraulic system to
the tubular diaphragms.

The major components shown on the simplified flow
diagram (Fig. 1) are:

(1)	Inlet Check Valve
(2) & (12)	Pressure Vessels
(3) & (11)	Tubular Diaphragms
(4) & (9)	Stroke Control Cylinders
(5) & (10)	Free Floating Pistons
(6)	Hydraulic Oil Reservoir
(7)	Motor and Hydraulic Pump
(8)	Four-Way Transfer Valve
(13)	Outlet Check Valve

A complete operating description has been previously published (Ref. 3).

The tubular diaphragm slurry pump enjoys several unique and noteworthy features.

1. No flush water system is required.
2. The high pressure slurry pump can be started up and operated against a closed line block valve.
3. There is no packing of any kind and consequently no possible leakage from this source.
4. The built-in internal relief system of the tubular diaphragm design can be set to a maximum desired system pressure above which oil will not be delivered to the hydraulic system.
5. External relief valves provide double protection against over-pressurization.

To accomplish the continuous pumping action, each pressure vessel alternately fills and pumps out once every 8 - 12 seconds. The slow cycle rate (5-7 cycles per minute) greatly prolongs check valve life. This reduced cyclical rate has resulted in check valve lives of up to one and one-half years in some installations.

However, depending on slurry solid characteristics, check valve life will normally range from several months to a year.

The major reason for the increased life is that even though the check valves are seeing the same material wear characteristics as in conventional pumping equipment, the cycle rate is so much reduced that it simply takes that much longer for wear to occur. It would be like comparing wear on a 5 - 10 RPM unit versus a 100 - 150 RPM unit. The wear characteristics are still there, only the rate is greatly reduced.

INSTALLATION

The general arrangement of the underground pumping facilities is illustrated in Fig. 2. The two sumps are elevated above the 25 level with water pumped from the underground workings to surface in one lift via 3 multistage centrifugal pumps (each .0315 m^3/s or 500 U.S. GPM at a total head of 670.6 m or 2200 ft.).

The sump underflow consists of gravel, wood fibre, and slimes with the slimes portion being heavier (and more abrasive) than originally planned. Density is controlled between 35% to 45% C_w prior to being charged into the tubular diaphragm high pressure slurry pump. Two parallel air diaphragm feed pumps are used for charging the high pressure slurry pump.

OPERATING EXPERIENCE

On Aug. 13, 1985, water was pumped from the 25 level to surface developing a pressure in excess of 69 bar (1000 psig).

The .0635 m (2-1/2") dedicated pipe to the surface began whipping rather violently at the surface due to a long unsupported section with 90° elbows. These were replaced with welded 45° angles to eliminate pipe vibration and movement and then anchored firmly. On the following day, actual slurry pumping began. Pumping pressures varied from 90 bar (1305 psig) to 103 bar (1495 psig)

depending on the C_w. After a few hours of pumping, the slurry check valves plugged with debris and the high pressure pump was shut down.

An analysis showed large particles plugging within the valves and in order to correct the situation, screen baskets with .00635 m (1/4") openings were fabricated and installed ahead of the mix tank to prevent large particles like wood slivers from plugging the pump valves.

The next area requiring attention was a means of preventing back pressure surges to the air diaphragm feed or charge pumps (Fig. 3). Whenever larger particle sizes would surge through the high pressure pump valves (due to screen baskets overflowing or splitting open and allowing large particle size passage), back pressure could blow back through ball checks momentarily unseated and rupture gasketing within the air diaphragm feed pumps. Corrective measures consisted of installing a 10.3 bar (150 psig), .0762 m (3") pressure relief valve in the recirculation line to syphon off any pressure surges back to the slurry mix tank. Also, replacement of .00635 m (1/4") screen baskets with a heavier duty .00953 m (3/8") mesh screen size helps to control the problem from developing (screen blinding and screen weakening allowing splits and cracks).

Slurry pumping requirements are only scheduled on a three day per week basis, one eight hour shift per day. General cleaning out of sumps and underground settling ponds may involve daily pump operation for up to a month at a time.

Per the operations group, maintenance since start-up in August, 1985, has been very minimal. Ribbed ball guides were replaced with non-ribbed guides in November, 1985 and since that date, there have been no major valve plug-ups. This allows passage of larger particle sizes by providing more open area around the valve balls (Ref. 3 & 4). On March 1, 1986 hydraulic oil was changed and pump components checked.

No valve or other spare parts usage has been required since startup 14 months ago. However, it must be recognized that the pump does not operate on a continual basis. Nevertheless, the operations group is appreciative of the fact that the relatively high pressure slurry pump operation has not required periodic check valve replacement as the valves are the usual area requiring periodic maintenance in conventional pumping equipment.

A further recommendation by the high pressure slurry pump manufacturer is installation of a check valve between the feed pumps and the relief valve for positive blowback protection (Fig. 3). However, after 1-1/4 years of operation, this has not proven necessary. If the relief valve operation would become erratic, then operations could install the check valve for additional protection of the feed pump system against back pressure surges.

Present operations utilize six trained operators who can review pumping details by viewing a 16 mm sound film which provides complete operations and maintenance details.

sump underflow is not only practical, it offers several benefits to underground mines.

A. By eliminating any solids portion inflow from mine dewatering centrifugal pumps, maintenance on the centrifugal pumps is reduced, often quite significantly. This can involve either many banks of pumps at various levels, or multistage pump units which are even more sensitive to solids of any kind.

B. Manual labor and shaft ramps for cleanout purposes can be eliminated.

C. Skip hoisting of solids is eliminated except for particle sizes too large to be pumped which in this installation comprise no great quantities of material.

Alternative systems can be considered. This paper briefly describes a dedicated pipe line for slurry handling alone. Another pumping strategy which is now in place at another mine injects the slurry solids into the clearwater line after the clearwater mine pumps and eliminates the need for a separate dedicated high pressure slurry pipe line. All that is required is pressure sufficient to allow injection of slurry into the clearwater line from the mine dewatering pumps. This allows a relatively small percentage of slurry flow to be transported by the larger volume of clear water to the surface. In multistage centrifugal installations, maintenance cost savings can be significant due to elimination of any solids contamination within the multistage pumping units.

REFERENCES

1) Westmin Opens $250 M Mine. THE NORTHERN MINER, September 30, 1985.

2) Zoborowski, M.E. and Radloff, D.E.: "Transportation of Slurried Minerals by a High Pressure Cylindrical Diaphragm Pump". Proc. 3rd Int. Conference on Slurry Transportation, Las Vegas, NV, U.S.A., (March 29-31, 1978). Slurry Transport Association, Washington, D.C., U.S.A.

3) Horn, A.C. and Zoborowski, M.E.: "Operating Experience of High Pressure Hydraulic Exchange Pumps in Pumping Ores and Industrial Minerals". Proc. Hydrotransport 7 Paper A1, Sendai, Japan (Nov. 4-6, 1980). Organized by BHRA Fluid Engineering in conjunction with the Slurry Transport Society of Japan.

4) Zoborowski, M.E.: "African Experiences with Pipeline Transportation of Rutile and Sand by Hydraulic Exchange Cylindrical Diaphragm Pump." Proc. Hydrotransport 8 Paper P3, Johannesburg, South Africa (Aug. 25-27, 1982). Organized by BHRA Fluid Engineering in conjunction with the South African Institution of Civil Engineers, The South African Institution of Mechanical Engineers, The South African Institute of Mining and Metallurgy and The Council of Scientific Research.

Photo A. High pressure pump—hydraulic side.

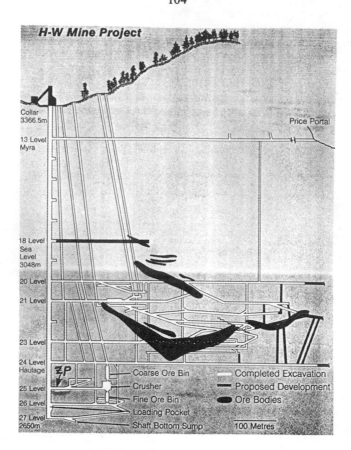

Schematic B.

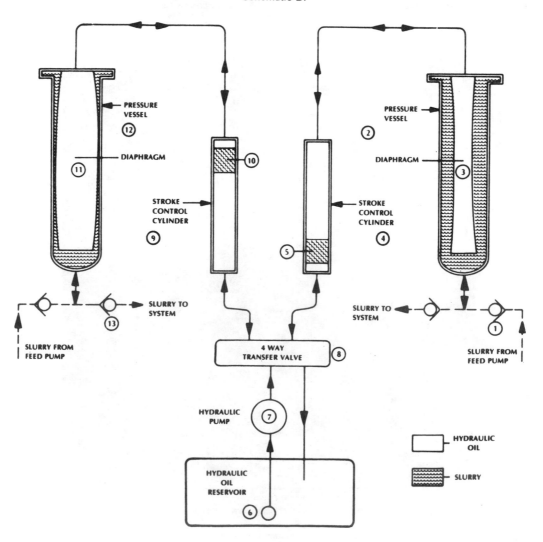

Fig. 1. High pressure pumps—simplified flow diagram.

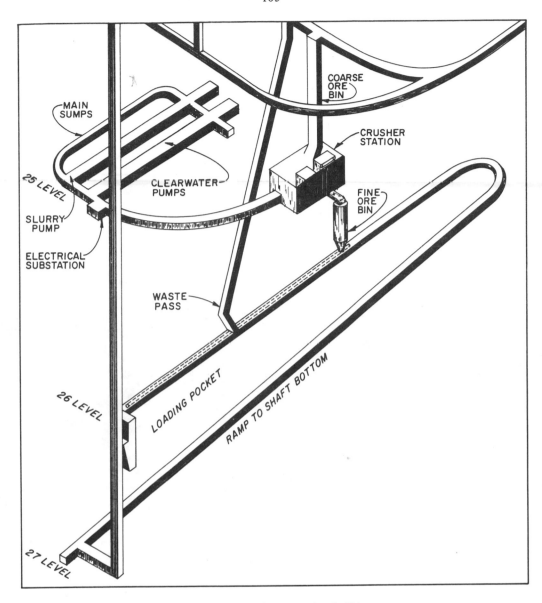

Fig. 2. Underground pumping facilities.

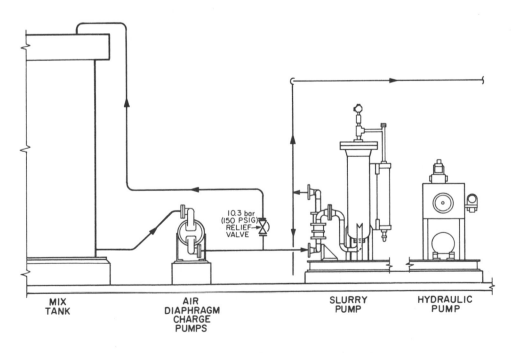

Fig. 3. High pressure slurry pump arrangement.

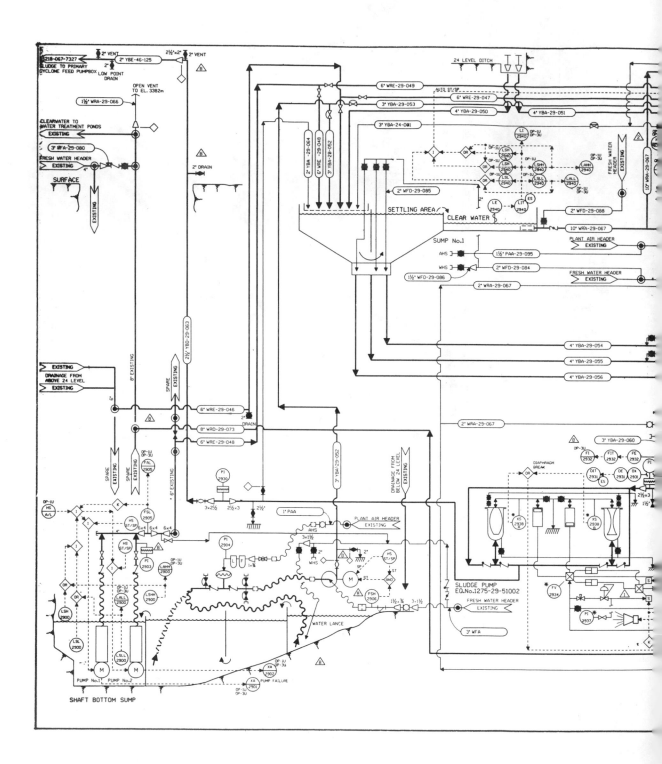

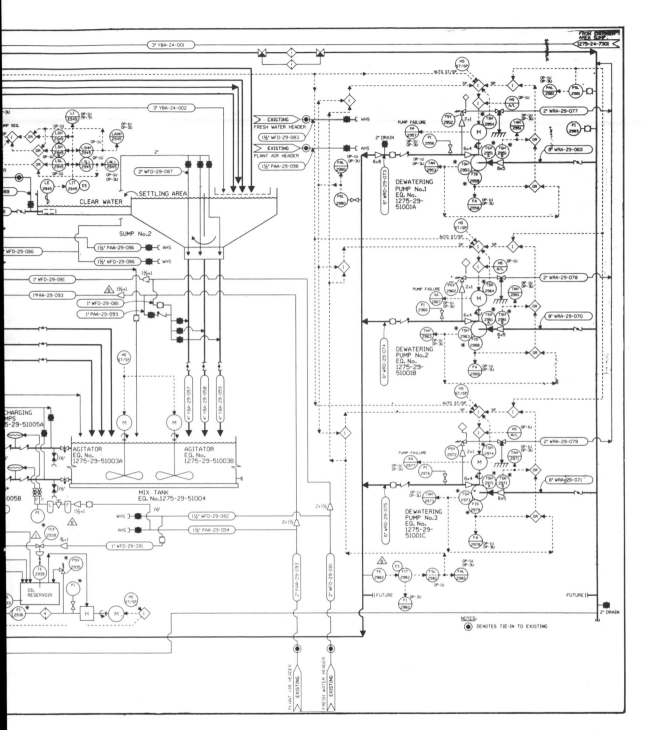

Fig. 4.   Mine dewatering system—piping and instrument diagram.

10th International Conference on the
Hydraulic Transport of Solids in Pipes

HYDROTRANSPORT 10

Innsbruck, Austria: 29-31 October, 1986

PAPER D2

APPLICATION OF THE AIR-LIFT PRINCIPLE TO SOLVE MAINTENANCE PROBLEMS

Etienne BERLEUR   Michel GIOT
Manager R & D   Professor
SBBM & Six Constructs, S.A.,   Université Catholique
Belgium.   de Louvain, Belgium.

1. INTRODUCTION - SCOPE OF THE STUDY

Since the early sixties siltation has been a source of trouble for many Port Authorities. Indeed, by deepening and widening of the approach channel the cost of maintenance dredging increases exponentially.

In many cases the ship channel is considerably deeper than the surroundings, which may cause a steady fluid mud flow from the sea or the estuary towards the deeper parts of the harbour.

On the other hand in harbours, along quay walls, under pontoons and in small marinas, the mud which enters with the flood, settles down at slack water and forms a mud layer of increasing thickness, which hinders the commercial exploitation of the harbour.

To avoid the difficult and expensive maintenance dredging at a harbour or dock entrance, in front of a navigation lock or under a landing-stage, it was suggested that a mud capture installation be built, provided with a fixed pump. The idea is basically to remove the mud by regular pumping, while it is still fluid, instead of dredging the consolitated mud. It has been developed by BERLEUR and al.([1] to [4]).

A mud trap is a fixed pumping plant, working according to simple rules and provided with automatic steering.

As pumping plant, it was therefore suggested to use the air-lift principle, which has the advantage of limiting heavy wear, as is the case for classical pumps, caused by the sediments passing through them.

The greater the length of the horizontal suction pipe and riser pipe, the greater is the wear of the pumps. Repair or exchange is very difficult and time consuming due to the difficult access to pumps immersed in the silted waters of the harbours and estuaries.

A possible arrangement of a airlift pump consists of a horizontal or inclined suction pipe connected to a vertical riser. The air injection port is located close to the lower end of the riser.

The pumping operation cost, in terms of energy absorbed by the air compressor is somewhat higher than it would be with a well designed dredge pump. However, this deficiency can be limited by optimizing the design of the system, as well as the air flow rate. Moreover the disadvantage of the energy consumption can be counterbalanced by the advantages of simplicity, reliability and low maintenance cost of the air-lift pump itself.

A theoretical and experimental study has been carried out at the Thermodynamic Department of the University of Louvain with the sponshorship of S.B.B.M. & SIX CONSTRUCT, in view of developing efficient air-lift pumping systems which could be used as mud trap installations in harbours.

2. CONCEPT OF AN AIR-LIFT PUMP OPERATING IN SHALLOW WATER

An air-lift pump mainly consists of a vertical pipe immersed in a liquid pool and supplied with compressed air at its lower end. As a result of the air supply, the fluid density is decreased inside the tube with respect to the density of the external liquid, and this creates a static pressure disequilibrium. Consequently, the mixture level starts to rise in the pipe above the liquid level until an equilibrium level is reached. If the upper end of the pipe is located below the equilibrium level, a discharge of the mixture occurs at a rate such that the friction and acceleration pressure losses in the pipe compensate the available static head.

According to GIBSON [5], the air-lift pump was invented by Carl LöSCHER at the end of the eighteenth century. Owing to its poor efficiency compared with other pumps, and its adaptability to many difficult cases of pumping, the air-lift pump is encountered today only in particular applications (GIOT, [6], [7]). Let us emphasize the lifting of solids in shaft and well drillings, the transport of coal in shafts and the pumping of manganese nodules from the bottom of the Pacific Ocean.

In the case of the maintenance of harbours, clay or sand sediments laying on the sea or river bed has to be raised over a short height which typically ranges between 5 and 30 meters. A possible arrangement (fig.1) consists of horizontal or inclined suction pipes connected to vertical risers. These suction pipes are ended with a suction head consisting of a perforated inlet like a well screen, covered by a cap. The air injection port is located close to the lower end of the riser in order to maximize the height of air-fluid mixture. The air injection device may consist of an annular chamber whose inner wall, having the same inner diameter as the riser, is perforated with holes with diameters of 2 to 5 mm. The air injection device is connected to an air compressor located at the surface, through a vertical air supply pipe, which is generally attached to the riser. The riser is ended with a 90° elbow and a horizontal pipe discharging the mixture either in the water under the water level, or in a settling basin. The water level can be subject to periodic variations according to the tide.

In view of the just defined concept, a typical geometrical configuration is adopted in this study (fig.2). For an easy understanding of the paper, we define the following characteristic points of the flow path :

A = a point located at the same level as the suction, and at such a horizontal distance from the pipe inlet that the flow velocity be negligible.

I = air injection

R = a point located at the same level as the discharge, and at such a horizontal distance from the pipe outlet that the flow velocity be negligible

N = the highest point of the riser

0,1,2 = respectively inlet, lower elbow and outlet sections of the pipe

The heights h_A, h_I and h_R are variable with the tide, the sign of h_R can even change.

3. ANALYTICAL MODELING

3.1. Momentum equations

In order to determine the flow rate of air $\dot{V}_G$ required to pump a given mixture flow rate $\dot{V}_m$ of water and solids, the following momentum balance equations are written :

inlet pressure losses :

The pressure losses at the inlet of the pipe are due to the acceleration of the mixture and to the singular pressure drop $\Delta p_{f,0}$ through the suction head :

$$p_0 - p_A = - \Delta p_{f,0} - \rho_m \left(\frac{\dot{V}_m}{A}\right)^2 \tag{1}$$

where ρ_m denotes the solid-liquid mixture density, $\dot{V}_m$ its volumetric flow rate and A the cross section area of the suction pipe. The singular pressure drop is determined by means of standard formulas or with the help of preliminary experiments.

- **pressure losses in the suction pipe**

The pressure losses in the suction pipe consist only of friction pressure losses :

$$\Delta p_{A-I} = f_m \frac{4 \, L_{A-I}}{D} \frac{\rho_m}{2} \left(\frac{\dot{V}_m}{A}\right)^2 \tag{2}$$

where L_{A-I} is the length of suction pipe plus an equivalent length to account for the elbow connecting the suction pipe to the riser. The friction factor f_m is calculated by means of the correlation proposed by CHURCHILL [8] :

$$f_m = 2 \left[\left(\frac{8}{Re_m}\right)^{12} + \frac{1}{(C + B)^{3/2}} \right]^{1/12} \tag{3}$$

where

$$C = \left[2.457 \, Ln \frac{1}{\left(\frac{7}{Re_m}\right)^{0.9} + 0.27 \frac{\epsilon}{D}} \right]^{16} \quad ; \quad B = \left(\frac{37530}{Re_m}\right)^{16}$$

and

$$Re_m = \frac{D}{\nu_m} \left(\frac{\dot{V}_m}{A}\right) \tag{4}$$

The kinematic viscosity of the mixture is derived from the expression proposed by RICHARDSON and ZAKI [9] :

$$\nu_m = \frac{\rho_L}{\rho_m} \nu_L \left(1 - \frac{\alpha_d}{0.62}\right)^{-1.55} \tag{5}$$

with

$$\alpha_d = \frac{\rho_m - \rho_L}{\rho_s - \rho_L}$$

In the above expression, ρ_L and ρ_s denote respectively the density of the water and of the solid phase, and ν_L is the kinematic viscosity of the water. Let us point out that equation (3) applied to the laminar flow regime is only valid for newtonian suspensions.

- **pressure losses in the riser**

The pressure loss Δp_I through the air injector device is approximated by :

$$\Delta p_I = p_{I+} - p_{I-} = \rho_m v_{m I-}^2 - \left[\alpha_G \rho_G v_G^2 + (1-\alpha_G)\rho_m v_m^2\right]_{I+} \tag{6}$$

where v_G and v_m denote respectively the velocity of the air and of the liquid-solid mixture, and α_G is the void fraction of the three-phase flow.

The riser is subdivided into n pipe sections, each of then having a length Δz. For each of these pipe sections, the pressure loss is calculated by means of the following equation $(1 \leqslant j \leqslant n)$:

$$p_{j+1} - p_j = - \Delta p_{f,j} - [\bar{\alpha}_G \bar{\rho}_G + (1-\bar{\alpha}_G)\rho_m]g\Delta z + [\alpha_G \rho_G v_G^2 + (1-\alpha_G)\rho_m v_m^2]_j$$
$$- [\alpha_G \rho_G v_G^2 + (1-\alpha_G)\rho_m v_m^2]_{j+1} \tag{7}$$

In order to take the slip of the air with respect to the mixture into account, the ARMAND parameter K is used to relate the void fraction to the volumetric quality β :

$$\alpha_G = K \, \beta \tag{8}$$

where

$$\beta = \frac{\dot{V}_G}{\dot{V}_m + \dot{V}_G}$$

For practical calculations, we have taken $K = 0.8$. Moreover, $\bar{\alpha}_G$ and $\bar{\rho}_G$ in equation (7) are calculated at the average pressure along the pipe section. The friction pressure drop is derived from the following correlation (GIOT [7]) :

$$\Delta p_{f,j} = (1 - \bar{\alpha}_G)^{-1.75} \, f_{m,0} \, \frac{4 \, \Delta z}{D} \, \frac{\rho_m}{2} \, (\frac{\dot{V}_m}{A})^2 \tag{9}$$

where $f_{m,0}$ is the friction factor of the mixture flowing alone in the pipe, i.è. at velocity V_m/A.

- pressure losses in the discharge line

Equations (7) to (9) can be used to predict the pressure losses through the discharge line, except that the gravity term of equation (7) vanishes. An equivalent length is used to predict the pressure losses through the elbow.

Assuming that pressure p_2 at the outlet of the pipe is equal to ambient pressure p_R, the correct value of the air flow rate is achieved when

$$p_2 = p_{atm} + \rho_L g h_R \quad \text{if the discharge level is below the free surface}$$
$$p_2 = p_{atm} \quad \quad \text{if the discharge level is above the free surface}$$

3.2. Air compressor power

Taking into account the isothermal efficiency η_{iso} of the compressor, and the air pressure losses Δp_{air} through the air supply pipe, the expression for the air compressor power is :

$$P_c = \frac{1}{\eta_{iso}} \, \rho_{G,atm} \, \dot{V}_{G,atm} \, RT \, Ln \, \frac{p_I + \Delta p_{air}}{p_{atm}} \tag{10}$$

3.3. Definition of the pumping efficiency

The useful work produced by the air-lift system to pump a unit of mass of mixture at the required flow rate is :

$$W_{us} = (1 - \frac{\rho_L}{\rho_m}) \, g \, (h_A - h_R) + \frac{1}{2} \, (\frac{\dot{V}_m}{A})^2 + W_{f,m} \tag{11}$$

where $W_{f,m}$ denotes the energy loss due to friction that would be obtained if the mixture was flowing alone through the system. The overall efficiency is then

$$\eta_c = \frac{\rho_m \, \dot{V}_m \, W_{us}}{P_c} \tag{12}$$

As equation (11) takes into account the real configuration of the circuit, efficiency η_c is system dependent.

4. EXPERIMENTAL SET-UP

4.1. Test loop

In order to validate the above model, tests were carried out by means of the experimental set-up shown in figure 3.

The mixture is stored in vessel R, whose useful capacity is about $6 m^3$ and in pipe T_2 having a diameter of 400 mm, by means of gate valve V_5. The mixture flow rate is measured by means of an electromagnetic flowmeter D located in the vertical downward flow. Air is supplied by a volumetric compressor through air injection I consisting of an annular chamber whose inner wall is perforated with 3 mm holes. The air flow rate is measured by means of venturi M_2, the air pressure and temperature being measured upstream of the venturi.

The three-phase flow is directed upwardly and discharged above the free surface through T_4, or diverted through tube T_3 and discharged below the free surface by means of two pneumatically actuated gate valves V_4 and V_3. A gammadensitometer SC is located in the upper part of the riser and enables the determination of the void fraction along a beam coïnciding with a diameter of the riser. The discharge pipe T_4, located above a transparent perspex section P, has the same inner diameter as the lower part of the riser, i.e. 205 mm, and can rotate around a vertical axis in order to divert momentarly the flow into a weighing tank placed inside the storage vessel R. This weighing tank is very useful when one deals with flow rate measurements of settling slurries.

Tube T_3 has an inner diameter of 255 mm. The height of the top of the U tube T_4 with respect to the floor of the laboratory is 10.940 m.

In order to determine accurate piezometric lines along the riser, pressure taps have been drilled in the riser wall at 10 different levels distant from each other by about 775 mm. At each of these levels, 4 pressure holes have been drilled. They are located at 90° from each other, and connected in a symmetric way to a set of electro-magnetic valves. These valves enable the connection of each pressure line to the HP or the LP side of a pressure transducer selected for small differential measurements. A second pressure transducer enables the measurement of the absolute pressure at the bottom of the riser.

4.2. Data acquisition system

The date logging and treatment is performed using a home-made micro-computer called MIDAS (Modulary Interactive Data Acquisition System) whose diagram is shown in figure 4. MIDAS consists of a Z80 microprocessor, a ROM memory of 8 kbytes, a 24 kRAM memory, an internal clock, an ADC converter and a few circuit boards specially developed for gammadensitometer acquisition. Two time bases were programmed in the clock : 1 s and 1/100 second.

MIDAS is equipped with the FORTH programming language, which is about 10 times quicker than the BASIC.

5. TEST RESULTS AND VALIDATION OF THE MODEL

5.1. Water pumping

The air-lift pump operating in shallow water is very sensitive to variations of the pressure at air injection due to pressure drops in the suction pipe. In view of the many singularities located in the test loop upstream of the air injector, it was decided to determine experimentally the total friction pressure drop upstream of the air injector. An equivalent length of suction pipe was deduced from these measurements made at 20 different water flow rates : L_{eq} = 23.6 m.

Solving the momentum equations presented in section 3, with the correct pressure at the injection, should enable to predict the piezometric line along the riser, as well as the required air flow rate. Figure 5 shows a typical piezometric line: it can be concluded that all data points lie along a straight line. The predictions give also a straight line, whose slope is mainly affected by the void fraction model, and, to a much smaller extent by the friction pressure drop correlation.

With the void fraction model given by equation (8) with K = 0,8, the slope of the predicted piezometric line is generally close to the slope of the experimental line. The agreement can still be improved by using the classical model proposed by ZUBER and FINDLAY [10] :

$$\alpha_G = \frac{\beta}{C_0 + \frac{v_{Gj}}{J}} \tag{13}$$

where $C_0 \approx 1$ at pressures close to the atmospheric pressure, J is the total volumetric flux :

$$J = \frac{\dot{V}_m + \dot{V}_G}{A}$$

and v_{Gj} is the drift velocity of bubbles in case of bubbly flow :

$$v_{Gj} = 1.41 \left[\frac{\sigma g(\rho_m - \rho_G)}{\rho_m^2}\right]^{1/4} \tag{14}$$

where σ is the surface tension. As v_{Gj} is flow pattern dependent, equation (14) can be replaced by another more appropriate correlation whenever needed. Here, we have simply divided v_{Gj} calculated with equation (14) by a factor of 3, in order to obtain a good agreement between the calculated piezometric lines and the data (figure 6).

The void fraction measurements show that the ARMAND void fraction model as well as the ZUBER and FINDLAY model predict too large void fractions at low water flow rates. Then, it is not surprising that they underestimate the air flow required for lifting the water. In order to improve the predictions, the void fraction model proposed by WEBER and al. [11] has also been tested :

$$\alpha_G = \cfrac{1}{\cfrac{1}{\alpha_{G,0}} + \cfrac{\dot{V}_m}{\dot{V}_G}} \qquad (15)$$

where $\alpha_{G,0}$ is the void fraction observed in a bubble column with no net flow of water. From the data reported by WEBER and al. referring to a 94 mm diameter pipe, we obtain the following correlation :

$$\alpha_{G,0} = -0.948(\frac{\dot{V}_G}{A})^{1/2} + 3.894(\frac{\dot{V}_g}{A})^{1/4} - 2.346(\frac{\dot{V}_G}{A})^{1/8} \qquad (16)$$

Equation (13) overestimates the void fraction at low water flow rates, whereas equation (15) overestimates the void fraction at high water flow rates.
This is the reason why an automatic selection of the lowest value at each step of the integration of the momentum balance equations has been introduced in the computer program. Figures 7 and 8 illustrate the results obtained in terms of water flow rate versus air flow rate respectively at low and high water flows. When the water discharge is achieved through pipe T_4 above the free surface of tank R, we have noted that the atmospheric pressure was reached at the top of the upper U bend, due to partial filling of this bend or phase separation. Figure 9 shows the pumping efficiency of our laboratory facility. It is interesting to note that a maximum efficiency is reached for some particular value of the water flow rate : at smaller flow rates the air slip increases, causing a loss of efficiency, whereas at larger flow rates the friction becomes the limiting factor.

5.2. Pumping of clay-water mixtures

Clay slurries at three different densities have been used, namely 1075, 1165 and 1211 kg/m³. The non-newtonian character of these mixtures is illustrated in figure 10 where some results of rheologic tests have been plotted in a shear-stress vs.shear-rate diagram. At least for the two mixtures having the largest densities, a BINGHAM plastic model seems to be adequate. The friction correlation proposed by LAZARUS [12] for yield pseudo-plastic fluids was not found successful in this case : using the equivalent length determined with water flows, the measured friction factor of the highest density mixture is 1.15 instead of the calculated value of 0.7.

Les us define a two phase friction multiplier :

$$\Phi^2 = \frac{\Delta p_{f \text{ clay+water}}}{\Delta p_{f \text{ water}}}$$

A preliminary correlation for this parameter is given by

$$\Phi^2 = 1 + a(X) - b(X)\dot{V}_m \qquad (17)$$

where

$$X = \rho_m - \rho_L \ .$$

$$a = -0.0226 \, X + 0.00252 \, X^2$$

$$b = -0.692 \, 10^{-3} + 0.0626 \, X^2$$

The three-phase flow friction factor is calculated by the following expression derived from gas-liquid flow considerations :

$$f_{3\Phi} = (1 - \bar{\alpha}_G)^{-1.75} \, \Phi^2 f_{m,0}$$

where Φ^2 is calculated at the velocity of the mixture in the three-phase flow.

A good agreement between the data and the calculations is obtained when the ARMAND void fraction model is used with a smaller value of parameter K than in air-water flow : K = 0.6 (fig.11).
Further investigations should concentrate on this important parameter.

6. CONCLUSIONS

A model for predicting the performance of an air-lift pump used to pump mud in shallow waters has been proposed. Tests carried out on a large test facility have contributed to the validation of the model.
The void fraction calculations affect the predictions to a very large extent, since the void fraction controls the gravity term in the momentum equation.

A substantial difference of void fractions, and consequently of air-liquid slip velocity, was found between air-water and air-mud flows. Further studies should relate this difference to the non-newtonian character of the mud.

The easiness of operation of the air-lift pump has been confirmed by the tests. Careful designs of practical installations can lead to better efficiencies than the test facility itself.

Acknowledgments

The authors are indebted to Mr.A.FOUARGE and Mr.J.F.REY for their important contribution to this study.

LIST OF SYMBOLS

A   cross section area
C_0 ZUBER and FINDLAY parameter, eq (13)
D   pipe inner diameter
J   total volumetric flux
K   ARMAND parameter, eq (8)
L   length
P   power

f   friction factor
g   gravity
h   height
p   pressure
v   velocity
z   axial coordinate

α void fraction or volumetric concentration
β volumetric quality
η_{iso} isothermal efficiency
ν kinematic viscosity
ρ density
σ surface tension
Φ friction multiplier

Subscripts

atm	atmospheric	L	liquid phase
d	dispersed	m	mixture
f	friction	S	solid phase
G	gas phase	3Φ	three-phase

REFERENCES

1. BERLAMONT J., THIENPONT M., VERNER B., VAN BRUWAENE A., NEYRINCK L., MAERTENS L. - CHURCHILL - 1983. Mud capture installation at the Sea Lock of Zeebrugge, Proceedings of the 8th International KVIV-Congress on Harbours, June 1983.

2. BERLAMONT J., VAN GOETHEM J., BERLEUR E., VAN BRUWAENE A. - 1985. A permanent Mud Pumping Installation as an alternative for Local Maintenance Dredging, Proc. of the 21st IAHR-Congress, Melbourne Australia, August 1985.

3. BERLAMONT J., VAN GOETHEM J., BERLEUR E. - 1986. Local Maintenance Dredging by means of a fixed mud pumping plant, accepted for Proc. Of the World Dredging Conference, Brighton, March 1986.

4. VERREET G., VAN GOETHEM J., VIAENE W., BERLAMONT J., HOUTHUYS R., BERLEUR E. - 1986. Relations between physico-chemical and rheological properties of fine-grained muds paper accepted for the Proceedings of the 3th International Symposium on River Sedimentation, Jackson, MS USA, March-April 1986.

5. GIBSON A.H. 1961. Hydraulics and its Applications, 5th Ed., Constable, London.

6. GIOT M. 1979. Le système de pompage air-lift et la remontée des nodules polymétalliques marins, La Houille Blanche, 6/7, 359-365.

7. GIOT M. 1982. Three-Phase Flow, Handbook of Multiphase Systems, G.Hetsroni, Ed. Hemisphere McGraw-Hill, Washington, ch.7.2, 7.29-7.45.

8. CHURCHILL - 1977. Chem.Eng.84, 24, 91.

9. RICHARDSON and ZAKI, 1954. Sedimentation and Fluidization, Part I, Trans.Inst.Chem.Engrs.32, 35-53.

10. ZUBER N. and FINDLAY J.A. 1965. Average Volumetric Concentration in Two-Phase Flow Systems, J.Heat Transfer, Trans.ASME, nov.1965, 453-468.

11. WEBER M., DEDEGIL Y. and FELDLE G. 1978. New Experimental Results regarding Extrem Operating Conditions in Air Lifting and Vertical Hydraulic Transport of Solids according to the Jet Lift Principle and its Applicability to Deep Sea Mining, Hydrotransport 5, B.H.R.A., England, F.7.11 - F.7.26.

12. LAZARUS J.H. 1980. Rheological Characterization for Optimising Specific Power Consumption of a Phosphate Ore Pipeline, Hydrotransport 7, B.H.R.A., Sendai, Japan, 133-142.

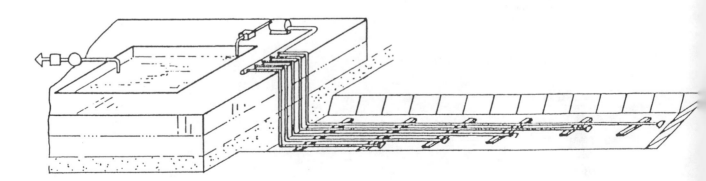

Fig. 1. Concept of an air-lift pumping system.

117

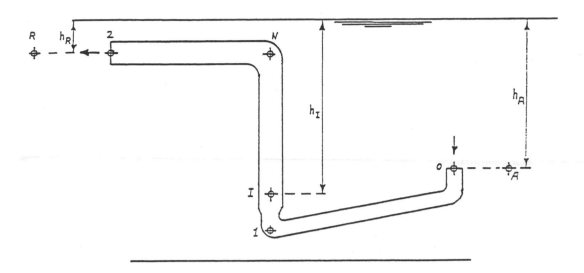

Fig. 2. Geometrical configuration of the air-lift pump.

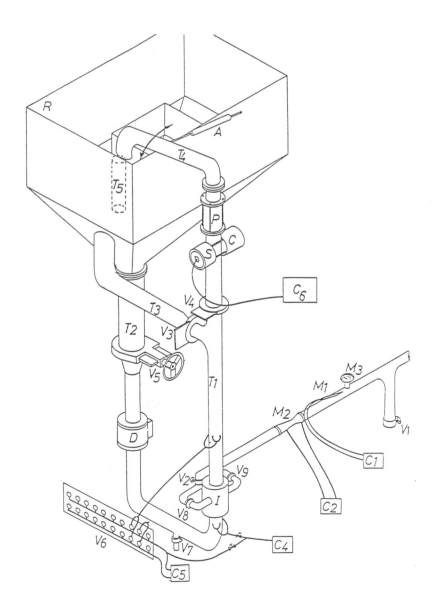

Fig. 3. Experimental set-up.

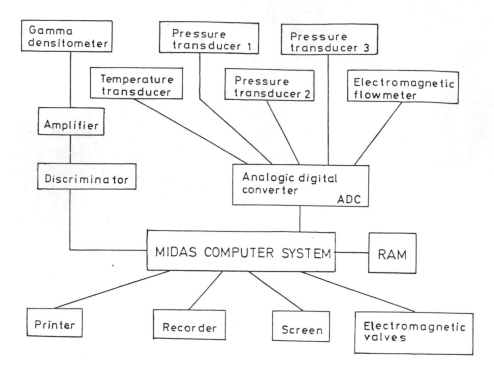

Fig. 4. Data acquisition system.

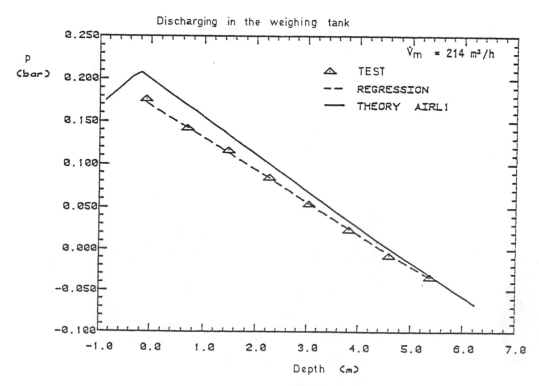

Fig. 5. Comparison between an experimental piezometric line and
the model with K = 0.8.

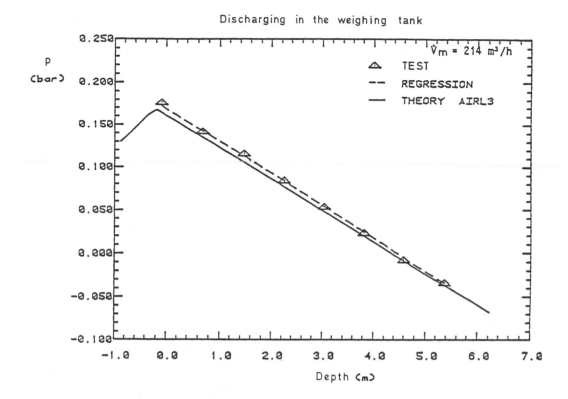

Fig. 6. *Comparison between an experimental piezometric line and the model with equation (13).*

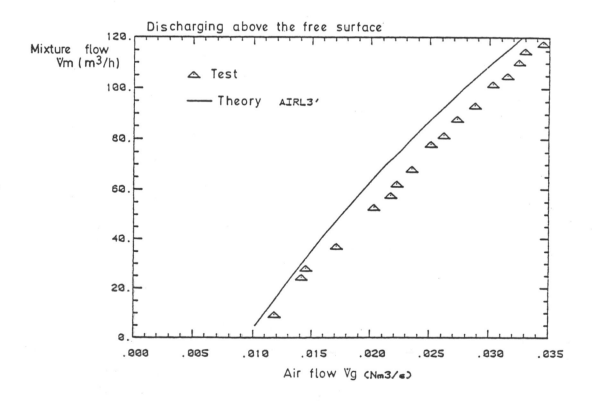

Fig. 7. *Water flow rate versus air flow rate: comparison between data and model.*

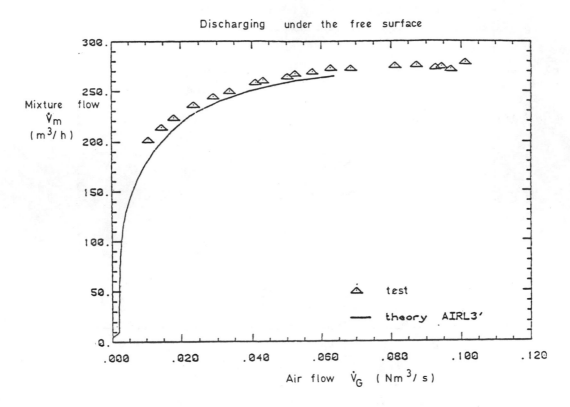

Fig. 8. *Water flow rate versus air flow rate: comparison between data and model.*

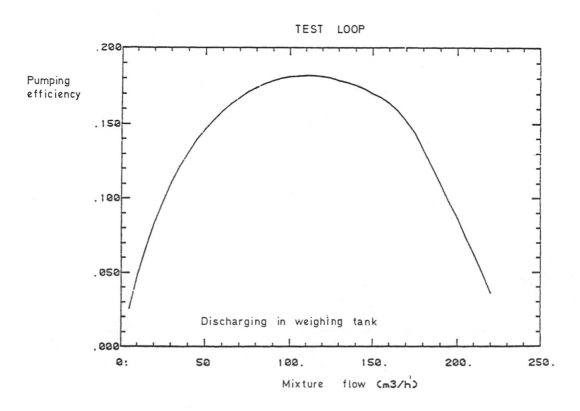

Fig. 9. *Pumping efficiency versus water flow rate.*

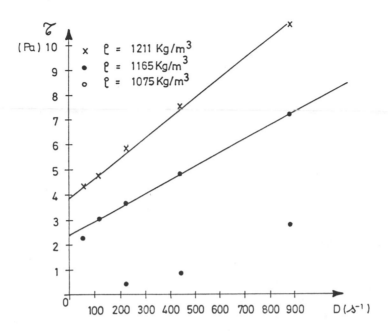

Fig. 10.   Shear stress versus shear rate of slurries at three different densities.

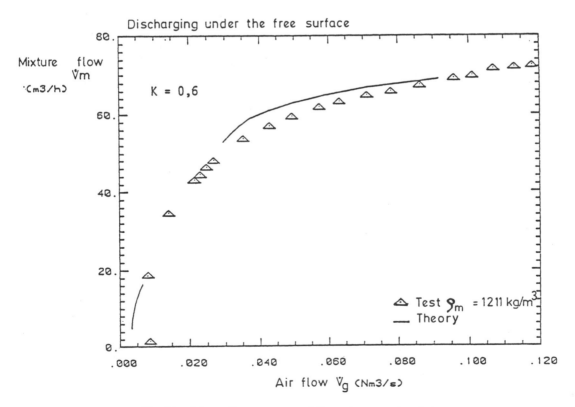

Fig. 11.   Mixture flow rate versus air flow rate: comparison between data and model.

10th International Conference on the
Hydraulic Transport of Solids in Pipes

HYDROTRANSPORT 10

Innsbruck, Austria: 29-31 October, 1986

PAPER D3

THE HANDLING OF SAND DEPOSITS AND LITTORAL DRIFT
BY SUBMERGED JET PUMP, WITH REFERENCE TO
THE NERANG RIVER PROJECT

A W Wakefield, B.Sc.(Eng.), Dipl.Civ.Eng., C.Eng.,
F.I.C.E., F.I.Mech.E., A.I.Q., M.Cons.E.

Partner of Wakefield and Imberg, Consulting and
Development Engineers

Summary

The paper reminds us of the need for beach and
sand bar management, describes a new technique
using jet pumps by which it may be achieved and
illustrates this with reference to the world's
first fixed sand by-passing scheme, at the Nerang
River mouth, Surfers' Paradise, Queensland,
Australia. Recent jet pump development is
described, in particular relating to the design of
non-blocking slurry systems, and the extension of
the technique to hybridisation with centrifugal
slurry pumps explained with reference to existing
sand quarry installations.

Symbols
_____

C   concentration by volume
H   head, metres
i   hydraulic gradient
     hydraulic efficiency
Q   volumetric flow, cubic metres per second
R   R value = (cross-sectional area of nozzle)/
     (cross-sectional area of mixing chamber)
V   velocity, metres per second

LITTORAL DRIFT

At almost any given point on the continental
margin one direction of ocean current
predominates. Particularly during storms, this
current drags at the coast line, moving it where
it can. Sandy margins thus travel at typical rates
of 10^5 to 10^6 tonnes per year.

Provided that nothing interferes with this
movement and that replenishment equals scour, the
effect is not obvious. However, any coast is a
sequence of discontinuities: headlands, groynes,
river mouths, harbours.

It is usual for sand accretion to occur to
the upstream of an obstruction and scour to the
downstream side, although sometimes local
circulations restore sand as quickly as it is
scoured. Figure 1 shows a typical case. Headland A
caused beach B to be denuded, so a rock groyne was
built at C. This robbed beach D which was restored
by groyne E which resulted in the stripping of
beach F.

Sand deflected out to sea may not rejoin the
coast for some kilometres. A river outlet will
push the sand out only to the depth where the
velocity is low enough for deposition to take
place, and a sandbar may be the result. At a
certain point the river may break through but such
locations tend to be unstable. Outlets often drift
in the direction of the current, occasionally
being brought back to the river at times of heavy
rainfall inland. Figure 2 shows a typical case.
The drift has been checked by a long training wall
but there is nothing to prevent waves and sand
from entering and heavy siltation upstream is
evident.

Harbour entrances are guarded by training
walls. Sand accretion often occurs against one of
these, beach loss adjacent to the other. As the
accretion reaches the end of the breakwater, it is
sometimes extended but this is not always an
acceptable or desirable solution. Figure 1 shows a
typical case of wall-extension. The sand accretion
has just about reached the end of the second wall.

As the accretion zone builds, deflecting sand
seaward, siltation occurs on the incoming tide and
continuous dredging may be necessary within the
harbour. Dredging by seagoing vessels may
additionally be required just offshore to
intercept the sand movement before it reaches the
harbour. Trailing suction hopper dredges typically
take a load of 1000 to 3000 cubic metres of sand
to a safe distance to the downstream side of the
entrance and dump it, returning for a fresh cargo
on a two-hour cycle.

Close inshore dredging is hazardous, many
rock walls being permanently adorned by the wrecks
of such dredges. And when they are most needed,
during storms, dredges cannot operate.

Major investment on-shore characterises
recreational stretches of coast and a good beach
is essential to the economic survival of such
regions. It is well known for a beach, which may
have been stable for as long as anyone remembers,
to disappear in the course of even a single
season. This may be the unpredicted and
unpredictable result of some coastal work tens of
kilometres distant, beyond the range of hydraulic
models. One small beach visited by the author in
February had had sand deposited on it by seagoing
dredger to the value of $2.5m the previous October
and November. Every trace of this sand had
disappeared.

Such developments often extend up-river into
marinas and water-front properties. For these,

clear access to the sea needs to be preserved for at least most of the year. Figure 3 is typical, and shows threatening siltation.

Clearly two elements are necessary in the stabilisation of the coastline. One is the construction of features according to the best hydraulic modelling techniques. The second, which has not been available until now, is a sand by-passing, beach management system which is safe, unobtrusive and reliable, which can be operated automatically in all weathers, and which offers sand handling at an acceptable unit cost.

The basic requirement for such a system is a sand pump which may be buried permanently in the sand deposit and turned on and off at will. The pump and pipeline will thus always contain sand. Restarting must be completely reliable.

It has been generally assumed that this goal is unattainable and little attempt seems to have been made to achieve it.

The present paper outlines the philosophy of development of a sand pump, a jet pump which satisfies these requirements.

THE DEVELOPMENT OF THE SUBMERGED JET PUMP

The author's first contact with jet pumps was in 1962. Then, the theory available was under-developed and not reliably applicable to pumps whose configuration was dictated by the requirements of a specific function, such as handling large particulate solids with minimum wear, long fibrous material without clogging, foodstuffs without damage, aeration of mine water with no installed motive power unit, and so on.

In such cases, the "black box" or empirical approach was unavoidable. What seemed likely to be good configuration was drawn, manufactured and tested. In the case of a product range, either each separate model had to be the subject of a trial or some method of assessing scale effect had to be arrived at. Measuring output resulting from a given input allowed a transfer function to be derived, but this implied and required little understanding of the jet pump. Available mathematical models were unreliable and it was the only approach open to the author in the pre-computer years prior to 1975.

Latterly a conscious attempt has been made to conduct design by mathematical model. Not only is the operation of the jet pump better understood, but the computer provides the means of manipulating the model to any desired end. Once a basic model, or equation, is written it becomes a matter of systematically examining all its elements and wherever and as soon as possible replacing the empirical with a true repesentation of what is happening. Gradually these elements yield to this treatment, allowing the scope of successful application of the jet pump to increase. Once a comprehensive model has been constructed, reversion to transfer functions for specific applications is advantageous in terms of computing speed.

This approach might have been followed earlier had its importance been realised, but it

was generally assumed that the jet pump was suitable only for a few well-defined applications one field impinged little on another. The power station man had small interest in tin mining; the problems of food processing were of no concern to the aeronautical engineer. In each single, specific application the empirical approach worked well enough.

Of recent years, however, the ability to manipulate a comprehensive model of pump and system together, afforded by the computer, has led to considerable development of the jet pump, but still in isolated fields. The author has attempted to extend the range of valid applicability into many fields and to relate all to a single model.

If one source of design information covering both compressible and incompressible flow and most configurations of jet pump is to be quoted, it must at the time of writing be the Data Item on Ejectors and Jet Pumps offered by the Engineering Sciences Data Unit. However, this limits itself to consideration of the device in isolation and it is the author's practice to model the entire system, permitting overall optimisation which is not otherwise easy.

First, reference is made to the author's papers presented at the First and Second Symposia on Jet Pumps and Ejectors, 1972 and 1975, in London and Cambridge respectively. Much water has passed through nozzles and mixing chambers since then and it would be appropriate to review these two papers before passing on to the latest phase of development.

The 1972 Paper: Practical Solids-Handling Jet Pumps

Development of Two Practical Configurations. This epitomises the empirical approach, now superseded.

Design of Characteristic. Superseded by the 1975 paper.

Determination of Operating Point. Still valid. The same method continues to be employed by the author for the estimation of pipeline friction loss, with good results. It should be noted that the accuracy of prediction declines with increasing approximation of the material grading curve, the use of the D50 particle alone often giving a wildly inaccurate, more usually an overestimated, value for the friction head. With computer assistance it is reasonable to input whatever particle data is available, allowing the computer to estimate a likely S-curve and abstract from this a grading with, say, twenty to one hundred discrete sizes. Arrival at the hypothetical single particle must be by averaging according to the effect on pipe friction.

Cavitation Suppression. The method given is still considered completely valid. Annular jets are now predictable, the identical test curve as drawn (Fig 9 in that paper) proving valid for annular jet pumps having R values in the range 0.28 to 0.55. Work with low-R jet pumps does not suggest any unexpected deviation.

External Disintegration. Many dredging applications employ cutter-suction dredges and it is assumed that a cutter is necessary in many cases when it is not. The rate at which solids are fed to a centrifugal pump must be controlled or the pump may stall owing to a combination of increased pipeline head and reduced pump efficiency. The principal function of a cutter is often only to achieve this, in which case it is a cost-inefficient device, imposing structural loads and a maintenance penalty which have to be paid for. Both mobilisation of the deposit and dilution of the incoming slurry may often more economically be furnished by external and internal low-pressure water jets. Within native permeable material, these increase the pore water pressure, jacking the individual particles apart and mobilising the deposit. Impermeable materials are not generally suitable for disintegration by water jet, and a mechanical cutter is necessary.

Output Control. All the methods described are valid and remain in use: control of the supply of motive water to the jet pump, introduction of water into the suction chamber, and regulation of the power available to the disintegration system.

Internal and External Blockage Formation and Avoidance.

1. In the pipeline. The mechanism of pipeline blockage and its avoidance are discussed below.

2. In the mixing chamber. This remains valid. That is, a blockage will only occur if the pattern of wear is such as to produce a convergent taper and if an elliptical stone having a minor axis just less than the mixer inlet diameter and a major axis less than the worn diameter but less than the mixer exit diameter should enter and rotate through a right angle during its passage - a remote chance, but it happens occasionally.

3. In the suction chamber. The same principle applies as at the suction orifice, discussed below.

4. In the suction duct. The required rate of expansion of the suction duct has been found to be only about half of that then predicted, for what appears to be the following reason. As material travels the expanding duct it spreads and slows, flattening the arch, the centre of which tends continuously to collapse. An effective included angle of divergence of about 7.5 degrees seems to be sufficient.

5. At the suction orifice. As previously briefly described, it has been found that a relationship exists between the rate of convergence of the streamlines, the particle size, the particle grading and the solids concentration. The relationship is necessarily complex and for the time being must remain "rule of thumb."

If both are correctly designed, the entry to the suction duct is more critical than the entry to the mixing chamber and following the original principles of internal configuration no blockage will occur internally if material has already successfully negotiated the intake orifice. If, however, an open configuration is used then a large sphere of packed solids can form, centred on the entrainment zone and stabilised by surrounding pipework or inlet grilles. The cure is simply to reduce the concentration of solids to below a critical value by liberating water externally and placing any intake grille at a radius exceeding

the critical.

Operation into a Pipeline. The paper offers a design procedure which, although un-mechanised and therefore dated, remains in principle valid and forms the basis of the detailed and extensive computer programs now used.

Wear and Impact. Most of this section of the paper remains valid, but much work has since been done on materials for the manufacture of nozzles. Although certain basic principles exist the selection of a suitable material is still something of a black art.

Spheroidal graphite (SG) iron machines well, is impact-resistant and its lack of ductility enables it to resist large particles in the motive supply, but it is not very corrosion resistant and has poor resistance to erosion due to the cavitation occurring in the bore of long, parallel nozzles. Stainless steel offers good resistance to cavitation erosion and is work-hardening so survives stone impact well, but its resistance to abrasive erosion is poor.

Nozzles may be lined with hard metal or ceramic or plastic materials, or the whole nozzle may be machined from plastic. Hot applications demand ceramic, and both silicon nitride and high-density alumina are successful, the latter only in non-alkaline applications. At the moment it is impossible to predict with certainty the best material in an abrasive application. In one quarry, the replacement of high-density alumina inserts by acetal copolymer extended the interval between replacement by a factor of three. In another, the reverse was the case.

The best solution in a permanent installation where the unit cost of production is scrutinised rigorously and cut to the minimum is to provide for easily replaceable nozzle inserts. Costs of production of sand have been reduced by up to 90% using jet pump dredging equipment, the cost of dredging and delivery to screens perhaps comprising much less than 5% of the price of sand at the weighbridge. Yet the process of attention to detail in further reducing these costs continues.

For mixing chambers the sensitivity of the material to work-hardening influences is important in relation to particle size and velocity. In high-head pumps it may be advantageous to start diffusion within the mixing chamber casting.

Relative Economics. The points made in this discussion remain valid. In summary, hydraulic efficiency must be balanced against operational efficiency and operating skills available when considering competing systems. Usually, a modern jet pump dredge will consume less energy in the excavation and delivery if a tonne of solid material than will a conventional cutter-suction dredge, and its capital cost will be lower.

The 1975 Paper: Performance of Solids-Handling Jet Pumps at Low Reynold's Numbers

Most of this paper was concerned with jet pumps in general, with particular reference to low Reynold's Numbers.

The relationships derived give good results, within the limits of commercial measurement. However, detailed consideration of individual losses can be expected to provide good results over a wider range of application and configuration - the principle of the model. Such a method is given in the ESDU Data Item.

The Supplementary Paper, under the heading of "Shut-Off," refers to low values obtained for first and second shut-off. Subsequent work has established that this relates to the mixing which occurs after the nozzle tip and before the mixing zone is confined by the bore of the mixing chamber. A correction factor of (1-0.3R) applied to the relationships under the Appendix I heading, "Design Data," provides the necessary correction. The factor is a multiplier in the equation for Nx, No and N and a divisor in the expression for M. A correction factor, constant at least between R values of 0.28 and 0.55, allows all these expressions to be applied to annular jet pumps, the value of the constant taking into account the angle of the nozzle to the axis of the jet pump and other loss factors resulting from the sudden injection of fluid into a confined space, and being derived by test.

CONSTRUCTION OF THE PUMP CHARACTERISTIC

Let us look at mechanisms operating in slurry transport, and consider what might represent an ideal pump characteristic and whether it might be attainable.

To recapitulate what will be familiar to most readers:

In a small pipe, at low speeds, a fluid travels in a parallel, orderly stream - "laminar" flow. If the diameter or speed is increased, following a period of "transition," the flow becomes "turbulent." Laminar flow characterises transport by stabilised slurry or thixotropic medium and turbulent flow by the traditional, heterogeneous, essentially Newtonian fluid, the latter concerning us here.

In the turbulence of high velocity, the solids are carried along mostly in suspension and touching the bottom of the pipe hardly more than the sides or top. As expected, a small reduction in velocity produces a small reduction in friction head. As velocity further decays, the particles spend longer near the bottom, increasingly impeded by friction with the walls and accordingly absorbing energy. Eventually increase in solids frictional energy loss exceeds reduction in pure fluid energy loss, and the total energy loss rises. As more solids spend more time at low velocity in the lower part of the cross-section, the fluid velocity over the top rises, improving the conveying properties and striking a balance.

However, as the slow-moving bed deepens it becomes unstable, forming dunes. Locally, a greatly reduced area is available to the overlying fluid and local losses becomes excessive. As head increases, flow decreases and more solids settle out, increasing bed depth. When the dune meets the top of the pipe a blockage occurs and the system stalls. This possibility is well-recognised and feared in all systems pumping coarse particulate solids.

Figure 4a shows the form of the water-only pipeline friction curve and 4b a typical solids curve superimposed. At first sight it appears advantageous to operate at the minimum head point A, but minimum power is to the left, at point B.

Figure 4c adds a typical centrifugal pump curve, intersecting the friction curve at A. All slurry systems are subject to fluctuation, of feed rate, particle size, pump speed or whatever. A small increase in flow will cause an increase in friction head, pushing the operating point back to A. But a small decrease in flow also causes an increase in friction head, resulting in a further reduction in flow, moving the operating point to the left, this process continuing in the absence of a stabilising influence. The presence of the second intersection C is a good indication that pipeline blockage will ensue.

To avoid this, we impart a dynamic characteristic to the pump with a variable speed motor or provide for automatic water-injection, or as is the most usual, to avoid complexity, we shift the basic operating point to D, as Figure 4d. Operation is off best-efficiency and the line velocity is high, increasing wear and consuming more power.

Now, the ideal characteristic. We would like a free choice of operating point, not necessarily A or B. B may be uneconomic as pipeline diameter and capital cost increase, but in 4e, B is indicated as the left-most point at which it is worth considering working.

The pump characteristic must be steep at this point, in practice a gradient providing an intercept on the i-axis at around twice the operating i appearing adequate and putting the pump more or less on its best-efficiency point. The pump curve to the right of B is irrelevant.

Then, to be quite sure we cannot have a blockage, it would be good to crank the curve upwards towards the axis to provide a high reserve of pressure to blast the top off any incipient dune.

Finally, to confer a further stability trend, either the the pump characteristic or pipeline characteristic should be dynamic. Let us (arbitrarily) take the latter. A head increase should cause a reduction in pipeline friction, as indicated in Figure 4f.

A jet pump is able to provide this characteristic, the model allowing us to deduce the physical parameters.

Nerang is the first major installation illustrating the validity of this principle, although many smaller schemes have been implemented as the technique has developed. Always, the pump can be turned on and off at will, full of solid material. Vertical and horizontal pipelines are equally feasible, a millscale line including a vertical leg of 32m having been operating some ten years and a horizontal sand line of 700m somewhat longer. The former has been demonstrated at 70% solids concentration by weight, although normally running at much less, and the latter operates at 40% by weight. Other lines of 300 to 450m working in the range 40 to

50% have been running for many years on abrasive solids, the longest since 1968. Owing to the low operating velocity, in no case has a pipeline had to be replaced or even turned.

In Figure 21a we have a set of pipeline friction curves for a 300mm level pipeline 500m long. Cross-plots of tons per hour of solids have been added, and also pipeline "efficiency," expressed as the tonnes per hour moved for each water-kilowatt. The best-efficiency line is drawn, the peak "efficiency" of 2.34t/h/WkW occurring at about 33% concentration. The material is sand of specific gravity 2.65, conveyed in sea water. The partical grading is continuous from 0.001mm to 100mm, with a D50 of 0.6mm. Allowance is made for a normal incidence of flanged joints and other features.

It is assumed that we wish to pump silt, sand and gravel at 400t/h. Abstracting the minimum information to Figure 21b, we see that if it is to work safely a centrifugal pump must operate at close to point A, at 12% concentration and 4.8m/s. The operating point is well to the right of the minimum head point and 15% above it, and the pump curve clears the pipeline friction curve at all points to the left of it, if not by very much. "Efficiency" is only about 1.75t/h/WkW and the leftward position of the operating point on the pump curve means that the pump will be running at only about 75% of its best efficiency and in a high-wear condition.

Point B is the operating point of a jet pump, chosen because it is close to the best pipeline "efficiency," in fact about 2.15t/h/WkW, and exactly on the best efficiency point of the jet pump. It will be seen that concentration is greater at 18%. Taking a part-worn hydraulic efficiency for the centrifugal pump of 65% and for the jet pump of 30%, the system efficiency is thus some (30%x2.15)/(65%x0.75x1.75) = 75% as high for the jet pump. In practice, the regulating effect of the jet pump can, particularly in a dredging situation, easily reverse the efficiency advantage. The line velocity of 3.2m/s can be expected to provide a pipeline life four or five times as long.

A progressive increase in head would move the operating point along the pump characteristic to B2, the point at which secondary flow ceases to be induced and beyond which the curve turns upwards. It is clear that the pump curve is everywhere well clear of the friction curve, even allowing for uncertainty as to the exact vertical position of the sliding bed plateau.

But note that as the increasing head reduces the induced solids, the friction head will eventually move to B1, well below B2.

However, part of the cost is that of pipeline replacement and it may be considered preferable to accept a lower pipeline "efficiency" and work at a higher concentration, say 24%, or point C. This time, the jet pump curve is close to the full-concentration friction curve at the shut-off point C2, but the friction value is on its way to C1.

HYBRIDISATION

One further development is worthy of note.

The belt, braces and piece of string described above are not always essential. Stopping the line may not be necessary and a compromise may be acceptable if the result is a power saving.

It can be shown that a combination of jet and centrifugal pumps save power as compared with either alone and maintenance as compared with a centrifugal pump alone. Referring to Figure 4g-i, the curves for centrifugal and jet pumps are added to give a combination proven completely blockage resistant provided that the power input to the jet pump system is at least one-third of the total.

Again, referring to Figure 21a,b, it will be clear that combining a centrifugal pump curve with a jet pump curve will allow operation at point B with complete safety, and potentially at a greater system efficiency than either pump could provide alone.

Figures 5 and 6 show typical hybrid dredges of this type and Figure 7 the component parts of a dredging jet pump, although disintegration arrangements and the form of the suction duct vary widely according to the deposit to be worked and the intended operating mode. A rotating cutter is sometimes fitted. Figure 8 shows a typical submerged jet pump, used for sand reclamation. A cross flow of water increases dilution to suit the following treatment plant.

Other multi-stage possibilities exist. Two jet pumps are sometimes combined in series, the first pressurising the second to suppress cavitation and permit greater secondary flow.

Less obvious is the advantage deriving from the combination of annular and central jet pumps, but these have found application subsea where the productivity per diver has been increased by a factor of between three and five as compared with the simple central jet pump types used previously. The lightweight diver-held annular jet pump controls the concentration fed to the central jet pump and converts an uncontrolled suction line into a controlled pressure line. See Figure 9.

THE NERANG RIVER SAND BY-PASSING SCHEME

The present paper is concerned only with the jet pumps themselves. For other details of the scheme the reader is directed to the sources quoted in the Reference section.

Appendix I contains extracts from the official literature.

A line of ten jet pumps at right angles to the direction of the littoral drift intercepts all of it. Each jet pump is deeply embedded in the sand. Any number may be run according to the drift pattern, activated individually by turning on the motive water supply.

Discharge is into a flushed gravity flume leading into a sump from which a conventional centrifugal slurry pump transports the sand under the inlet to the beach the other side. Automatic flushing is provided if pressure rises or flow decreases outside set limits.

Technical Notes

The ten pumps are of the shrouded open-body type, with no inlet grille. Close to minimum safe fluidisation is installed and the induced flow is thus very near to the maximum possible.

The estimated native bulk density is 1700 kg/cbm, with a material specific gravity of 2.7, that is a volumetric solids concentration in-situ of 63%. To ensure that flow is able to take place, the facility of continous fluidising water is provided, and only a small flow has so far proved necessary. So, at least for the time being, the reduction of concentration must be purely nominal and a figure of 60% by volume, 80% by weight, might reasonably be assumed.

If the mass flow ratio is about 1.5, then the delivered concentration is 26% by volume, 48% by weight. The observed trial at the demonstration was running at reduced discharge head probably giving a mass flow ratio of at least 2.0, resulting in a delivered concentration of about 30% by volume, 53% by weight.

The whole installation was modelled on the computer and a single mixing chamber diameter derived which would give 140 tons/hour of sand delivery against each of the various static lift and delivery distances, simply by fitting different nozzles. The common mixing chamber diameter is 102mm, so ellipsoid particles of minor axis 100mm and major axis 250mm can be passed. Nothing approaching this size is expected. Weed will go through without difficulty.

Any number of pumps up to a maximum of seven, limited by the motive water supply, can be operated at once, according to demand, as the pattern of littoral drift is sensed, but it is envisaged that either any four or any seven will run at one time. Operation is automatic under the control of a programmable logic controller, with continuous readout of flow rates and total sand movement. A nucleonic sensor detects concentration and a doppler-shift flowmeter the flowrate.

A conventional Warman centrifugal pump booster, driven through a speed-reducing gearbox and scoop-controlled fluid coupling, passes the sand under to the entrance and on to the feeder beach. Provision is made for automatic clean-water flushing on accidental shut-down. Whilst the jet pumps may safely be started up and shut down at any time with pumps and pipework full of sand, this is not the case with the centrifugal booster.

The jet pumps deliver to a regulating sump where a surrounding launder permits water to be weired either in or out to regulate the concentration of solids induced by the booster.

A pair of motive pumpsets supplies water at 1100kPa for motive and fluidising purposes. On placing the jet pumps, the flow of fluidising water is intended to be turned off or greatly reduced. According to the number of jet pumps activated, either one or both the motive pumpsets will run.

It is likely that subsequent sand by-passing schemes will use hybrid jet pump/centrifugal pump systems. Several are currently under active consideration in various parts of the world.

Project Status

At the time of writing, May 1986, the ten jet pumps are placed at depths to eleven metres below sand level. For about two months they were not taken deeper than six metres, allowing time to establish that the ultimate angle of repose, or slope of the sides of the sand trap trench, would not be such as to risk undermining of the training wall. Had the slopes extended to anywhere near the rock wall, greater depth would not have been needed.

The first pump was lowered into position on 9 February 1986, using its own fluidising nozzles for penetration. The motive water was then turned off and the pump was left for some hours. It was restarted without difficulty, a high production rate of sand was noted and the pump was turned off again.

On the occasion of the author's visit to site on 14 February, at a seminar and demonstration organised by the GENFLO agents, the first pump restarted without hesitation. After running for some time it was shut down, left for a while and restarted. The second pump had meanwhile been carried by mobile crane along the jetty and positioned in its slideway. This pump was then connected to the motive main and the supply turned on. The pump was seen to descend smoothly to its designated depth of six metres where it was secured. The time required to collect the pump and its pipework from the storage yard, convey it to near the end of the jetty, fit it into the slideway, connect the hoses and pump it down to six metres was about ninety minutes.

The slurry density was generally estimated at something approaching 60% by weight and a test the next day established a production rate from the one pump of between 210 and 220 tonnes per hour.

The official opening was by Sir Joh Bjelke-Petersen, in the first week in April, 1986. All pumps are now in operation, having produced sand on test at a rate in excess of specification, and the job has been handed over by the contractors. Production capacity will increase slightly with wear before declining eventually to the specified rate, when replacement of wearing items will be indicated.

Acknowledgments

Acknowledgment is gratefully made to the following:

The Gold Coast Waterways Authority, constituted for the purpose of the Nerang scheme

The Beach Protection Authority, Queensland

McConnell Dowell Constructors Limited, the prime contractors

Slurry Systems Proprietary Limited

Simax Consolidated Limited, UK agents of GENFLO Pumps Limited, licensees of Wakefield and Imberg, and Kirrawee Engineering Services Proprietary

Limited who supplied jet pump equipment and matrix computer system predictions by Wakefield and Imberg to the project

GENFLO Pumps Limited who manufactured and supplied the jet pumps from the UK.

References

The Nerang River Project, June 1985, by the Gold Coast Waterways Authority

Nerang River Entrance Stabilisation, 1986, by the Beach Protection Authority, Queensland

Nerang River Entrance Sand Bypassing System - a Unique Technology Step, 1986, Slurry Systems Pty Ltd

Practical Solids-Handling Jet Pumps, A W Wakefield, BHRA First Jet Pumps and Ejectors Symposium, London, 1972

Performance of Solids-Handling Jet Pumps at Low Reynold's Numbers, A W Wakefield, BHRA Second Jet Pumps and Ejectors Symposium, Cambridge, 1975

Data Item 85032, Ejector and Jet Pumps, Design and Performance for Incompressible Liquid Flow, December, 1985

APPENDIX

THE NERANG SAND BY-PASSING SCHEME
EXTRACTS FROM THE OFFICIAL LITERATURE

"Queensland's Nerang River, with the Broadwater and the adjoining waters of Moreton Bay and the South Pacific, together form one of the World's most attractive boating and fishing environments.

"Ever since the white man came to what we now call the Gold Coast, people in boats have sought the haven of the Nerang River. But the Bar has always proved an obstacle, and a dangerous and fickle one at that ... to their safe passage.

"For decades, people interested in the marine environment of the Gold Coast Region have sought ways of controlling the moods of the Southport Bar.

"But more than that, as the years progressed, it became evident that the location of the River mouth was continually moving North." Figure 10 illustrates the trend.

"The Nerang River flows to the sea through a broad, shallow tidal estuary known as the Broadwater and meets the ocean via an entrance channel between the southern end of South Stradbroke Island and the Southport Spit."

The littoral drift is "estimated to be 500,000 cubic metres per annum."

"In essence, the Nerang scheme calls for the construction of two massive training walls shielding a new passage located some 700 metres south of the existing (1985) channel. The walls are aligned to extend approximately 15 degrees North of East.

"The eastern Australian coast features a natural northerly movement of sand. The function of the Sand Bypassing System is to collect the northward-moving sand as it reaches the Southern Training Wall, and then to pump it via a conduit located under the passage, to be deposited on the beach on South Stradbroke Island. This effectively prevents sand entering the Broadwater where it would form a bar. Instead, the maximum amount of sand will reach the Island to prevent further erosion.

"Recreational and fishing craft will for the first time enjoy a stable, deep and safe passage between the Pacific Ocean and the Broadwater. The navigation channel will be 170 metres wide with a low-water depth of 4.5 metres.

"Wave Break Island has been reclaimed immediately inside the new mouth. Comprising over a million cubic metres of sand, this island will protect the beaches of Labrador from the effects of waves penetrating the new entrance."

Delft Hydraulics Laboratory, commissioned by the Department of Harbours and Marine, advised in 1976 that "Sand bypassing would be required to prevent massive erosion of South Stradbroke Island."

"The Queensland Government in 1983 decided that the (Gold Coast) Waterways Authority should proceed with the project. The layout of the final entrance training scheme is shown in Figure 11."

The Delft report "gave a conceptual design for sand bypassing of the trained entrance. In practice, no comparable bypassing system was operating anywhere in the world and the design of such a facility was not clear cut."

"Tenders for the sand bypassing works were called for in September 1983. Ten tenders were received offering a number of alternative systems. The tenders were assessed by Cardno and Davies Australia Proprietary Limited with expert coastal engineering input from the Beach Protection Authority and the Department of Harbours and Marine." The Delft Hydraulics Laboratory were again consulted.

"The tender of McConnell Dowell Constructors Limited in the sum of A$5.3 million was accepted for the design and construction of the sand bypassing system similar to the one recommended in the Delft Report. The key components of the system are ten jet pumps suspended from and spaced about 30 metres apart along a jetty extending about 490 metres. The jetty is located some 250 metres south of the southern training wall. The jet pumps are submerged approximately 10 metres below the normal sea bed and their operation creates a trench or sand trap at right angles to the beach. The sand moving from the south gathers in the trench and is pumped through a pipeline under the entrance to the northern beach. Water to power the jet pumps is drawn from the Broadwater through a series of pumps. A schematic diagram illustrating this operation is shown in Figure 12.

"The system has the capacity to pump sand at rates of either 335 or 585 cubic metres per hour. The system is expected to operate 5 days per week and can operate unattended overnight. Under storm conditions, the system would work continuously at peak capacity."

"From a technical viewpoint the project highlights the level of expertise and skills being employed by coastal engineers in this State today.

The following facts reflect this situation:-

...... The use of a network of jet pump intakes along a fixed trestle structure is the first time such a system has been used anywhere in the world for bypassing sand past a trained river entrance."

"The Gold Coast Waterways Authority is proud of its role in bringing this exciting development to fruition."

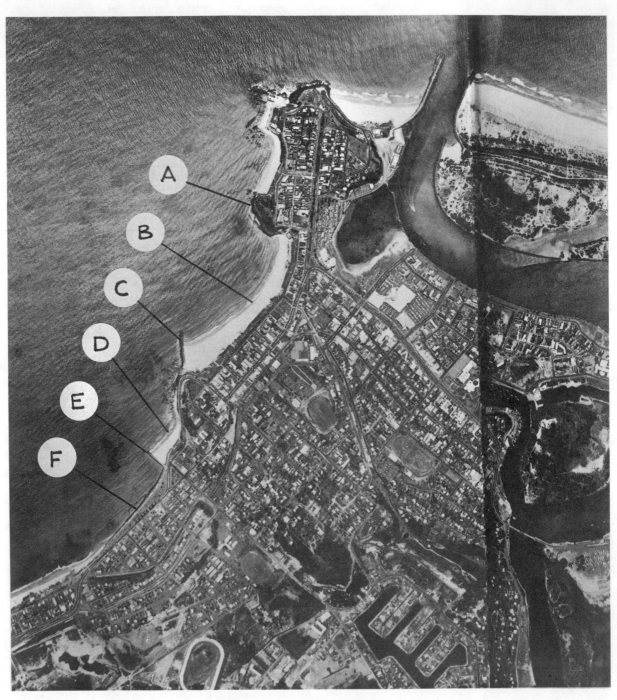

Fig. 1. Accretion and scour: extended training wall.

Fig. 2. Drifting entrance.

Fig. 3. Waterfront development threatened by siltation.

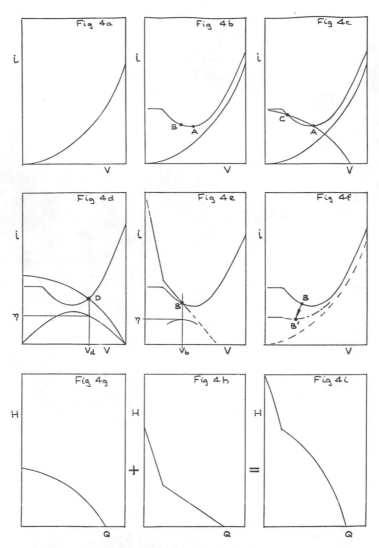

Fig. 4. Characteristic curves.

Fig. 5. Dredge designed for Joseph Adshead Ltd, subsequently
developed into hybrid for Tarmac Ltd.

Fig. 6. Dredge developed into hybrid for George Garside Sand Ltd
(ECC Quarries Ltd).

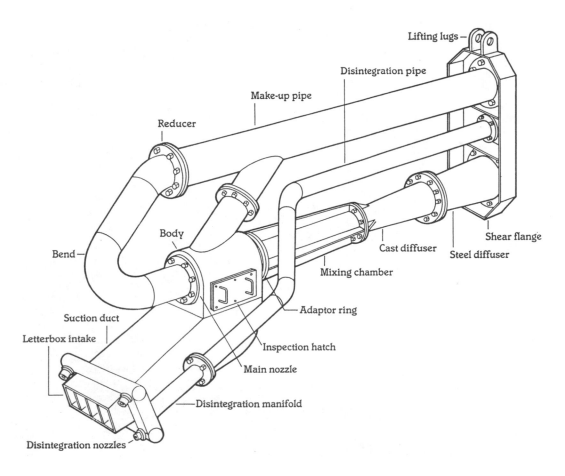

Fig. 7. The component parts of a typical dredging jet pump.

Fig. 8. *A typical submerged jet pump (sand reclamation, Lough Neagh).*

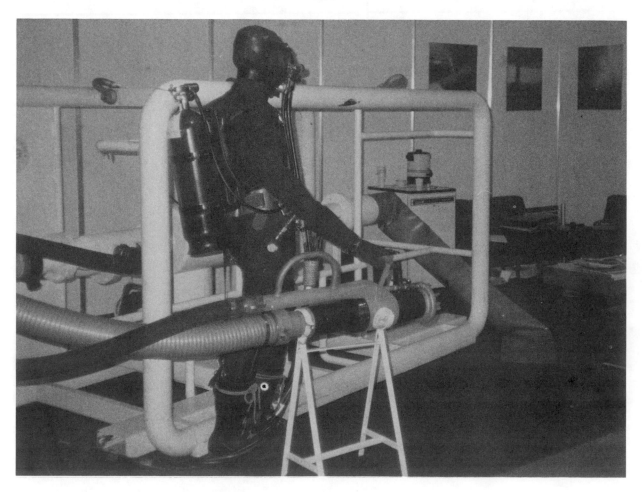

Fig. 9. *Diver-operated two-stage jet pump excavation system.*

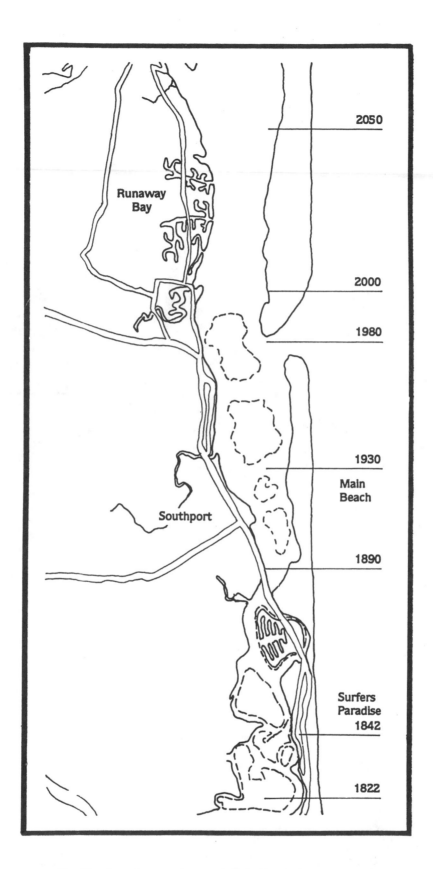

Fig. 10.   Over the years the mouth of the Nerang River has moved
steadily northwards. Without the Nerang Training Scheme, it is
estimated that by the year 2050 it would be north of Runaway Bay.

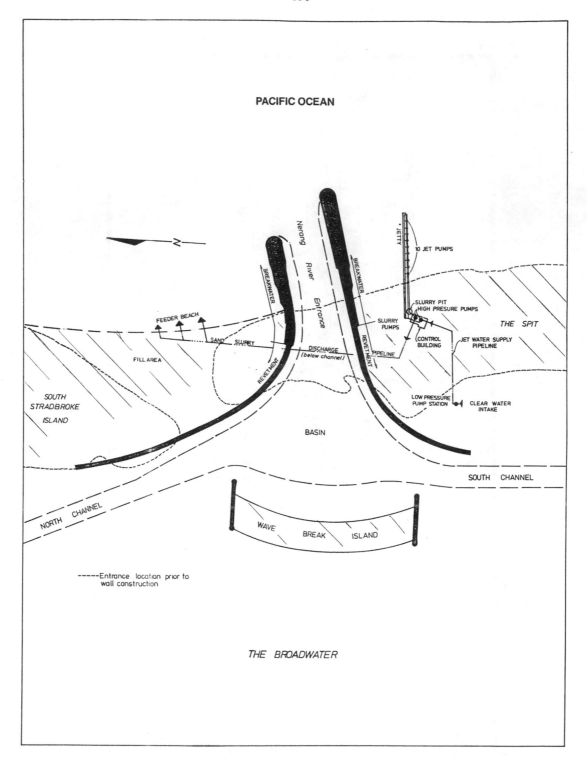

Fig. 11. Nerang River entrance—general arrangement.

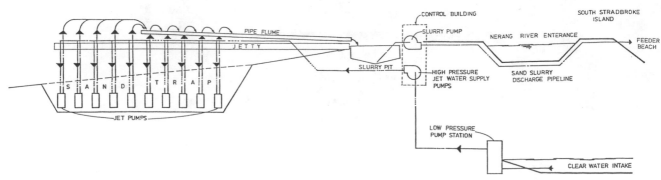

Fig. 12. Schematic diagram of sand bypassing system.

Fig. 13.   Nerang—jetty from the seaward end.

Fig. 14.   Nerang—jetty from the landward end.

Fig. 15. Nerang—motive pumpsets.

Fig. 16. Nerang—jet pump.

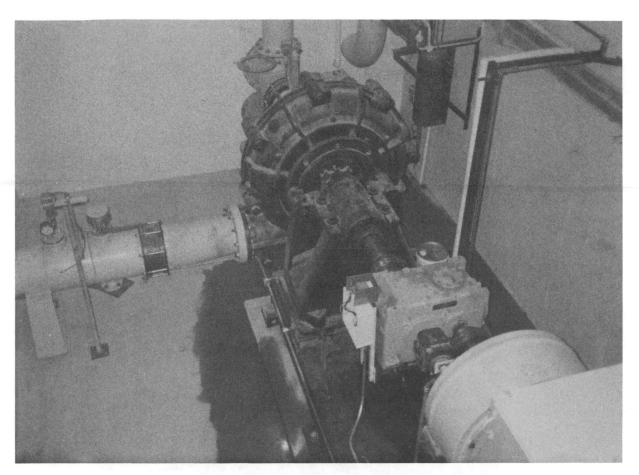

Fig. 17. Nerang—centrifugal booster.

Fig. 18. Nerang—jet pump and vertical pipework on way to installation.

140

Fig. 19. Nerang—jet pump and vertical pipework installed.

*Fig. 20. Nerang—jet pump discharging to waste during com-
missioning.*

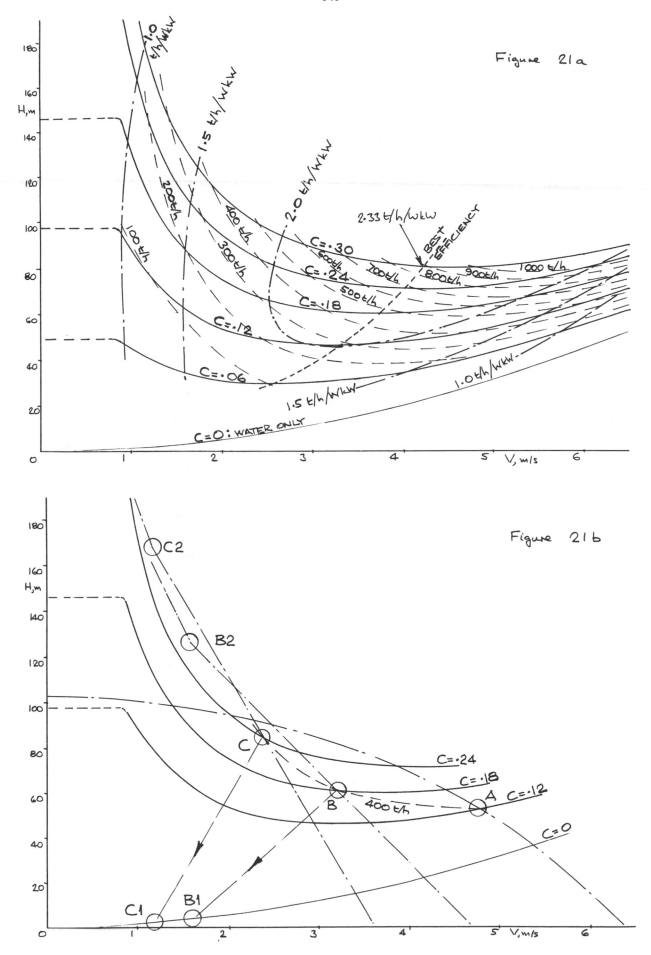

Fig. 21. Worked example.

10th International Conference on the
Hydraulic Transport of Solids in Pipes

HYDROTRANSPORT 10

Innsbruck, Austria: 29-31 October, 1986

PAPER E1

SLURRY FLUME DESIGN

Robert R. Faddick

Dr. Faddick is Professor of Civil Engineering in the Engineering Department at the Colorado School of Mines in Golden, Colorado, USA

SYNOPSIS

An analytical method is provided for the design of a slurry flume that includes consideration of optimum channel geometry, a supercritical flow regime, and flow resistance criteria for the selection of a channel slope that ensures a velocity greater than the deposition velocity for the coarsest particles.

The composite design equation is a combination of the continuity equation with solids throughput, Manning equation, Froude Number, and the turbulent velocity distribution for open channel flow. An example is given with field data to illustrate its use.

INTRODUCTION

Flumes are man-made open channels used to transport liquids or solids-liquid mixtures called slurries. Historically, water flumes go back to the origins of irrigation. Flumes have been used to transport various solids as well, including sugar cane in Hawaii and pulpwood logs in Canada. In the minerals industries, flumes or launders as they are often called, are used to transport ore and tailings and trace their origins to the turn of the century.

The largest slurry flumes in the world are believed to be the 140-km total length of wood and concrete channels of El Teniente in Chile. The newest is designed to carry ultimately 180,000 tonnes per day of copper tailings.

The advantages of flumes include low capital and operating costs. Their disadvantage is the necessity to maintain a minimum channel slope or gradient, which in turn controls its length. Slurry flows are also subject to solids deposition (plugging) with flow overtopping the channel sidewalls if the slope is insufficient to create a velocity to maintain solids in suspension. Erosive wear is usually not a significant problem.

The literature is relatively sparse on slurry flumes. Lau's 1979 Literature Survey (Ref. 1) lists only 30 references from 1945. Other references include works by Blench, Galay, and Peterson, (Ref. 2), Wood (Ref. 3), and Kuhn (Ref. 4). Green, Lamb, and Taylor (Ref. 5) in 1977 presented the most comprehensive slurry flume design. However, a computer code is recommended to handle the two-dozen calculation steps required to develop the slurry flume. The channel cross-sections that can be used include circular, U-shaped, and rectangular.

This paper seeks to reduce the design effort for slurry flumes by working with just the rectangular shape since it is by far the most common, and including a means for ensuring a local velocity high enough to maintain the largest solids in suspension.

DESIGN CRITERIA

The flow resistance equation used most often for single-phase flows in open channels is the Manning Equation given as:

$$V = Rh^{2/3} \, S^{1/2} /n \qquad \ldots (1)$$

where V = mean velocity of flow, m/s
Rh = hydraulic radius = A/P, m
A = cross-section of flow, m
P = wetted perimeter, m
S = channel bed slope, dimensionless
n = Manning's roughness coefficient

The Manning equation is valid for steady, uniform, and fully rough flow, i.e. the flow does not vary with time and space, and the Reynolds number is sufficiently high that the laminar boundary layer is penetrated by the roughness elements of the channel lining. These conditions are not always encountered in natural channels but are usually obtainable in artificial channels. Furthermore, these conditions can be measured in artificial channels with reasonable ease, and thus can be verified. Note that the Manning equation is not dimensionally correct unless n takes on peculiar units. For this reason, the Manning equation is considered to be semi-empirical.

Combining Eq. 1 with the equation of continuity

$$Q = VA \qquad \ldots (2)$$

where Q = flowrate, m^3/s

gives

$$Q = A \, Rh^{2/3} \, S^{1/2} /n \qquad \ldots (3)$$

Thus the Manning equation as given by Eq. 3 embodies continuity, flume geometry, and roughness.

For two-phase flows involving granular solids (mineral tailings or ores) conveyed in water, the Manning equation is still used.

However, the design criteria become more complicated because of several considerations:

1) the continuity equation involves a second-phase, the throughput of solids,
2) the heavy solids will settle out of suspension if the carrying velocity is too low.
3) A third consideration follows from the latter consideration; a higher velocity of flow is required to maintain the solids in suspension; but if too high, can lead to erosive wear on the channel or flume lining.

Before delving into a two-phase flow analysis, a brief look at the Manning equation for single-phase flows is worthwhile. Eq. 3 shows that flowrate varies with hydraulic radius to the power of 2/3, or that maximum flow occurs with minimum wetted perimeter, P, all else being equal. In fact, for a lined channel, a minimum value of P maximizes flow and minimizes lining costs. Standard textbooks on open channel flow give the optimum values of Rh for various channel cross-sections as follows:

General Shape	Optimized Shape	A	P	Rh
circle	semi-circle	$\pi D^2/8$	$\pi D/2$	$D/4$
rectangle	rectangular	$2y^2$	$4y$	$y/2$

(width = twice depth)

trapezoid	trapezoid inscribed	$A=y^2 (\csc\theta - \cot\theta)$
	by a circle	$P=2y (\csc\theta - \cot\theta)$

θ = exterior side angle $Rh=y/2$

By far the most common design for slurry flumes is a rectangular channel. Occasionally, a fillet is placed in each corner for various reasons: to strengthen the channel lining, reduce leadage, or to improve the flow pattern.

FLOW REGIME

In open channel flow, the flow regime can be either subcritical or supercritical flow. Critical flow is to be avoided because it is inherently unstable, producing wave action and fluctuating depths. From the depth-specific energy diagram (see Fig. 1), it can be seen that, although the specific energy E, is a minimum at critical depth, a small change in E by a small change in flume geometry, roughness, or flow rate can yield a substantial change in depth.

With slurry flow, the most important operation factor becomes the velocity. A minimum velocity must be ensured to keep the solids in suspension and not allow deposition to occur. While occasional small beds of solids are not harmful, if allowed to build, they tend to increase the local bed roughness, reduce the flow capacity of the channel, and when the flow is increased, lead to local increases in solids concentration. It must also be remembered that the deposition velocity varies with the solids concentration.

Because of the importance of velocity, a good design criterion is to maintain supercritical flow, i.e. shallow depth with high velocity. Several advantages accrue:

(i) the lower flow depth means less sidewall height and less channel lining.

(ii) the higher velocities generally exceed the deposition velocity of the coarser solids by a generous amount so that deposition is not a problem over a reasonable range of flows.

(iii) a supercritical flow requires by definition a steep slope. In hilly terrain, a steep channel bed slope requires a shorter length of flume and therefore reduces capital costs.

The flow regime design criterion becomes:

$$Fr = V / (gy)^{1/2} = Q / (gyA^2)^{1/2} =$$

$$Q / (4g\ y^5)^{1/2} \qquad \ldots(4)$$

for an optimized rectangular channel.

If the critical flow regime is to be avoided (Fr = 1.0), and for safety from depth fluctuations a range of 0.8 to 1.2 is avoided, than a Froude number of say, 1.25 could be assumed, giving

$$Q = 2.5\ (gy^5)^{1/2} \qquad \ldots(5)$$

To keep the design more general for the moment,

$$Q = 2Fr\ (gy^5)^{1/2} \qquad \ldots(6)$$

It can be shown that a Froude Number of 1.25 corresponds to a flow depth 86% of the critical depth for a given flowrate. This is a sufficient safety factor to avoid the inherent instability of the critical flow regime.

CONTINUITY-SOLIDS THROUGHPUT

For an optimized rectangular flume the continuity equation becomes

$$Q = VA = V2y^2 \qquad \ldots(7)$$

The solids loading or throughput of dry solids is given by

$$T = Q\ pm\ Cw\ 3600/1000$$
$$= 3.6\ Q\ pm\ Cw \qquad \ldots(8)$$

where T = dry solids throughput in tonnes per hour

Q = slurry flowrate, m^3/s

pm = slurry density, kg/m^3

Cw = solids concentration by weight, decimal fraction.

Normally, T, pm, and Cw are known so that the continuity equation is recast as

$$Q = T / 3.6\ pm\ Cw \qquad \ldots(9)$$

VELOCITY PROFILE

Vanoni (Ref. 6) showed that the Prandtl universal-logarithmic velocity-distribution law for pipes also applied to two-dimensional open channels. An open channel is said to be a wide channel and hence two-dimensional when the width is at least thirty times the depth. Few open channels can meet this requirement so in essence must be treated as three-dimensional channels complete with sidewall effects. However, the sidewall effects on the velocity distribution are not well established so it will be assumed for expediency that Prandtl's velocity distribution holds for an optimized rectangular channel in which the width is only twice the depth. The velocity distribution equation is given by

$$u = V + \frac{1}{X} \; (g \; Rh \; S)^{1/2} \; (1 + \ln(y'/y)) \quad \ldots(10)$$

where
- u = local velocity at depth y' from the channel floor
- V = mean velocity, m/s
- X = von Karman universal constant = 0.38 for clear water
- y = depth of flow, m
- Rh = hydraulic radius, m
- S = channel bed slope = hydraulic gradeline = energy gradeline for uniform steady flow, decimal fraction

The literature (Ref. 7) shows that in a controlled laboratory experiment X decreased to 0.2 for a sediment concentration of 16 gm/liter. For slurry transport in open channels, this is an extremely low concentration. The writer's experience with field data on sand tailings has shown that X does decrease with solids concentration but it is extremely difficult to obtain accurate measurements of local velocity and solids concentration at varying flow depths to prove X's variation with solids loading and flow depth. Furthermore a sensitivity analysis shows that a nearly 50% reduction in X from 0.38 to 0.20 with all the other variables held constant, increases the channel slope by only 10%. Since the error incurred in sampling local velocities and concentrations with Pitot tubes and tube samplers, for example, is probably in this range of 10%, a value of $X = 0.20$ is used in Eq. 10 to give a conservative computation for channel slope.

If indeed Eq. 10 is a universal velocity distribution valid for slurry transport in open channel flow, then it can be used as a design equation by taking a depth y' close to the floor of the channel (say 1/10 of the total depth) where the local velocity is u, and is sufficient to keep the largest particle $dmax$, in suspension. To keep the procedure simple, assume that the horizontal local velocity u, is a multiple of the settling velocity Vs, of the largest particle. A review of limited field data gives this multiple as 35 although more extensive field data and experience might suggest a higher or lower value.

Then

$$u = 35 \; Vs \quad \ldots(11)$$

Verification of this approach with field data is difficult due to limited instrumentation for measuring local velocities and sampling solids concentration.

FLOW RESISTANCE

The Manning equation will be used. It assumes steady, uniform flow and fully turbulent flow, i.e. the resistance to flow is independent of Reynolds number but depends on boundary roughness only where the roughness is Manning's n value.

Henderson (Ref. 8) gives a test for fully rough flow as

$$n^6 \; (Rh \; S)^{1/2} \geq 1.9 \times 10^{-13} \quad \ldots(12)$$

where Rh is in feet.

The Manning equation for an optimized rectangular channel becomes

$$v = 1.0 \; Rh^{2/3} \; (S)^{1/2} / n =$$
$$(y/2)^{2/3} \; (S)^{1/2} / n \quad \ldots(13)$$

COMPOSITE DESIGN EQUATION

Substitute Eq. 13 into Eq. 10 to obtain

$$u = (y/2)^{2/3} \; (S)^{1/2} / n +$$
$$(g \; y \; S/2)^{1/2} \; (1 + \ln y'/y)/X \quad \ldots(14)$$

Combine Eqs. 6 and 9

$$Q = 2Fr \; (gy^5)^{1/2} = T / 3.6 \; pm \; Cw$$

to obtain

$$y = (T/3.6 \; pm \; Cw \; 2Fr \; (g)^{1/2})^{2/5} \quad \ldots(15)$$

Let $W = T/7.2 \; pm \; Cw \; Fr \; (g)^{1/2}$

As noted before, set $Fr = 1.25$ giving

$$W = T / 28.19 \; pm \; Cw \quad \ldots(16)$$

Then

$$u = W^{4/15} \; (S)^{1/2} / n \; 2^{2/3} +$$
$$(g \; S/2)^{1/2} \; W^{1/5} \; (1 + \ln y'/y)/X \quad \ldots(17)$$

Let $u = 35Vs$, set $y'/y = 1/10$, $X = 0.2$ and solve for S as a percentage.

$$S,\% = \frac{100 \; (35Vs)^2}{(0.63W^{4/15} / n - 14.42 \; W^{1/5})^2} \quad \ldots(18)$$

The design procedure can be summarized as follows:

1. Solve Eq. 16 for W knowing dry solids throughput T, in tonnes per hour, solids concentration Cw as a decimal fraction, and slurry density pm, kg/m³.
2. Solve Eq. 18 for minimum channel slope in percent.
3. Solve Eq. 9 for flowrate, Q, m³/s.

4. Solve Manning equation, Eq. 13 knowing channel slope S as a decimal fraction, and roughness n. Solve for uniform depth y. For optimized rectangular cross section, the channel width is twice the flow depth.

5. Solve Eq. 7 for velocity. Verify Eq. 4 for Froude No = 1.25.

Fig. 2 is a nomograph giving channel slope S as a function of the variable W for maximum sand particle sizes and a channel roughness of 0.012. The sand specific gravity is 2.65 and the von Karman universal constant was set at X = 0.2. The local velocity u, was set at 35 times the settling velocity of the maximum particle at a channel depth ratio of 1/10.

EXAMPLE

A tailings slurry (S = 2.74) of 38.4% by weight (Sm = 1.32) is transported at 2536 tonnes per hour in a rectangular flume with a concrete floor and wooden walls (n = 0.010). The maximum particle size is 0.4 mm with a settling velocity of 0.06 m/s. Find the channel slope to keep the solids in suspension without any bed formation. Let the local velocity be 35 times the settling velocity of the maximum particle size at 1/10 depth of flow.

Solution.

$$W = T/28.19 \text{ pm } Cw$$
$$= 2536/(28.19 \times 1320 \times 0.384) = 0.1775$$

$$S = \frac{100 \ (35 \times 0.06)^2}{(63 \ W^{4/15} - 14.42 \ W^{1/5})^2}$$

$$= 0.506\% \text{ or } 0.00506$$

$$Q = T/(3.6 \text{ pm } Cw)$$
$$= 2536/(3.6 \times 1320 \times 0.384) = 1.39 \text{ m}^3/\text{s}$$

$$y = (Q \times 2^{2/3} \times n \ /2 \ S) = 0.497 \text{ m}$$

$$b = 0.994 \text{ m}$$

$$V = 2.81 \text{ m/s and Fr} = 1.27$$

Actual channel conditions were:

$$S = 0.0047 \quad y = 0.474 \text{ m} \quad V = 2.69 \text{ m/s}$$
$$Fr = 1.25$$

FURTHER WORK

Eq. 10 as noted earlier may not be valid for a narrow channel nor for high solids throughput. In fact Graf (Ref. 9) suggests that for sediment transportation in open channels a more proper velocity distribution may be obtained if the exchange mechanism of the solid particles is included. Some of these empirical velocity distributions will be examined in the future with field data. Also the settling velocity of the largest particle needs to be reduced by the presence of finer particles developing a heavy medium or viscous effect.

ACKNOWLEDGEMENTS

The writer wishes to acknowledge the United Nations Industrial Development Organization for sponsoring his visit to Chile during June 1985 to serve as a technical advisor on slurry technology to the copper and coal industries through the Pontificia Universidad Catolica de Chile.

REFERENCES

1. Lau, H.H., "A Literature Survey on Slurry Transport by Flumes," BHRA Fluid Engineering TN 1564, August, 1979.

2. Blench, T., V.J. Galay, and A.W. Peterson, "Steady Fluid-Solid Flow in Flumes," BHRA Fluid Engineering, Hydrotransport 7, Paper C1 Sendai, Japan, November 4-6, 1980.

3. Wood, P.A., "Optimization of Flume Geometry for Open Channel Transport," BHRA Fluid Engineering, Hydrotransport 7, Paper C2, Sendai, Japan, November 4-6, 1980.

4. Kuhn, M., "Hydraulic Transport of Solids in Flumes in the Mining Industry," BHRA Fluid Engineering, Hydrotransport 7, Paper C3, Sendai, Japan, November 4-6, 1980.

5. Green, H.R., D.M. Lamb, and A.D. Taylor, "A New Launder Design Procedure," ASME Paper No. 78-B-16, Denver, CO, November, 1977.

6. Vanoni, V., "Transportation of Suspended Sediment by Water," Trans. Am. Soc. Civil Engrs. Vol III, 1946.

7. ASCE Task Committee, "Sediment Transportation Mechanics: Suspension of Sediment," Proc. Am. Soc. Civil Engrs., Vol 89, No. HY5, September, 1963.

8. Henderson, F.M., "Open Channel Flow," MacMillan Co., 1966.

9. Graf, W.H., "Hydraulics of Sediment Transport," McGraw-Hill, 1971.

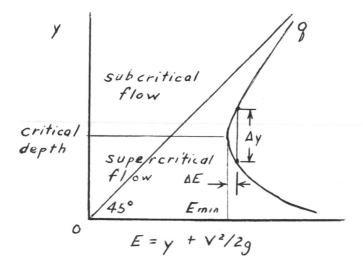

Fig. 1.   Specific energy.

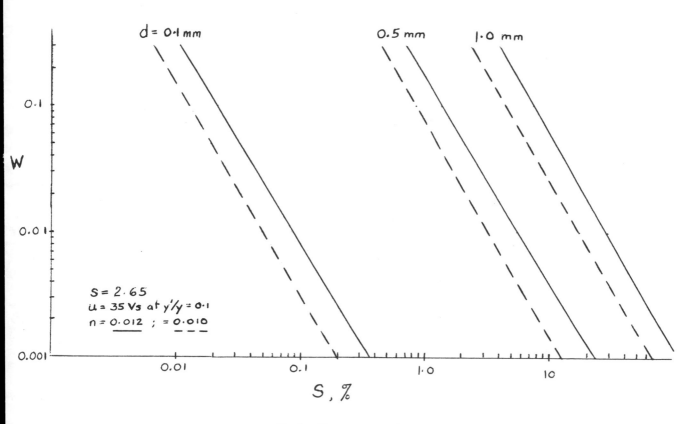

Fig. 2.   Slope nomograph.

10th International Conference on the
Hydraulic Transport of Solids in Pipes

HYDROTRANSPORT 10

Innsbruck, Austria: 29-31 October, 1986

PAPER E2

A COMPARISON OF SOME GENERALISED CORRELATIONS FOR THE HEAD LOSS GRADIENT OF MIXED REGIME SLURRIES

A W Sive and J H Lazarus

A W Sive is a Senior Research Officer and PhD student under the supervision of Professor J H Lazarus, both of the Hydrotransport Research Unit, Department of Civil Engineering, University of Cape Town, Republic of South Africa.

SYNOPSIS

Milled slurries generally exhibit well graded and sometime bi-modal particle size distributions. The fine particles constitute a 'vehicle' which carry the coarse particles as a suspended load. The mechanisms involved in increased carrying capacity for mixed regime slurries are investigated. The generalised correlations of four researchers are compared with data collected by the Hydrotransport Research Unit and the Saskatchewan Research Council. The mathematical techniques used in the presented correlations are discussed and conclusions drawn from the results produced.

NOMENCLATURE

A	internal area of pipe	(m^2)
a_{bed}	area of bed in pipe of radius $r = 1$	
C_a	concentration at reference level a	(%)
C_b	concentration of solids in loose-packed bed	(%)
C_D	particle drag coefficient	
C_r	concentration ratio C_{vb}/C_b	
C_v	volumetric concentration	(%)
C_{vb}	concentration of solids in contact load	(%)
C_w	weight or mass concentration	(%)
d	particle size	(m)
d_n	particle size at which n% of the particles are smaller than d_n	(m)
d_{Lmax}	particle size for $R_{ep} = 1$	(m)
d_{Tmin}	particle size for $R_{ep} = 1000$	(m)
D	internal pipe diameter	(m)
f	friction factor	
fn	function of ...	
g	gravitational constant	(m/s^2)
ΔH	head loss	(m)
i_m	head loss gradient in units of water per unit length of pipeline	(m/m)
i_{mcal}	calculated value of i_m	(m/m)
i_{mobs}	observed value of i_m	(m/m)
j_m	head loss gradient in units of mixture per unit length of pipeline	(m/m)
k	mean size of pipe roughness element	(m)
L	length of pipeline measuring section	(m)
M_s	solid mass flow rate	(kg/s)
M_m	mixture mass flow rate	(kg/s)
ΔP	pressure drop	(Pa)
Q	volumetric flow rate	(m^3/s)
R	ratio of mass of fines to total mass of particles	
R_e	Reynolds number	
R_{ep}	particle Reynolds number	
S	root mean square deviation of log i_m	
S_f	particle shape factor	
S_m	relative density of mixture	
SPC	specific power consumption	(Ws/Nm)
S_s	relative density of solid	
v	local velocity	(m/s)
V_{dep}	critical deposit velocity	(m/s)
V_m	mean mixture velocity	(m/s)
V_{mt}	value of V_m at threshold of turbulent suspension	(m/s)
V_t	particle settling velocity	(m/s)
V'_t	hindered particle settling velocity	(m/s)

V^*	friction velocity $V_m\sqrt{f/2}$	(m/s)
w_i	weight fraction of i-th particle diameter	
τ	shear stress	(Pa)
τ_o	wall shear stress	(Pa)
τ_y	yield shear stress	(Pa)
δ	sub-layer thickness	(m)
μ	dynamic viscosity	(Pas)
ρ	density	(kg/m^3)
ϕ	$(i_m-i_w)/(C_{vd}\, i_w)$	
α,β,σ	non-dimensional parameters	
β	angle defining bed load portion	

SUBSCRIPTS

m,s,w	mixture, solid, water
o	pipe wall
c,f	coarse, fine

1.   INTRODUCTION

The objective of considering mixed regime slurries is to find a generalised theory which includes all particulate slurries. The investigation of a unified theory is one of the research projects in progress at the Hydrotransport Research Unit (HTRU) of the University of Cape Town.

An appraisal of some of the existing correlations for the analysis of mixed regime slurries is presented. Test data generated by the HTRU and the Saskatchewan Research Council (SRC)[1] for mixed regime slurries is used to confirm the accuracy of the correlations presented.

2.   DEFINITION OF MIXED REGIME FLOW

Mixed regime slurries, by definition, display heterogeneous and homogeneous flow phenomena simultaneously. Particle size distribution,

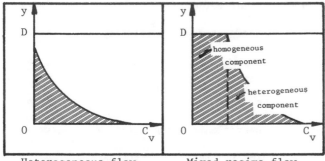

Fig. 1 : Definition of mixed regime flow according to the concentration profile at constant mean velocity and particle size distribution.

volumetric concentration and mean flow velocity influence the regime ratio and hence energy gradient. Wasp et al[2] describes mixed regime flow as a two phase transporting medium (called the vehicle) with suspended solids carried in this medium. The concentration and particle size distribution at the top of a horizontal pipeline defines the vehicle for a particular flow velocity. The assymetrical part of the pipeline concentration profile is ascribed to the heterogeneously transported coarse suspension (see Fig. 1).

3.   BENEFIT OF MIXED REGIME FLOW

Mixed regime flow results in a decreased energy gradient and a higher total solids concentration than an equivalent mixture of unsized particles. The increased concentration depends on particle size distribution. The increased carrying capacity results in a decrease in Specific Power Consumption where

$$SPC = \frac{i_m}{C_v S_s}\quad Ws/Nm \qquad [1]$$

This leads to an overall increase in the transport system efficiency. Abrasive wear and particle attrition are also reduced[3].

4.   MECHANISMS DECREASING ENERGY GRADIENT

The mechanisms for decreasing the energy gradient for equivalent concentrations of mixed regime slurries compared with unsized slurries are predominant at low velocities when the flow would be heterogeneous[4].

The mechanisms causing this decrease are listed below.

4.1   The fines content results in an increase in the apparent viscosity and a damping of turbulent eddies[5,6].

4.2   There is a decrease in density differential between vehicle and suspended load[6].

4.3   Mixed regimes eliminate, or significantly reduce coarse particle settling resulting in a decrease in the relative slip velocity of the coarse component and a decrease in turbulent suspension requirement[5,6,7]. The vehicle may exhibit non-Newtonian behaviour with a yield stress which results in a central core containing coarse solids[7].

The increase in the mixture energy gradient over the carrier fluid (water) is due to the homogeneous non-Newtonian vehicle with large particles being transported with little settling requiring low additional pressure loss to keep them in suspension. Cheng and Whittaker[8] exploited this fact in analysing mixed regime flow as a non-Newtonian fluid if a specified settling interface criteria could be met.

A stabilised slurry with $\tau_y \geqslant A\, gd\, (\rho_s-\rho_m)$ is able to support particles due to the increased yield stress at concentrations above $C_v = 30\%$.

where	A = 2/3π	(ref. 11)
	A = 0,092	(ref. 9)
	A = 0,083 - 0,10	(ref. 12)

4.4   There is a reduction in pipewall roughness due to the entrapment of fines in the pipewall irregularities.

4.5   Plate and needle like fine particle (d < 10μm) cause a lubricating effect.

4.6   The ratio of particle size to the viscous sublayer thickness, $\delta = 5\mu/\rho V*$ causes a pressure reduction if $d/\delta \geqslant 5$ for pseudo-homogeneous flow which will occur if $V_t/V* \leqslant 0,11$ [9].

Examples of the pressure reducing phenomena are:

1.   At C_v = 15,7% and V_m = 1,5 m/s

For sand only i_m = 0,012 m/m
For sand and bentonite i_m = 0,08m/m [4].

2.   An increase of up to 15% by volume of large particles in fines has little influence on hydraulic gradient. An increase of 20% of fines in a coarse mixture significantly influences hydraulic gradient [13].

3.   A clay content of 7% decreases the hydraulic gradient by 10% at $V_m \simeq V_{crit}$ and $C_v \simeq 20\%$ [13].

4.   Klose [14] has reported a decrease in hydraulic gradient caused by the attrition of coal particles. Figs. 2 and 3 show this decrease.

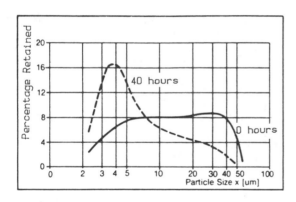

Fig. 2 Change in PSD for untra-fine coal (ref. 14)

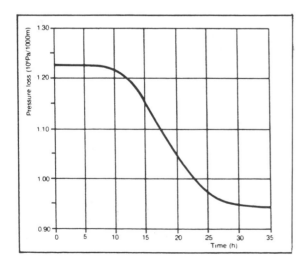

Fig. 3 Change in pressure loss for coal slurry with time (C_v = 41%, V_m = 1,75 m/s, D = 400mm)

5.   CORRELATIONS FOR HYDRAULIC GRADIENT

Table 1 presents some correlations for mixed regime flow. The correlations used for analysis are marked by an asterisk.

Hanks [6] noted that the choice of a "favourite mean diameter" would not sufficiently describe the Particle Size Distribution (PSD) of any slurry. The only way to describe the PSD is by taking points along the distribution and actually using these in the analysis.

Type of Correlation	Author		Reference
1. Empirical non-dimensional	Bonnington,	+	15
	Durand & Condolis	+	16
	Zandi & Govatos		17
2. Semi-empirical non-dimensional	Clift et al		18
	Vocadlo & Charles	+	19
	Smoldyrev		13
3. Two component	Faddick	*+	20
	Hanks		6
	Newitt et al	*+	21
	Wasp et al	*	2,5,22
4. Two layer	Wilson	*	24 - 26
5. Two layer, multi-component	Lazarus		27
6. Multi layer, multi-component	Roco & Shook		28 - 30

TABLE 1 : Some correlations for mixed regime flow.

* - correlations used to analyse data in this paper

+ - correlations compared by Wani et al(ref.31)

5.1   CORRELATIONS NOT USED FOR COMPARISON IN THIS PAPER

5.1.1   The correlations of 14 authors for mixed regime flow, indicated thus (+), are compared by Wani et al[31]. The correlations selected required the use of a weighted mean particle diameter (d_w) for the best agreement between experimental and calculated values.

$$d_w = \frac{\Sigma w_i d_i}{\Sigma w_i} \qquad [2]$$

5.1.2   Clift et al[18] presented a scaling technique which requires a set of experimental data points for the material being considered. The correlation therefore requires no assumed particle diameter and has since been combined with Wilson[26] in the form of nomograms which are not used for comparison since a representative particle size is chosen in the analysis rather than a particle size distribution.

5.1.3 Smoldyrev [13] presented an equation of the form

$$i_m = 2f_f \frac{V_m^2 C_{v1}}{gD} + 2f_w \frac{V_m^2 a^* C_{v2}}{gD}$$

$$+ Aa^* C_{v3} \frac{V_t \sqrt{D/d_w}}{V_m} \qquad [3]$$

where C_{v1} = concentration of particles
$\qquad$ < 0,04 mm

$\qquad C_{v2}$ = concentration of particles
$\qquad$ 0,04 - 0,2 mm

$\qquad C_{v3}$ = concentration of particles
$\qquad$ 0,2 - 2,5 mm

$\qquad a^*$ = $(S_s - (1 + a C_{v1}))/(1 + aC_{v1})$

$\qquad a$ = $(\rho_s - \rho_w)/\rho_w$

$\qquad A$ = constant

The concentrations of each particle size are arbitrarily separated with the variable a* being used to modify the heterogeneous component.

5.1.4 Lazarus [27] is not used for comparison because the method is still in the process of being updated. This method involves a mechanistic model consisting of the vehicle portion or slow settling component which is generally non-Newtonian and rheologically active, the suspended load portion and the bed load portion travelling as a sliding bed. The analysis is iterative and each particle size fraction is considered separately using a criterion for distinguishing the vehicle portion from the suspended portion and a further criterion for distinguishing the suspended portion from the bed load portion.

5.1.5 Roco and Shook's [28-30] models are not used for comparison; firstly, because empirical coefficients are still required in a method which is rather complex to apply and involves an averaging approach specifically for multi component turbulent flow; secondly, although the University of Cape Town Applied Mechanics Research Unit (AMRU) are leaders in finite element analysis, it was not possible to include Roco and Shook's model in time for this paper.

5.2 CORRELATIONS COMPARED IN THIS PAPER

5.2.1 Faddick [20] as modified by Streicher [32] required the calculation of two characteristic diameters.

$$d_{Lmax} = 2,62 \left(v^2/g(S_s - 1) \right)^{1/3}$$

$\qquad$ maximum particle diameter settling in water according to Stokes' law

$$d_{Tmin} = 26,5 \ d_{Lmax}$$

$\qquad$ minimum particle diameter settling in water according to Newton's law

$\qquad R$ = Ratio of fines for $d < d_{Lmax}$ to total solid

$$S_{mf} = \frac{S_w + C_{vd} R(S_s - S_w)}{1 - C_{vd}(1 - R)} \quad \text{(ref. 32)} \qquad [4]$$

$$C_{vf} = \frac{S_{mf} - S_w}{S_s - S_w} \qquad [5]$$

$$\mu_f = \mu_w \left(1 - \frac{C_{vf}}{0,6}\right)^{-2,5} \qquad [6]$$

$$C_{vc} = C_v - C_{vf} \qquad [7]$$

$$C_D' = \frac{3 S_w C_D}{4 S_f (S_s - S_w)}$$

$$\frac{i_m - i_{mf}}{C_{vc} \ i_{mf}} = 180 \left[\frac{V_m^2 \sqrt{C_D'}}{gD (S_s - 1)} \right]^{-1,5} \qquad [8]$$

i_{mf} $\quad$ vehicle hydraulic gradient calculated from Colebrook White equation with viscosity and density of the mixture

C_D $\quad$ drag coefficient for $d = d_{Tmin}$

5.2.2 Newitt [21] presented a correlation for a mixture of two discrete particle sizes. The settling velocity used in the correlation is replaced by a variable given by

$$B = (1 - X) V_{tf}' + X (1 - (1-X)C_v) V_{tc}' \qquad [9]$$

X = Ratio of coarse fraction to total solid.

V_t' = hindered settling velocity
$\quad$ = $V_t S_f (1 - C_v)^\alpha$ $\quad$ (ref. 33) $\qquad [33]$

The data analysed was arbitrarily divided into a d_{30} particle size representing the fine component and d_{70} representing the coarse component.

5.2.3 Wasp et al [2] presented an iterative method for calculating the rheologically active part of a mixture. The procedure used follows:

1. Guess initial value of C_{vf} (say $C_{vf} = C_v$)

2. $S_m = S_w + C_{vf} (S_s - S_w)$

$\qquad \mu_m = (1 + C_{vf}/C_b)^{-2.5}$

3. Calculate f_m from Colebrook-White

$\qquad$ equation for $R_e = \frac{\rho_m V_m D}{\mu_m}$

4. Calculate $V_t(I)$ the settling velocity for each particle diameter for S_m and μ_m

5. Calculate $\log \left(\frac{C}{C_a}\right)_i = -1.8 \dfrac{V_t(I)}{k\beta V_m \sqrt{f_m/2}}$ [10]

6. Sum all the values obtained in 5 as follows:

$$C_{vf} = \frac{\Sigma \left(\frac{C}{C_a}\right)_i}{N} \quad [11]$$

N = number of data points in P S D

7. Compare C_{vf} in Step 6 to C_{vf} in Step 1 and if C_{vf} (from 6) - C_{vf} (from 1) > 0,001 then recalculate C from Step 2 using new C_{vf} from step 6.

8. Calculate;

$$i_{mf} = \frac{2f_m S_m V_m^2}{gD} \quad [12]$$

9. $C_{vc}(I) = C_v(I)(1-C/C_a)$ [13]

$C_v(I)$ = percentage of each particle size retained

10. $\psi'(I) = C_{vc}(I)\, 150 \left(\dfrac{gD(S_s-1)}{V_m^2 \sqrt{C_D}}\right)^{-1.5}$ [14]

11. $\psi = \Sigma \phi'(I)$ [15]

12. $i_m = i_{mf}(1 + \psi)$ [16]

5.2.4 Wilson[24-26] presented a mechanistic model for the analysis of sliding bed flow. The analysis has an inherent method for stationary bed classification. Wilson presented nomographic charts to alleviate the need for a computer analysis. The analysis presented here is for a computer and required considerable interpretation of Wilson's papers.

1. Calculate the bed load concentration C_{vb} by averaging the equation

$$\frac{C_{vb}}{C_v} = \left(\frac{V_{mt}}{V_m}\right)^\alpha \quad [17]$$

This was interpreted for this paper as

$$C_{vb} = C_v \left[\frac{\Sigma \left[\frac{V_{mt}(I)}{V_m}\right]^2}{N}\right] \quad \text{for} \quad \frac{V_{mt}(I)}{V_m} < 1 \quad [18]$$

$V_{mt}(I) = 0,6\, V_t(I)(8/f_w)^{1/2} \exp \dfrac{45d(I)}{D}$
for each particle size I [19]

2. $C_r = \dfrac{C_{vb}}{C_b}$ [20]

3. $a_{bed} = \pi C_r$ (bed load area) [21]

4. Calculate β for a given bed load angle using an iterative method where

$$a_{bed} = \beta - \cos\beta \sin\beta \quad [22]$$

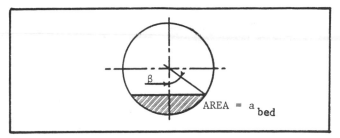

Fig. 4 : Definition of terms

5. Determine the friction factor associated with the interface. The value of d_{30} is an interpretation for this paper as a result of trying various values of d .

$$f_b = \left(4 \log\left(\frac{3,7D}{d_{30}}\right)\right)^{-2} \quad [23]$$

6. Calculate the hydraulic gradient for the suspended load portion (last term in equation 24 was an interpretive addition for this paper).

$$i_{susp} = ((S_s-1)(C_v-C_{vb}) + 1)\, i_w \quad [24]$$

7. Calculate the hydraulic gradient for plug flow of solids from;

$$i_p = 2\,\mu_s\,(S_s-1)\,C_b \quad [25]$$

8. Calculate the ratio (ξ) from

$$\xi = 2\, f_b/f_w \quad [26]$$

9. Wilson defined nondimensional hydraulic gradient parameters

$$Y = \frac{i}{i_p} = \frac{\sin\beta - \beta\cos\beta}{1 - a + \dfrac{a\xi\,\sin\beta}{\pi-\beta+\xi\sin\beta}} \quad [27]$$

and $\quad X = \dfrac{i_w}{i_p} = \dfrac{f_w V_m^2}{\mu_s gD(S_s-1)C_b}$ [28]

10. By definition (interpreted for this paper)

$$i_b = (Y-X)\, i_p \quad [29]$$

11. The total hydraulic gradient is calculated from the sum of equations 24 and 29 and the head loss gradient for clear water

$$i_m = i_b + i_w + i_{susp} \quad [30]$$

12. The value of X at the bed slip point is calculated from

$$X_s = \frac{\pi(1-a)^2 \dfrac{\sin\beta - \beta\cos\beta}{\pi}}{\bar{r}\,\sin\beta} \qquad [31]$$

If $X < X_s$ then a stationary bed exists.

6. EXPERIMENTAL DATA

Experimental Data derived from the Saskatchewan Research Council (SRC) and the Hydrotransport Research Unit (HTRU) are used to compare the correlations discussed. Table 2 is a list of the data used.

N	D [mm]	C_v [%]	S_s	S_w	μ_w [Pas]	T [°C]	Ref.	Material
1	158,5	14	2,65	0,9991	$1,14.10^{-3}$	-	1,p.47	Sand
2	158,5	30	2,65	0,9991	$1,14.10^{-3}$	-	1,p.48	Sand
3	107,57	13,5	2,65	0,9991	$1,14.10^{-3}$	-	1,p.44	Sand
4	107,57	30	2,65	0,9991	$1.14.10^{-3}$	-	1,p.45-46	Sand
5	45,42	21	2,96	-	-	40	HTRU	UT
6	57,96	19	2,96	-	-	40	HTRU	UT
7	141,7	15	2,27	-	-	30	HTRU	Fly Ash
8	141,7	23	2,27	-	-	30	HTRU	Fly Ash
9	80,89	22	2,27	-	-	33	HTRU	Fly Ash

UT = Uranium Tailings

TABLE 2 : Experimental data used to compare correlations tested

7. COMPARISON OF CORRELATION ERRORS

The correlations are compared numerically using a log standard error where

$$S_{error} = \sqrt{\frac{\Sigma\,(\log i_m - \log i_{mcal})^2}{n - 1}}$$

where n = number of data points

S_{error} = root mean square deviation of the log of observed points from the log of calculated points.

DATA BANK NUMBER	CORRELATION				
	$S_m \ast i_w$	Faddick	Newitt	Wasp	Wilson
1	0,1087	0,2045	**0,0692**	0,5078	0,1650
2	0,1547	0,4068	**0,0686**	0,5813	0,3760
3	0,1392	0,1584	**0,0843**	0,4852	0,1810
4	0,1424	0,3990	**0,0556**	0,5137	0,4715
5	0,5526	0,1235	0,3798	0,2151	**0,0511**
6	0,3314	0,1630	0,2354	0,2863	**0,0501**
7	0,1344	0,2504	0,1409	0,3902	**0,0740**
8	0,1439	0,1125	0,1418	0,1583	**0,0518**
9	0,1043	0,0636	0,1019	0,1023	**0,0300**

TABLE 3 : Log standard errors for calculated hydraulic gradients of mixture data. The minimum values are shown in bold type.

Hydraulic gradient versus mean velocity graphs are presented for each correlation.

Quadratic least squares curves are drawn through the points generated by a correlation. This was done because each point is calculated for a unique concentration.

8. DISCUSSION AND SELECTION OF CORRELATIONS FOR ENERGY GRADIENT PREDICTION

The data presented, although always demonstrating a well graded particle size distribution, is essentially of two types

1. sand mixture with minimum particle sizes settling above the Stoke's region
2. mixtures with minimum particle sizes settling well within the Stoke's region (laminar settling).

8.1 PSEUDO-FLUID CORRELATION

The pseudo-fluid correlation is presented as a datum by which to assess the more complex correlations. Mixed regime flow is apparent from the large underestimations of i_{mcal} for the pseudo-fluid correlation shown in the graphs.

8.2 FADDICK CORRELATION

This correlation underpredicts pseudo-homogeneous flow and overpredicts heterogeneous flow components. For the sand data i_{mcal} increases with decreasing velocity. For the Hydrotransport Research Unit data i_{mcal} increases (and is too high) only at velocities below the deposit velocity. These velocities are outside the defined range of the correlation.

8.3 NEWITT ET AL CORRELATION

This correlation was developed in a 45mm diameter pipeline for poorly graded mixtures. The predictions for the sand mixtures (SRC data) show good correlation. This is expected since the mixture has a wide size distribution and approximates a mixture of two sand sizes. The arbitrary choice of a d_{30} and d_{70} particle size to define the fine and coarse components respectively could be further investigated to describe the sensitivity of this correlation to this size of selection.

The HTRU data predictions are all low since the correlation assumes the carrier fluid to be water and not a non-Newtonian fluid.

The equation used for pseudo-homogeneous flow is

$$\frac{i_{mcal} - i_w}{C_v i_w} = K(S_s - 1)$$

K = 0,6 from Newitt.

Lazarus[34] showed that the value of K used by Newitt et al was too low and presented a graph of K versus concentration (C_v) with relative density as parameter.

The correlation could be improved by assuming

particle settling in heavy media and an increase in the value of K from Figure 5.

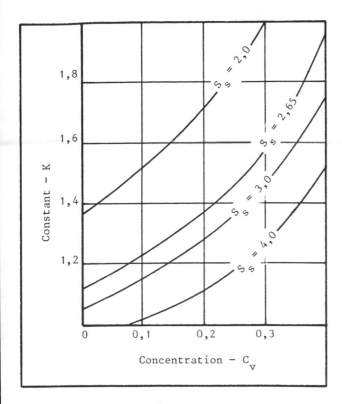

Fig. 5 Newitt et al constant K as a function of concentration with solid relative density as parameter.

8.4 WASP ET AL CORRELATION

This correlation shows a significant over-prediction of the heterogeneous energy gradient. The results obtained for the first four sets of data (sand mixtures) are significantly higher than the experimental results. For the Hydrotransport Research Unit data the prediction is low at high velocities (pseudo-homogeneous flow) and high at low velocities indicating the heterogeneous flow over prediction.

8.5 WILSON CORRELATION

This correlation is adapted from a correlation for heterogeneous flow of unsized particles. For the sand data (SRC) the correlation over-predicts. Increasing concentration causes an increase in the overprediction. For the mixed regime slurries (HRTU) this correlation produced very good results. The data produced by the HTRU, although of two different types of solid, shows distinctive mixed regime characteristics. No significant bias to either the heterogeneous or homogeneous component is apparent. It must be stressed that significant interpretation of the correlation was required to achieve the results shown. From Table 3 it can be seen that the Wilson correlation test predicts these results.

9. CONCLUSIONS

1. Slurry pipelines operated in mixed regime flow show significant advantages over other transport regimes.

2. The Wilson correlation is an effective tool for modelling mixed regime slurries containing significant quantities of fines.

3. The Newitt et al correlation best describes coarse particle mixtures although the particles may be well graded. The effect of choosing different representative diameters needs further investigation.

4. None of the above correlations is universally applicable for modelling slurries with well graded particle size distributions.

10. ACKNOWLEDGEMENTS

The authors wish to acknowledge the support received from the following organizations -

1. Council for Scientific and Industrial Research

2. Chamber of Mines

3. Rössing Uranium

4. University of Cape Town

5. Mather and Platt, S.A.

6. Georgia Iron Works, U.S.A.

References

1. SCHRIEK, W.; SMITH, L.G.; HAAS, D.B.; HUSBAND, W.H.W.; Report VII (E73-21), Engineering Division, Saskatchewan Research Council, Nov. 1973.

2. WASP. E.J.; REGAN, T.J.; WITHERS, J.; COOK, P.A.C.; Pipeline News, July 1963, pp 20-28.

3. CHARLES, M.E.; CHARLES, R.A.; Advances in Solid Liquid Flow, Edited by Zandi, I., Pergamon, 1971, pp 187-197.

4. KAZANSKIJ, I.; BRUHL, H.; HINSCH, J.; Paper D2, Hydrotransport 3, Colorado, BHRA, May 1977, pp D2.11-21.

5. WASP, E.J.; AUDE, T.C.; KENNY, J.P.; SEITER, R.H.; JACQUES, R.B.; Paper H4, Hydrotransport 1, Warwick, BHRA, Sept. 1970, pp H4.53-76.

6. HANKS, R.W.; R.W. Hanks Association Inc., Orem, Utah, 1981, Chapter 8.

7. STEPANOFF, A.J.; Mechanical Engineering, ASME, Vol.86, Sept. 1964, pp 29-35.

8. CHENG, D.C.-H.; WHITTAKER, W.; Paper C3, Hydrotransport 2, Warwick, BHRA, Sept. 1972, pp C3.21-39.

9. THOMAS, A.D.; Paper D5, Hydrotransport 5, Hannovr, BHRA, May 1978, pp D5.63-78.

10. PARZONKA, W.; KENCHINGTON, J.M.; CHARLES, M.E.; The Canadian Journal of Chemical Engineering, Vol.59, June 1981, pp 291-296.

11. DUCKWORTH, R.A.; PULLUM, L.; LOCKYEAR, C.F.; C.S.I.R.O. Report, Division of Mineral Engineering, Australia.

12. TRAYNIS, V.V.; Terraspace Inc., Rochville, USA., 1977, Translated from Russian.

13. SMOLDYREV, A.Ye.; Terraspace Inc., Rochville, USA., 1982, Translated from Russian.

14. KLOSE, R.B.; 7th International Tech. Conf. on Slurry Transportation, Lake Tahoe, STA, Mar. 1982, pp 61-64.

15. BONNINGTON, S.T.; BHRA Report No. RR637, Oct. 1959.

16. DURAND, R.; CONDOLIS, E.; Proc. of a Colloquim on the Hyd. Trans. of Coal. Paper 4, Nat. Coal Board, London, 1952.

17. ZANDI, I.; GOVATOS, G.; J. Hydr. Div. ASCE, HY3, Vol.93, 1967, pp 145-159.

18. CLIFT, R.; WILSON, K.C.; ADDIE, G.R.; CARSTENS, M.R.; Paper B1, Hydrotransport 8, Johannesburg, BHRA, Aug. 1982, pp 91-101.

19. VOCADLO, J.J.; CHARLES, M.E.; Paper C1, Hydrotransport 2, Warwick, BHRA, Sept. 1972, pp C1.1-12.

20. FADDICK, R.R.; Short Course on Hydraulic Transport, 8, Johannesburg, Aug. 1982, Chapter 6.

21. NEWITT, D.M.; RICHARDSON, J.F.; ABBOT, M.; TURTLE, R.B.; Trans. Instn. Chem. Engrs., Vol.33, 1955, pp 93-113.

22. WASP, E.J.; AUDE, T.C.; SEITER, R.H.; THOMPSON, T.L.; Advances in Solid-Liquid Flow, Edited by Zandi, I., Pergamon, 1971, pp 119-210.

23. WILSON, K.C.; Paper E1, Hydrotransport 3, Colorado, BHRA, May 1976, pp A1.1-13.

24. WILSON, K.C.; Paper A1, Hydrotransport 4, Alberta, BHRA, May 1976, pp A1.1-16.

25. WILSON, K.C.; JUDGE, D.G.; Paper A1, Hydrotransport 5, Hannover, BHRA, May 1978, pp A1.1-12.

26. WILSON, K.C.; Paper A1, Hydrotransport 6, Kent, BHRA, Sept. 1979, pp 1-12.

27. LAZARUS, J.H.; Hydrotransport Research Unit Report, Uranium Tailings, phase 1., UCT, Cape Town, July 1985.

28. ROCO, M.C.; SHOOK, C.A.; 7th International Tech. Conf. on Slurry Transportation, Lake Idahoe, STA, Mar. 1982, pp 175-192.

29. ROCO, M.C.; SHOOK, C.A.; Journal of Pipelines, Vol.4, Elsevier, 1984, pp 3-13.

30. ROCO, M.C.; SHOOK, C.A.; AIChE Journal, Vol.31, No.8. Aug. 1985, pp 1401-1403.

31. WANI, G.A.; SARKAR, M.K.; PITCHUMANI, B.; Journal of Pipelines, Vol.3, No.1, Elsevier, 1982, pp 23-33.

32. STREICHER, D.J.; M.Sc. Thesis, Dept. Civil Engineering, U.C.T., 1984, p 66.

33. RICHARDSON, J.F.; ZAKI, W.N.; Chem. Eng. Sci. Vol.3, 1954, p 65.

34. LAZARUS, J.H., Course Notes on Hydraulic Transport of Solids, Chapter 9, University of Cape Town, 1985.

Data Bank Number 1

S.R.C. Sand Mix 1	(Ref. 1,p.47)
Diameter : 158.50 mm	Relative Density : 2.65

Percentage Passing	Particle Size [mm]
20	0.034
30	0.080
40	0.149
50	0.200
60	0.285
70	0.395
80	0.505
90	0.701
100	2.380

Vm [m/s]	Cv [%]	im [m/100m]	im=Sm≭iw		Faddick		Newitt		Wasp		Wilson	
			im	Error	im	Error	im	Error	im	Error	im	Error
3.52	13.7	7.03	6.45	-8.3	9.27	31.9	6.46	-8.1	15.43	119.5	9.56	36.0
3.22	14.1	6.28	5.52	-12.1	8.92	42.0	5.52	-12.1	15.85	152.4	9.03	43.8
3.25	13.8	6.31	5.59	-11.4	8.86	40.4	5.60	-11.3	15.75	149.6	8.98	42.3
3.09	13.9	5.97	5.11	-14.4	8.69	45.6	5.12	-14.2	16.07	169.2	8.75	46.6
3.01	14.1	5.73	4.88	-14.8	8.68	51.5	4.97	-13.3	16.30	184.5	8.71	52.0
2.69	13.9	5.03	3.97	-21.1	8.41	67.2	4.36	-13.3	17.30	243.9	6.75	34.2
2.54	14.2	4.80	3.59	-25.2	8.52	77.5	4.15	-13.5	18.06	276.3	6.65	38.5
2.33	14.2	4.86	3.07	-36.8	8.63	77.6	3.86	-20.6	19.31	297.3	6.53	34.4

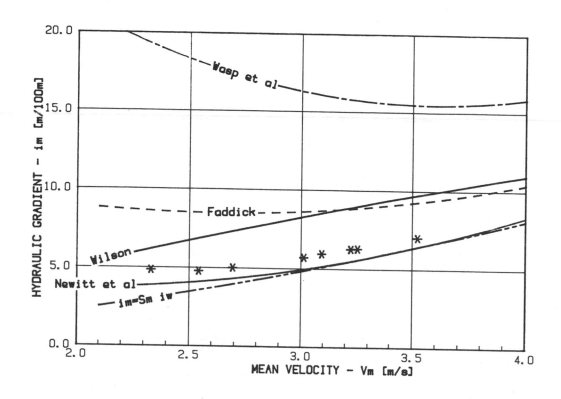

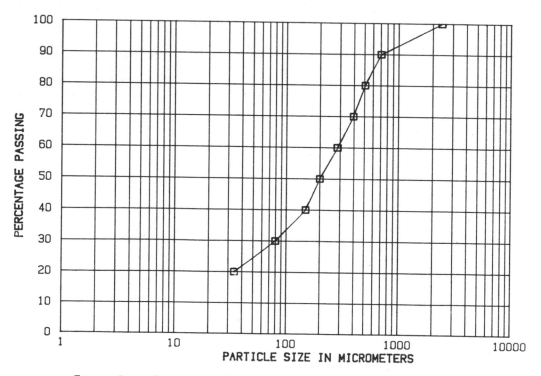

Fig.   6   Data Bank Number 1. Comparison of correlations
              and Particle Size Distribution.

Data Bank Number 2

S.R.C. Sand Mix 1	(Ref. 1,p.48)
Diameter : 158.50 mm	Relative Density : 2.65

Percentage Passing	Particle Size [mm]
20	0.060
30	0.160
40	0.195
50	0.260
60	0.350
70	0.485
80	0.600
90	0.798
100	2.380

Vm [m/s]	Cv [%]	im [m/100m]	im=Sm*iw im	Error	Faddick im	Error	Newitt im	Error	Wasp im	Error	Wilson im	Error
3.44	29.9	8.48	7.54	-11.1	14.92	75.9	8.12	-4.2	20.39	140.4	15.46	82.3
3.27	30.0	8.00	6.89	-13.9	14.98	87.3	7.87	-1.6	20.88	161.0	15.37	92.1
3.10	30.3	7.53	6.27	-16.7	15.22	102.1	7.69	2.1	21.50	185.5	15.56	106.6
2.91	30.5	7.08	5.60	-20.9	15.56	119.8	7.53	6.4	22.42	216.7	16.03	126.4
2.78	30.7	6.75	5.17	-23.4	15.91	135.7	7.47	10.7	23.21	243.9	16.62	146.2
2.63	30.3	6.43	4.65	-27.7	16.13	150.9	7.36	14.5	24.12	275.1	17.24	168.1
2.43	31.1	6.11	4.07	-33.4	17.28	182.8	7.48	22.4	26.00	325.5	13.48	120.6
2.28	30.9	5.90	3.61	-38.8	17.92	203.7	7.53	27.6	27.49	365.9	14.16	140.0
2.20	30.9	5.90	3.39	-42.5	18.40	211.9	7.60	28.8	28.43	381.9	14.73	149.7

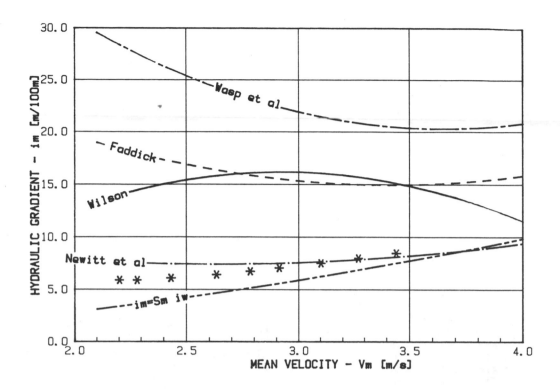

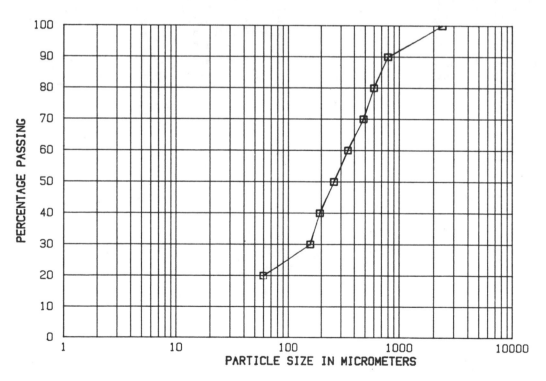

Fig. 7   Data Bank Number 2. Comparison of correlations
            and Particle Size Distribution.

Data Bank Number 3

S.R.C. Sand Mix 1	(Ref. 1,p.44)
Diameter : 107.57 mm	Relative Density : 2.65

Percentage Passing	Particle Size [mm]
30	0.061
40	0.150
50	0.205
60	0.301
70	0.408
80	0.510
90	0.680
100	2.380

Vm [m/s]	Cv [%]	im [m/100m]	im=Sm*iw im	Error	Faddick im	Error	Newitt im	Error	Wasp im	Error	Wilson im	Error
3.11	13.5	8.75	8.14	-7.0	10.66	21.8	8.15	-6.9	16.91	93.3	12.89	47.3
2.82	13.4	7.74	6.81	-12.0	9.85	27.3	6.82	-11.9	17.10	120.9	10.20	31.8
2.48	12.8	6.54	5.35	-18.2	8.97	37.2	5.59	-14.5	17.76	171.6	9.10	39.1
2.51	12.9	6.70	5.48	-18.2	9.06	35.2	5.68	-15.2	17.69	164.0	9.20	37.3
2.51	13.3	6.68	5.51	-17.5	9.21	37.9	5.71	-14.5	17.77	166.0	9.36	40.1
2.37	13.2	6.25	4.96	-20.6	8.98	43.7	5.34	-14.6	18.26	192.2	9.11	45.8
2.19	14.1	6.01	4.35	-27.6	9.21	53.2	5.01	-16.6	19.35	222.0	9.47	57.6
2.20	13.4	5.93	4.35	-26.6	8.92	50.4	4.97	-16.2	19.13	222.6	9.12	53.8
2.01	14.7	5.80	3.76	-35.2	9.51	64.0	4.73	-18.4	20.73	257.4	10.00	72.4
1.86	10.0	5.51	3.06	-44.5	7.34	33.2	3.94	-28.5	20.79	277.3	8.00	45.2

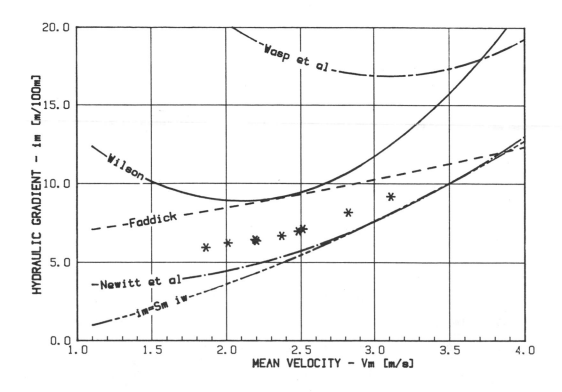

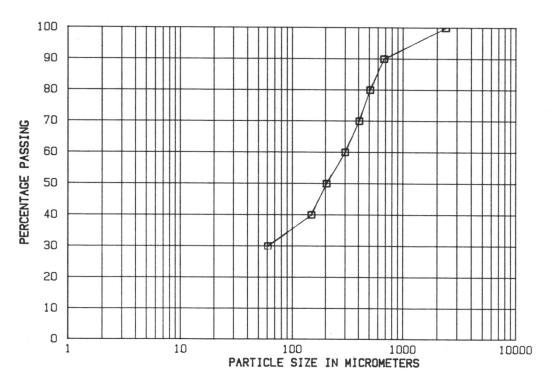

Fig. 8   Data Bank Number 3. Comparison of correlations
and Particle Size Distribution.

Data Bank Number 4

S.R.C. Sand Mix 1	(Ref. 1,p.45-46)
Diameter : 107.57 mm	Relative Density : 2.65

Percentage Passing	Particle Size [mm]
30	0.068
40	0.147
50	0.180
60	0.210
70	0.310
80	0.480
90	0.640
100	2.380

Vm [m/s]	Cv [%]	im [m/100m]	im=Sm*iw im	Error	Faddick im	Error	Newitt im	Error	Wasp im	Error	Wilson im	Error
3.14	30.0	10.61	10.13	-4.5	16.48	55.3	10.14	-4.4	18.92	78.3	23.09	117.6
2.80	30.6	9.30	8.29	-10.9	16.27	74.9	8.29	-10.9	19.25	107.0	18.59	99.9
2.65	30.4	8.56	7.49	-12.5	16.15	88.7	7.49	-12.5	19.56	128.5	18.66	118.0
2.54	29.9	7.92	6.90	-12.9	15.98	101.8	6.90	-12.9	19.81	150.1	18.70	136.1
2.34	30.7	7.41	6.00	-19.0	16.59	123.9	6.17	-16.7	20.86	181.5	20.51	176.8
2.20	29.7	6.80	5.31	-21.9	16.45	141.9	5.86	-13.8	21.46	215.6	21.25	212.5
2.03	30.6	6.40	4.64	-27.5	17.53	173.9	5.67	-11.4	22.99	259.2	25.12	292.5
1.88	31.3	5.96	4.07	-31.7	18.74	214.4	5.57	-6.5	24.68	314.1	20.90	250.7
1.89	30.3	5.90	4.06	-31.2	18.10	206.8	5.51	-6.6	24.29	311.7	19.88	236.9
2.05	29.6	6.32	4.67	-26.1	16.91	167.6	5.63	-10.9	22.56	257.0	23.60	273.4
1.73	29.7	5.96	3.44	-42.3	18.86	216.4	5.37	-9.9	26.10	337.9	22.68	280.5
1.65	28.0	5.90	3.10	-47.5	18.45	212.7	5.21	-11.7	26.73	353.1	13.37	126.6
2.53	29.2	7.85	6.79	-13.5	15.68	99.7	6.80	-13.4	19.69	150.8	18.31	133.2
2.54	30.1	8.01	6.91	-13.7	16.07	100.6	6.92	-13.6	19.85	147.8	18.82	135.0

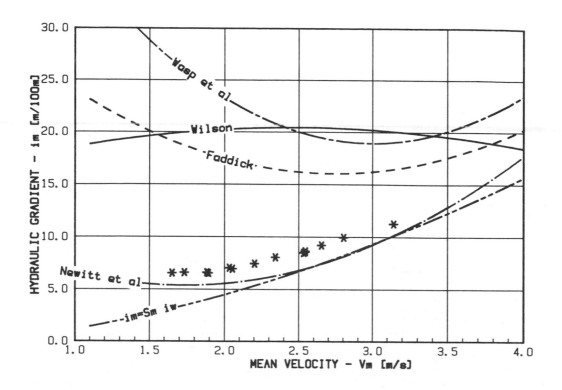

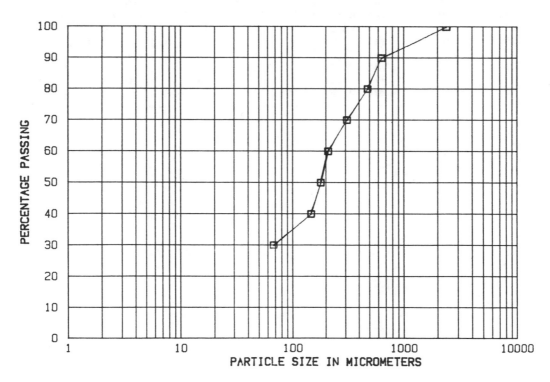

Fig. 9   Data Bank Number 4. Comparison of correlations
              and Particle Size Distribution.

Data Bank Number 5

H.T.R.U.	Uranium Tailings
Diameter : 45.42 mm	Relative Density : 2.96

Percentage Passing	Particle Size [mm]
10	0.006
20	0.008
30	0.012
40	0.019
50	0.031
60	0.050
70	0.080
80	0.110
90	0.180
100	0.600

Vm [m/s]	Cv [%]	im [m/100m]	im=Sm*iw im	im=Sm*iw Error	Faddick im	Faddick Error	Newitt im	Newitt Error	Wasp im	Wasp Error	Wilson im	Wilson Error
3.09	21.6	35.54	23.41	-34.1	24.64	-30.7	23.59	-33.6	23.49	-33.9	30.81	-13.3
2.05	21.7	17.72	11.24	-36.6	14.52	-18.1	11.32	-36.1	15.76	-11.1	15.84	-10.6
1.76	21.4	14.69	8.46	-42.4	12.57	-14.4	8.53	-41.9	14.74	.3	12.75	-13.2
.77	21.0	12.05	1.94	-83.9	13.79	14.4	3.14	-73.9	23.12	91.9	3.56	-70.5
.51	20.4	13.10	.93	-92.9	19.99	52.6	3.50	-73.3	36.21	176.4	3.41	-74.0
1.12	20.6	13.25	3.76	-71.6	10.98	-17.1	3.78	-71.5	16.38	23.6	6.52	-50.8
2.50	21.2	21.77	15.95	-26.7	18.12	-16.8	16.07	-26.2	18.22	-16.3	21.35	-1.9
2.87	19.9	28.65	19.99	-30.2	21.34	-25.5	20.14	-29.7	20.67	-27.9	26.02	-9.2
3.24	20.8	35.60	25.20	-29.2	26.10	-26.7	25.39	-28.7	24.64	-30.8	32.82	-7.8

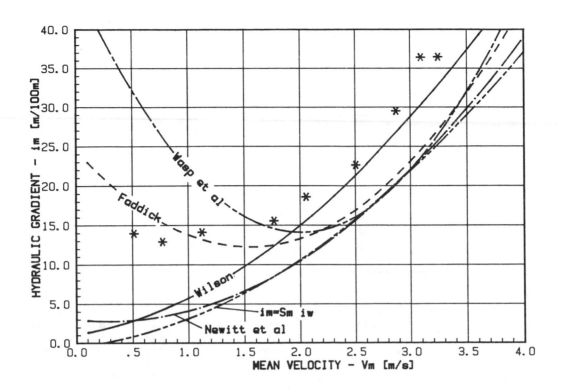

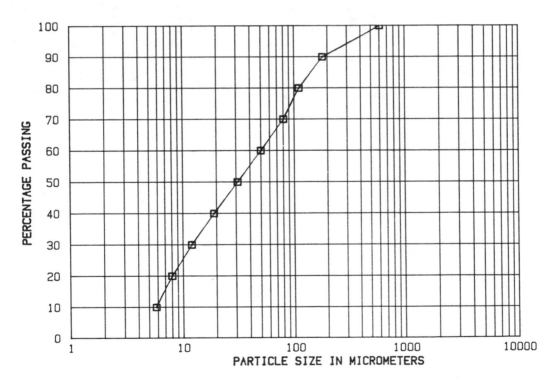

Fig. 10   Data Bank Number 5. Comparison of correlations
and Particle Size Distribution.

Data Bank Number 6

H.T.R.U.	Uranium Tailings
Diameter : 57.21 mm	Relative Density : 2.96
Percentage Passing	Particle Size [mm]
10	0.006
20	0.008
30	0.012
40	0.019
50	0.031
60	0.050
70	0.080
80	0.110
90	0.180
100	0.600

Vm [m/s]	Cv [%]	im [m/100m]	im=Sm*iw im	Error	Faddick im	Error	Newitt im	Error	Wasp im	Error	Wilson im	Error
3.36	18.5	30.24	19.82	-34.5	20.74	-31.4	19.98	-33.9	19.88	-34.3	25.36	-16.1
2.56	18.9	16.99	12.23	-28.0	14.30	-15.8	12.32	-27.5	14.87	-12.5	16.09	-5.3
1.81	19.0	8.32	6.51	-21.8	10.26	23.3	6.56	-21.2	12.80	53.8	9.55	14.8
1.04	18.9	6.87	2.43	-64.6	10.15	47.7	2.84	-58.7	16.90	146.0	4.57	-33.5
.68	18.1	6.43	1.10	-82.9	13.72	113.4	2.55	-60.3	26.19	307.3	2.35	-63.5
1.41	18.8	7.13	4.16	-41.7	9.39	31.7	4.19	-41.2	13.59	90.6	7.47	4.8
2.18	18.8	12.04	9.13	-24.2	11.89	-1.2	9.20	-23.6	13.35	10.9	12.40	3.0
2.97	18.3	21.56	15.83	-26.6	17.20	-20.2	15.95	-26.0	17.02	-21.1	20.34	-5.7
3.46	18.5	30.25	20.90	-30.9	21.69	-28.3	21.06	-30.4	20.67	-31.7	26.70	-11.7
3.52	18.7	30.14	21.66	-28.1	22.40	-25.7	21.83	-27.6	21.25	-29.5	27.70	-8.1

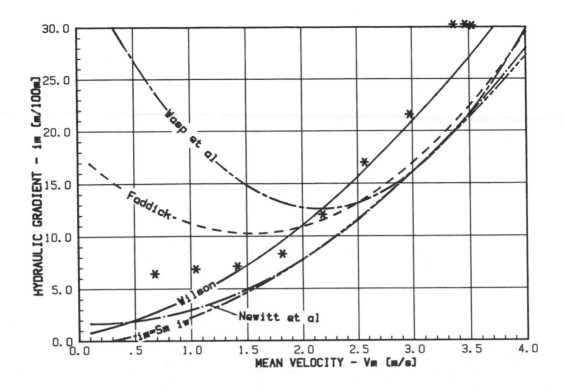

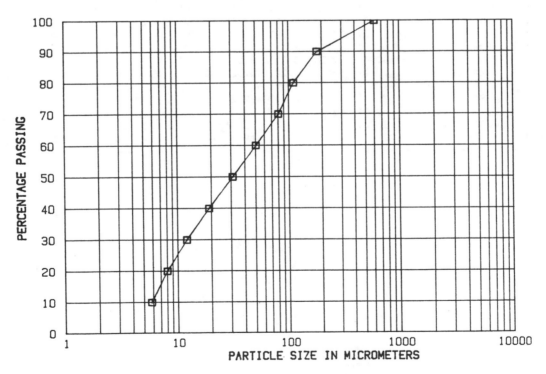

Fig. 11   Data Bank Number 6. Comparison of correlations
and Particle Size Distribution.

Data Bank Number 7

H.T.R.U.		Field 1 Fly Ash	
Diameter : 57.21 mm		Relative Density : 2.96	
Percentage Passing		Particle Size [mm]	
10		0.006	
20		0.008	
30		0.012	
40		0.019	
50		0.031	
60		0.050	
70		0.080	
80		0.110	
90		0.180	
100		0.600	

Vm [m/s]	Cv [%]	im [m/100m]	im=Sm*iw im	Error	Faddick im	Error	Newitt im	Error	Wasp im	Error	Wilson im	Error
6.55	14.8	25.30	20.53	-18.9	20.56	-18.7	20.65	-18.4	20.28	-19.8	23.83	-5.8
6.08	14.6	21.70	17.88	-17.6	17.98	-17.1	17.98	-17.1	17.91	-17.5	20.72	-4.5
5.37	15.0	18.00	14.31	-20.5	14.57	-19.1	14.39	-20.1	14.79	-17.8	16.66	-7.4
4.64	16.5	13.90	11.13	-19.9	11.62	-16.4	11.19	-19.5	12.21	-12.2	13.16	-5.3
4.22	15.7	14.20	9.28	-34.6	9.84	-30.7	9.33	-34.3	10.68	-24.8	10.93	-23.0
3.46	16.8	8.20	6.54	-20.2	7.41	-9.6	6.57	-19.9	8.82	7.6	7.86	-4.1
2.87	16.5	7.40	4.64	-37.3	5.76	-22.2	4.66	-37.0	7.82	5.7	5.67	-23.4
2.10	15.7	2.60	2.61	.4	4.20	61.5	2.62	.8	7.69	195.8	3.44	32.3
1.46	17.0	.70	1.37	95.7	4.09	484.3	1.51	115.7	9.99	.	2.49	255.7
2.11	16.9	2.70	2.66	-1.5	4.38	62.2	2.68	-.7	7.84	190.4	3.56	31.9
3.43	15.7	7.00	6.36	-9.1	7.17	2.4	6.39	-8.7	8.63	23.3	7.57	8.1
4.76	16.8	14.40	11.70	-18.8	12.17	-15.5	11.76	-18.3	12.70	-11.8	13.86	-3.8
6.64	14.3	23.60	20.94	-11.3	20.93	-11.3	21.06	-10.8	20.64	-12.5	24.19	2.5

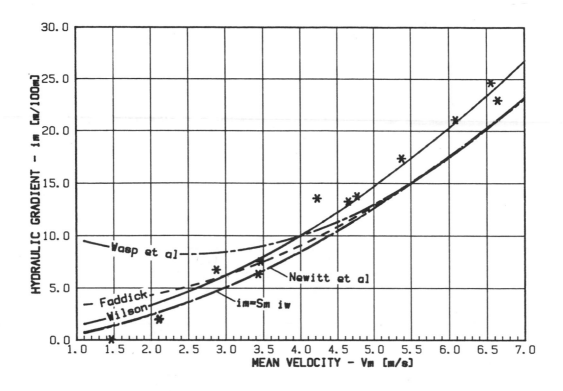

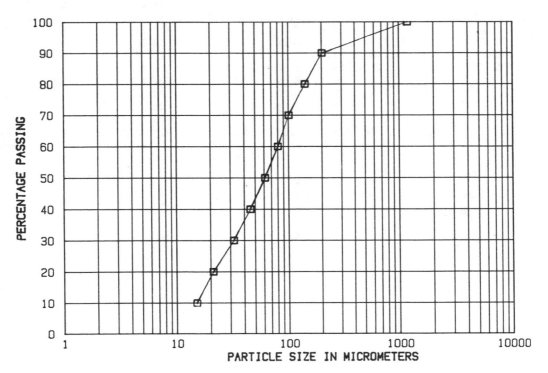

Fig. 12   Data Bank Number 7. Comparison of correlations
and Particle Size Distribution.

Data Bank Number 8

H.T.R.U.	Field 1 Fly Ash
Diameter : 141.70 mm	Relative Density : 2.27

Percentage Passing	Particle Size [mm]
10	0.015
20	0.021
30	0.032
40	0.045
50	0.061
60	0.080
70	0.100
80	0.140
90	0.200
100	1.180

Vm [m/s]	Cv [%]	im [m/100m]	im=Sm‡iw im	Error	Faddick im	Error	Newitt im	Error	Wasp im	Error	Wilson im	Error
5.96	23.0	28.70	19.01	-33.8	19.55	-31.9	19.10	-33.4	19.92	-30.6	23.62	-17.7
5.42	23.1	23.30	15.99	-31.4	16.65	-28.5	16.07	-31.0	17.21	-26.1	19.91	-14.5
4.77	22.4	18.20	12.58	-30.9	13.35	-26.6	12.64	-30.5	14.07	-22.7	15.61	-14.2
4.06	23.8	13.20	9.50	-28.0	10.59	-19.8	9.55	-27.7	11.87	-10.1	11.99	-9.2
3.43	23.2	9.40	6.95	-26.1	8.26	-12.1	6.98	-25.7	9.91	5.4	8.83	-6.1
2.76	23.1	6.40	4.67	-27.0	6.40	0.0	4.69	-26.7	8.78	37.2	6.12	-4.4
2.11	22.7	3.90	2.86	-26.7	5.20	33.3	2.87	-26.4	8.77	124.9	4.08	4.6
2.74	22.9	5.10	4.60	-9.8	6.32	23.9	4.62	-9.4	8.71	70.8	6.02	18.0
3.52	23.2	10.20	7.28	-28.6	8.56	-16.1	7.32	-28.2	10.13	-.7	9.23	-9.5
3.98	22.7	11.40	9.06	-20.5	10.11	-11.3	9.11	-20.1	11.29	-1.0	11.35	-.4
4.87	23.0	17.90	13.14	-26.6	13.93	-22.2	13.20	-26.3	14.68	-18.0	16.38	-8.5
6.43	23.1	29.60	21.86	-26.1	22.31	-24.6	21.97	-25.8	22.59	-23.7	27.16	-8.2

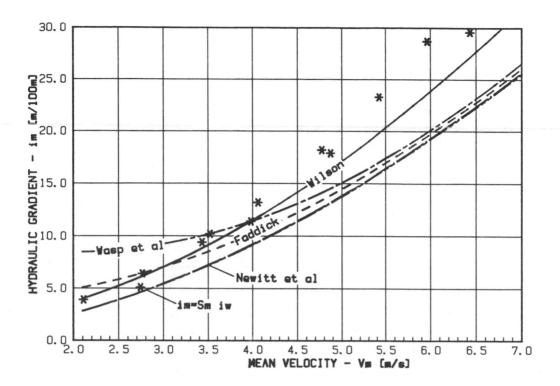

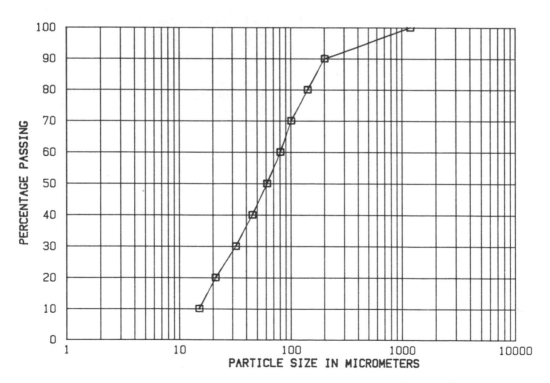

Fig. 13   Data Bank Number 8. Comparison of correlations
and Particle Size Distribution.

Data Bank Number 9

H.T.R.U.	Field 1 Fly Ash
Diameter : 80.89 mm	Relative Density : 2.27

Percentage Passing	Particle Size [mm]
10	0.015
20	0.021
30	0.032
40	0.045
50	0.061
60	0.080
70	0.100
80	0.140
90	0.200
100	1.180

Vm [m/s]	Cv [%]	im [m/100m]	im=Sm*iw im	im=Sm*iw Error	Faddick im	Faddick Error	Newitt im	Newitt Error	Wasp im	Wasp Error	Wilson im	Wilson Error
5.56	23.5	37.40	32.15	-14.0	32.65	-12.7	32.33	-13.6	33.14	-11.4	40.00	7.0
5.15	22.2	33.10	27.60	-16.6	28.09	-15.1	27.76	-16.1	28.27	-14.6	34.02	2.8
4.68	21.3	28.40	22.98	-19.1	23.53	-17.1	23.11	-18.6	23.75	-16.4	28.17	-.8
4.30	23.0	24.70	20.02	-18.9	20.83	-15.7	20.15	-18.4	21.57	-12.7	24.94	1.0
3.79	22.8	20.30	15.90	-21.7	16.85	-17.0	15.99	-21.2	17.83	-12.2	19.84	-2.3
3.28	22.5	16.20	12.20	-24.7	13.33	-17.7	12.27	-24.3	14.65	-9.6	15.32	-5.4
2.87	22.4	12.30	9.57	-22.2	10.91	-11.3	9.62	-21.8	12.64	2.8	12.18	-1.0
1.90	21.4	6.60	4.49	-32.0	6.63	.5	4.52	-31.5	10.02	51.8	6.47	-2.0
1.90	22.3	5.90	4.53	-23.2	6.78	14.9	4.56	-22.7	10.24	73.6	6.61	12.0
2.80	22.3	11.40	9.14	-19.8	10.52	-7.7	9.19	-19.4	12.31	8.0	11.67	2.4
4.68	22.6	26.40	23.28	-11.8	23.93	-9.4	23.41	-11.3	24.41	-7.5	28.83	9.2
5.99	23.8	39.70	36.93	-7.0	37.37	-5.9	37.13	-6.5	37.94	-4.4	46.04	16.0

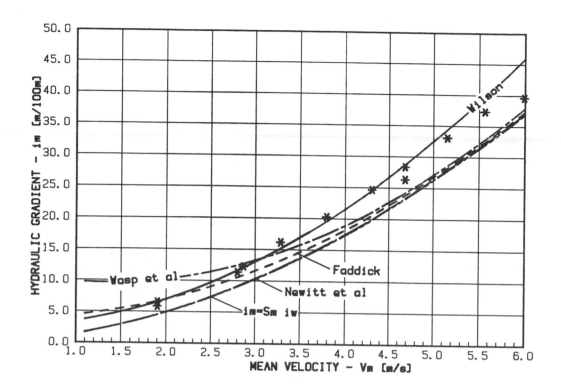

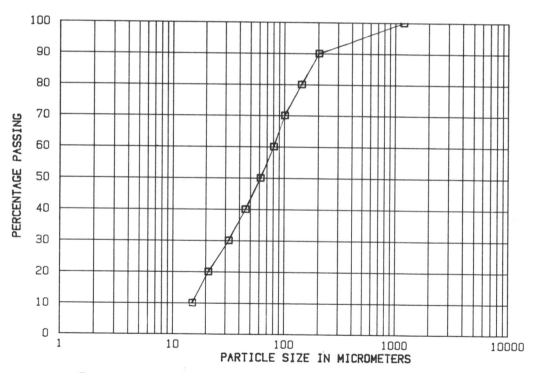

Fig. 14   Data Bank Number 9. Comparison of correlations
and Particle Size Distribution.

10th International Conference on the
Hydraulic Transport of Solids in Pipes

HYDROTRANSPORT 10

Innsbruck, Austria: 29-31 October, 1986

PAPER E3

INVESTIGATION OF OPTIMAL GRAIN-DISTRIBUTION
FOR TRANSPORT WITH HIGH CONCENTRATION

H.C.Hou

Prof.Guangdong Res.Inst.Hydraulic Eng.,
Guangzhou,China

SYNOPSIS

This paper deals with optimization of grain-size
in pipe-transportation of solid on the theoretical
background. The condition required to form and
sustain a high concentration slurry have been
identified; a slurry containing a given amount of
fines, will support large particles only if there
exists a continuous particle size distribution.
Theoretical predictions are in good agreement with
experimental results.

INTRODUCTION

During the last two decades, long-distance
hydraulic conveying of solids has found widespread
application in the transport of minerals,
particularly in the mining of coal.
Transportation with successful selection of some
parameters of the solid mixture the transport
efficiency was attained to such a high
concentration, by which one ton of coal could be
carried away by only one ton of water.

Hydrotransport with high efficiency could be
obtained in case of transport with very high
concentration, ie "hyperconcentration." Two-phase
flow with hypoerconcentration with fine particles
is such a high concentration, by which the
Newtonian system should transform to a Non-
Newtonian one, in particular, to Bingham fluid.
The physical-mechanical properties of the liquid
with hyperconcentration are different from those
of a Newtonian and there are a lot of problems
remaining unknown to date.

Hydraulic conveying at high solids concentration
is very attractive and a large amount of
experimental data has been accumulated to date.
In the literature, however, the approach remains
purely empirical.
In this concern, many fundamental problems still
need to be answered, such as: What
is the distinguishing feature of energy
dissipation between the dilute and high
concentration in hydrotransport? What is the
condition for the formation of hyperconcentration? With what value of concentration the
transport should became the transport with
hyperconcentration? Under what circumstance
should exist both the hyperconcentration
transport and the lower energy expenditure?
This paper is even dealt with these problems
on the theoretical background.

TWO FORMS OF TRANSPORT

As the density of solid particle is higher
than the water, they would settle in quiescent
fluid under gravitational force. However, it
could be suspended in flowing water. There are
two quite different conditions, by which the
solid particle could sustain in suspension:

The first case is the case of dilute concentration
with fine and/or coarse materials, in which the
solid particles are suspended individually in the
flowing water. In this case the energy
expenditure for suspension of particle is
compensated by the turbulence energy, which is
taken from the kinetic flow field with the result
that the von Karman constant for flow with
sediment suspension should be less than that for
the pure water. For a given particle-composition,
the higher the value of concentration, the lower
the value of von Karman constant. The reduction
of von Karman's constant with the presence of
sediment means that the increase of dissipated
energy directly from the mean flow field, as a
result it needs increase the energy-supply from
outside. Therefore, hydraulic conveying of low
concentrations is accompanied with high energy
consumption. Therefore, the hydraulic conveying
of solids by means of Newtonian fluids is not
economic.

Another form of hydrotransport is the transport
with hyperconcentration including some of fine
particles in the range $D \leqslant 0.02 \sim 0.04mm)$
This forms a Non-Newtonian system which is complex
and requires thorough analysis.

Recently, many experiments concerned with
the rheological property of slurry with
fine particles showed that the system would
transform from the Newtonian fluid to that of
Bingham's one, when the value of concentration
should exceed some critical value. In this
case, the shear stress τ can be expressed by

$$\tau = \tau_B + \eta \frac{du}{dy} \qquad (1)$$

here τ_B is Bingham yield stress, and η is the
plastic viscocity.

Some current experiments showed that the
critical volume-concentration $(C_v)_{cr}$, by
which the system-transfer should be occured,
was at about [1]

$$(C_v)_{cr} \approx 6\%$$

For simplicity to analysis it is assumed that
the particles are distributed geometrically
uniformly in space(Fig.1), then the volume-
concentration can be expressed by the relative
distance between the neighboring particles:

$$\left(\frac{D}{1}\right)_v = 1.24c_v^{1/3} \ , \ \frac{R}{1} = \overline{R} = 0.62c_v^{1/3} \qquad (2)$$

where R is radius of particle. Substituting above critical concentration into (2), we get the critical relative distance as

$$\left(\frac{D}{1}\right)_{cr} \approx 0.48 \ , \quad \left(\frac{R}{1}\right)_{cr} = \overline{R}_{cr} \approx 0.24 \qquad (3)$$

or, the critical relative gap $(S/D)_{cr}$ should approach to

$$\left(\frac{S}{D}\right)_{cr} \approx 1 \qquad (4)$$

However, the above description is based on geometry only. From the micromechanical point of view, the Bingham stress is caused by the formation of space-network of coagulation. The higher the concentration of fine particles, the stronger the rigidity of network-structure of coagulum. This network by fine sediments, not only could sustain the constituent particles themselves in suspension against the gravity force but also could sustain and then could carry some part of coarse particles without excessive additional energy requirement.

The space-network of coagulation is caused by the attracting force between the particles. A theoretical expression has been derived by Hamaker (2) about half a century ago, and its attractive potential V_A was expressed by

$$V_A = -\frac{1}{6}\left[\frac{2R^2}{1^2-4R^2} + \frac{2R^2}{1^2} + \ln\frac{1^2-4R^2}{1^2}\right] =$$

$$= -\frac{1}{6}\left[\frac{2\overline{R}^2}{1-4\overline{R}^2} + 2\overline{R}^2 + \ln(1-4\overline{R}^2)\right] \qquad (5)$$

Differentiating V_A with respect to the relative distance $\overline{R}$, we get the non-dimensional attractive force $\overline{F}$ as

$$\overline{F} = -\frac{\partial V_A}{\partial R} = \frac{2}{3}\ \overline{R}\left[\frac{1}{(1-4\overline{R}^2)} - 1\right]^2 \qquad (6)$$

It shows that when $\overline{R} \rightarrow 0$(zero or very dilute concentration), then $\overline{F} \rightarrow 0$; but when $\overline{R} \rightarrow 1/2$ (neighbouring particles come into contact, ie maximum concentration), then $\overline{F} \rightarrow \infty$.

Assuming that the critical concentration, corresponding to the Bingham yield stress, coincides with the critical attracting distance, then the effective attracting force can be found by substituting (3) into (6), ie

$$\overline{F}_o = \frac{2}{3}\ 0.24\left[\frac{1}{(1-4\times0.24)^2} - 1\right]^2 \approx 92.16 \qquad (7)$$

It is interesting to note that several authors (3) report that the Bingham stress and the plastic viscosity vs concentration of fine particles coincide and are equal in value to the result given by eqn (6). In other words between $\overline{F}_o$ and τ_B, η existed some direct proportionality. The examples for τ_B, η vs C_v are shown in Fig.2.
When the concentration apporoaches to its own ultimate value, the values of τ_B, η increase

infinitely. If we assume that the Bingham stress is caused by the Van der Waals attractive force, then the infinite increase of the attractive force should correspond to the infinite increase of the yield stress τ_B and the plastic viscosity η.

The family of curves in Fig.2 suggests that the coarse material in the slurry is not coagulated. If this were the case, then they ought to unify into one single curve.

MAXIMUM DIAMETER OF PARTICLE, CAPABLE SUSTAINED IN BINGHAM SLURRY

Consider a coarse particle with specific weight settling in a Bingham slurry (muddy water) with specific weight γ_m $(\gamma_s > \gamma_m)$.
Its movement is resisted by the shear stress at the surface on its lower semi-sphere. When equilibrium is established its settling velocity should equal to zero, therefore $du/dy=0$, in this case, the shear stress would be purely the Bingham stress; from (1), we have

$$\tau = \tau_B \qquad (8)$$

Under equilibrium condition the weight of particle W should be balanced by the resultant upward resistance $T_{B,y}$, which is sumed up from on the lower semi-sphere-surface of the particle:

$$W = T_{B,y} \qquad (9)$$

The effective weight of particle is equal to

$$W = \frac{1}{6}\pi D^3(\gamma_s - \gamma_m) \qquad (10)$$

With the aid of the spherical coordinates we can readily derive $T_{B,y}$ by the elementary calculus. Let the tangential shear stress τ_B subjects on an elementary surface with area(Fig.3):

$$Rd\phi R \sin\phi\, d\theta = R^2\sin\phi\, d\phi\, d\theta$$

the effective upward resistance on an unit area is equal to

$$\tau_B\sin\phi$$

and the effective resistance on this elementary surface is equal to

$$R^2\sin\phi\, d\phi\, d\theta\, \tau_B\sin\phi = \tau_B R^2\sin^2\phi\, d\phi\, d\theta$$

Therefore, the resultant resistance $T_{B,y}$ can be expressed by

$$T_{B,y} = \int_0^\pi\int_{-\pi}^\pi \tau_B R^2 \sin^2\phi\, d\phi\, d\theta =$$

$$= \pi^2 R^2\ \tau_B = \frac{\pi^2}{4}\ D^2\ \tau_B \qquad (11)$$

Substituting (10),(11) into (9), we get

$$D(\gamma_s - \gamma_m) = \frac{6}{4}\pi\ \tau_B \approx 4.712\ \tau_B$$

therefore, the theoretical maximum diameter of particle, which could be sustained in Bingham fluid, can be determined by this formula:

$$D_{max} \approx 4.712\left(\frac{\tau_B}{\gamma_s - \gamma_m}\right) \qquad (12)$$

equation (12) was verified with some current experiments (4). Fig. 4 compares the predicted with the experimental results showing a good agreement.

Eq.(12) shows that the diameter depends on $\tau_B + \gamma_m$. For the two-phase flow, both τ_B and γ_m are determined by the value of concentration with fine particles. Anyway, the higher the concentration, the higher the value of τ_B and γ_m and then, the larger the sustained diameter of suspended particles.

ULTIMATE CONCENTRATION AND MAXIMUM CONCENTRATION

Some size-distributions of coal slurries used in hydrotransport are shown in Fig.5 and its characteristic values are summed up in Tab.1. With each specified size-distribution, the concentration of coal will approach to its own highest value. This highest value of concentration is called the ultimate concentration. The ultimate concentration for any giving size-distribution can be determined only by experiment. Tab.1. illustrates the influence of size-distribution on its ultimate value of concentration. It also reveals that particle size-distribution is a crucial factor regarding the efficiency of hydraulic conveying.

Furthermore, we define the optimal size-distribution as distribution by which the particles could most densely occupy all the space within the coagulation-network. In another words, the optimal distribution is corresponding to maximum concentration. It is necessary to distinguish the maximum concentration with the ultimate concentration, the later is corresponding to some specified size-distribution, but the former is corresponding to an ideal distribution, which is to be find out in the following paragraph.

OPTIMAL SIZE-DISTRIBUTION AND MAXIMUM CONCENTRATION

Kolmogorov (10) had investigated the mathematical model for weathering process of rock, he showed that when the rock should randomly divided infinitely, its size distribution would approach to log-normal distribution. Namely, its density distribution function f(D) should be

$$f(D) = \frac{1}{\sigma\sqrt{2\pi}}\exp\left[-\frac{(\ln D - \ln D_m)^2}{2\sigma^2}\right]$$

in which D_m is the mean diameter of particle, σ is its variance.

Our model is conversed to this weathering model. We imagine that the disintegrated by weathering gravels is being pasted up again into the rock, then the space should be occupy by the material with this distribution. If we fill the space within the network with the material of this distribution, then we would obtain a concentration, theoretically speaking, up to $C_v=100\%$! Therefore, the log-normal distribution is the distribution by which the most economic transport could be established.

Of course, the form of every solid particle differs significantly; furthermore, some place in the network space would be occupied by the contained water also, therefore it may be impossible for the concentration to approach 100%. But whatever, the distribution with log-normal form must be still belong to an optimal size distribution among anothers.

Now the remaining questions are: concerning to the real size-distributions in Fig.5, are they close or far away from the optimal size distribution? What is their representative optimal size-distribution?

For calculating each corresponding theoretical size-distribution it is necessary to give the upper and lower limit values of diameter. The actual mean diameter and maximum diameter in Tab.1. are used for the mathematical mean, and the upper limit of diameters respectively. However, the upper limit of diameter for the theoretical distribution is meaningless, for the upper limit would approach infinity, when the cumulative percentage P should approach 100%. The difficulty was eliminated by using

$$P_{D=D_{max}} \approx 0.99$$

Though all the size-distribution of coal, used in the hydrotransport with hyperconcentration, were quite different in form, all the distribution lines are concentrated to P=15~18%, its mean value is at about

$$P_{D=0.04mm} \approx 16\%$$

This, perhaps, is the least percentage of the fine particles, by which the two-phase system could form the Bingham system, and only after the system-transformation the hydrotransport could realize the transport with hyperconcentration and lower energy consumption, as already mentioned above. Therefore, the above value of P is accepted as another boundary condition for our case.

For simplicity of script let

$$x = \ln D \ , \ x_m = \ln D_m$$

then the theoretical cumulative curve of the size-distribution is expressed in Gaussian form:

$$P(a<x<b) = \int_a^b f(x)dx = \frac{1}{\sigma\sqrt{2\pi}}\int_a^b \exp\left[-\frac{(x-x_m)^2}{2\sigma^2}\right]dx$$

$$= \frac{1}{\sqrt{\pi}}\int_{\frac{a-x_m}{\sigma\sqrt{2}}}^{\frac{b-x_m}{\sigma\sqrt{2}}} \exp(-t^2)dt$$

here the integral is trnsformed through

$$\frac{x-x_m}{\sigma\sqrt{2}} = t$$

or shortly

$$P(a<x<b) = \frac{1}{2}\left[\text{erf}\ \frac{b-x_m}{\sigma\sqrt{2}} - \text{erf}\ \frac{a-x_m}{\sigma\sqrt{2}}\right] \quad (13)$$

where

$$erf(x) = \frac{2}{\sqrt{\pi}} \int_0^x \exp(-t^2)dt$$

In our case,

$$a \sim \ln(D_{0.04}) \ , \ b \sim \ln(D_{max}) \ , \ x_m \sim \ln(D_m)$$

and the percentage P should vary in this range

$$0.16 \leq P \leq 0.99$$

the values of D_{max}, D_m are listed in Tab.1. With known values of $a=\ln(0.04)$, $b=\ln(D_{max})$ and $x_m=\ln(D_m)$, the theoretical numerical variance σ^2, corresponding to any actual distribution of Tab.1, can be calculated; when b represents the logarithm of an unknown diameter, ln D, then the cumulative curve can be calculated from eq.(13) finally. The numerical results of four theoretical cumulative curves are shown in Fig.6 as the dash-lines. The corresponding real size is represented by the solid lines. It seems that all the real cumulative curves deviate to the left of the theoretical ones; the theoretical size curves seem almost finer than that of the real ones. The existence of differences between thw two sets of curves suggests that the rational selection of particle size-distribution leads to a direct increase of the transport-efficiency.

OPTIMIZATION OF HYDROTRANSPORT OF SOLID

The efficiency of solid-transport in the pipe line is defined as the ratio of output power over input power, ie

$$\eta = \frac{P_{out}}{P_{in}} \qquad (14)$$

For the solid-transport in pipe flow, the input of pipe-line can be expressed by

$$P_{in} = \gamma_m Q \Delta H = \left[\gamma + (\gamma_s - \gamma)C_v\right]Q\Delta H \qquad (15)$$

where γ_m, γ, γ_s are the specific weight of slurry, pure water and particle respectively Q is discharge, and ΔH is the water head loss through the pipe-line length L. The effective output for the pipe-line P_{out} can be expressed by

$$P_{out} = QC_v \gamma_s L \qquad (16)$$

substituting (15),(16) into (14),we get

$$\eta = \frac{C_v \gamma_s}{\left[\gamma + (\gamma_s - \gamma)C_v\right] J} \qquad (17)$$

here L is the length, $J=\Delta H/L$, is the hydraulic gradient. It is inversely proportional to the hydraulic gradient: the higher the value of J the lower the value of η, but it is direct proportional to the concentration also. When $C_v \to 0$, then $\eta \to 0$; when $C_v \to 1$, then η approach to an ultimate value as

$$\eta_{max} \to \frac{1}{J}$$

though the hydraulic gradient is a function of concentration also.

A useful form of efficiency is the efficiency for the whole pipe-line system(efficiency of installation). Furthermore, the output is

calculated by the electric energy expenditure in this case, the dimension for the efficiency should be T-Km/Kw.h.

The pipe-transport of coal in Black Mesa,U.S. was brought into operation since 1970, its characteristic values [5,6] are listed in Tab.2 Based upon these values we can calculate the active efficiency of the Black Mesa system as

$$\eta = \frac{500 \times 400 \times 10^4}{4 \times 2 \times 1300 \times 0.99 \times 24 \times 365} \ \text{T-Km/Kw.h}$$

$$= 24.36 \text{T-Km/Kw.h}$$

or,in inverse form

$$\frac{1}{\eta} = 0.041 \ \text{Kw.h/T-Km}$$

Besides, some experiments [11] for determining the active efficiency of pipe-transport with fixed concentration of coarse material (D 0.05mm, C =20,30%) the active efficiency should change by the adjustment of concentration of fine material (Fig.7). It is interesting to note that when the concentration (including fine particles) is about $C_v \sim 20 \sim 30\%$, then $1/\eta$ is a minimum or is a maximum. This optimum concentration corresponds to the most economic transport.

The physical essence here is quite evident: Parallel to increasing the concentration, the physical viscocity of the system also increases which in turn increases the energy consumption.

From Fig.7, when concentration of $C \sim 30\%$ is attained, then the efficiency reduces to

$$\left(\frac{1}{\eta}\right)_{max} \sim 4 \times 10^{-4} \ \text{Kw.h/T-Km}$$

The reasons for this overwhelming difference are not clear but particle size distribution is almost certainly the centre point to the answer.

CONCLUSION AND DISCUSSION

The optimization of grain-size in hydrotransport is a complex and profound problem. First of all the target object is to find out the critical condition, by which the Newtonian system should be transformed to the Bingham system. To perform this transformation it is necessary to adjust the concentration with the fine particle. Meanwhile, to achieve economic transport it is necessary to adjust the concentration of fines also.

The maximum diameter, which could be sustained in suspension, is directly determined by the Bingham yield stress, and is indirectly determined by the concentration of fine particles also.

Although the above constraints are interrelated the optimal size-distribution can be obtained by the log-normal distribution theory.

REFERENCES

1. Hou,H.C.,An analysis on some problems of pipeline transport of mineral slurry of heavy concentration,J.Sediment Res.,No.4, 1981,p.50.
2. Hamaker,H.C.,The London-Van Der Waals attraction between spherical particles, Physica,Vol.4,1937,p.1058.

3. Jiang,S.Q.,Calculation of sediment transport in pipes at hyperconcentration,J. Sediment Res.,No.2,1982,p.45.
4. Chien,N. and Wan,Z.H.,Mechanics of Sediment Transport,Sci.Press,1983.
5. Hale,D.W.,Slurry pipelines in North America,Past,Present and Future, Hydrotransport 4,1976,paper G2.
6. Wasp,E.J.,Solid-Liquid Flow Slurry Pipeline Transportation,Trans.Tech.Publ., Germany,1977.
7. Rigby,G.R. and Callcott,T.G.,A system for the transportation,cleaning and recovery of Australian cooking coal,Hydrotransport 5,1978,paper E5.
8. Snock,P.E. et al.,Utilization of pipeline delivered coal,Hydrotransport 4,1976, paper E1.
9. Haas,D.B. and Husband,W.H.W.,The development of hydraulic transport of large sized coal in Canada-Phase I,Hydrotransport 5, 1978,paper H1.
10. Kolmogorov,A.N.,Logarithmically normal distribution of fragmentary particle sizes, Dokl.Akad.Nauk SSSR,31,No.2,1941,pp99-101.
11. Sakamoto,M.et al.,A hydraulic transport study of coarse materials including fine particles with hydrohoist,Hydrotransport 5, 1978,paper D6.

SYMBOLS

f	Distribution function of Probability
l	Distance between neighbouring particle
C_v	Concentration by volume
$(C_v)_{cr}$	Critical concentration by volume, corresponding to initial transfer from Newtonian to Bingham
D	Diameter of solid particle
D_{max}	Maximum diameter of particle,sustained in Bingham system
$\bar{F}$	Nondimensional attracting force of solid particles
$\bar{F}_o$	Nondimensional attracting force,corresponding to initial transfer of system
ΔH	Water head losses
J	Hydraulic gradient
P	Cumulative percentage
P_{in}	Input-Power
P_{out}	Output-Power
Q	Discharge
R	Radius of particle
$\bar{R}$	Nondimensional radius of particle
$\bar{R}_{cr}$	Critical value of R,corresponding to initial system-transfer
S	Gap between neighbouring particles
$T_{B,y}$	Resultant resistant force along the vertical direction,caused by the Bingham yield stress
$\bar{V}_A$	Nondimensional attraction potential of solid particles
W	Weight of particle
γ	Specific weight
γ_m	Specific weight of slurry
γ_s	Specific weight of solid particle
η	Bingham plastic viscocity or Efficiency
σ	Variance of normal distribution
τ	Shear stress
τ_B	Bingham yield stress

Tab.1

Location	Diameter(mm)		Ultimate concentrationCw (%)	No
	max	mean		
McIntyre Canada	20	1.0	45	(1)
Sheerness Canada	20	2.7	45	(2)
Lignite Canada	20	5	33	(3)
Black Mesa	3.0	0.27	?	(4)
Kemblg Australia	8	0.71	?	(5)

Tab.2

Inner diameter(mm)	457
Distance of transport (Km)	440
Annual amount (10^6 T)	5
Amount of pump stations	4
Capacity of each station(Kw)	3 1300
Utility factor	99%
Number of pump in operation	2

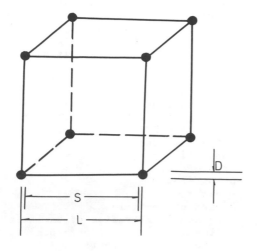

Fig. 1.   Distribution of particles.

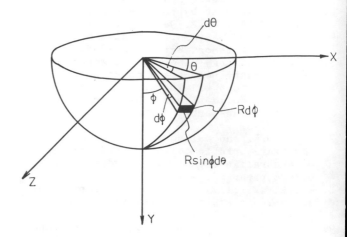

Fig. 3.   Coordinates of semi-sphere.

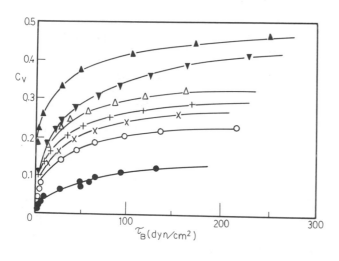

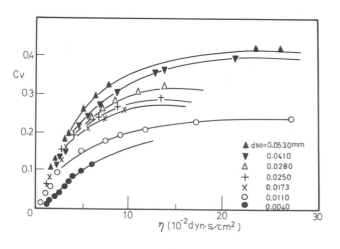

Fig. 2. τ_B, η vs concentration and diameter.

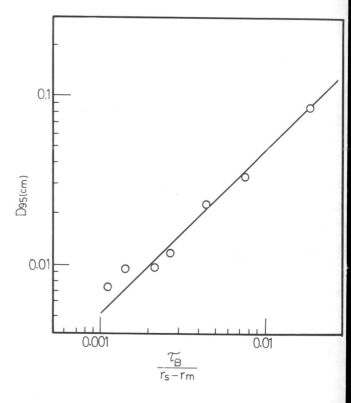

Fig. 4.   Maximum diameter vs Bingham stress: ◯, experiments; —, theory.

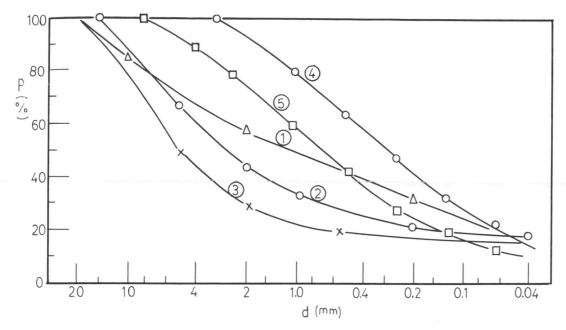

Fig. 5. Size-distribution of coal: (1) McIntyre; (2) Sheerness;
(3) Lignite; (4) Black Mesa; (5) Kemblg.

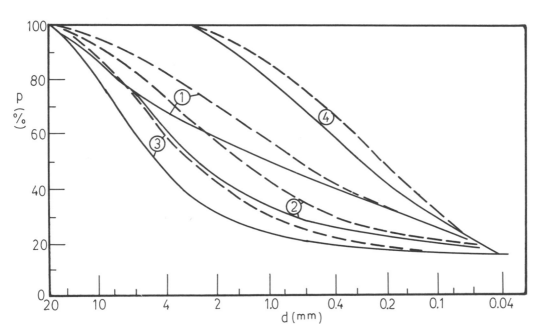

Fig. 6. Comparison of size-distribution: dashed-line, theoretical
log-normal distribution; solid line, real cumulative curve.

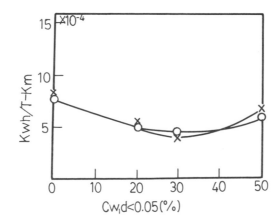

Fig. 7. Efficiency vs concentration.

10th International Conference on the
Hydraulic Transport of Solids in Pipes

HYDROTRANSPORT 10

Innsbruck, Austria: 29-31 October, 1986

PAPER E4

THE EFFECTS OF PIPELINE TURBULENCE ON THE BEHAVIOUR OF CONVEYED SOLIDS

R.S. NEVE

Senior Lecturer in Fluid Mechanics
Department of Mechanical Engineering,
The City University, London, EC1V 0HB.

SYNOPSIS

In cases where conveyed solids are large enough to be subject to settling, the critical velocity involved and the excess pressure drop due to the presence of solids are both dependent on the particles' drag coefficient C_D. In the past, designers have far too frequently resorted to "standard curve" values for sphere C_D versus Reynolds No., even though experimental evidence has long shown that this curve is modified considerably by turbulence in the fluid.

This paper reviews past results on the effects of turbulence intensity on sphere drag and shows, in addition, some startling new results obtained by the author indicating that scale as well as intensity should be considered when estimating a typical C_D value. In some cases the correct combination of scale and intensity can produce values of C_D almost as high as those for flat plates perpendicular to a stream, which is about as high as one can experience anywhere.

An indication is given of what particular zones of particle:pipe diameter ratio are likely to be subject to these problems.

NOMENCLATURE

a	pipe radius
C	particle concentration
C_D	drag coefficient $[=\text{Drag}/\frac{1}{2}\rho u_s^2.(\pi d^2/4)]$
d	particle diameter
D	pipe diameter [=2a]
I	relative turbulence intensity $[=u'_{rms}/u_s]$
k	Nikuradse equivalent sand roughness
L_x	turbulence longitudinal macroscale
Re	Reynolds No. $[=u_s d/\nu]$
u	fluid axial velocity
u_s	axial slip velocity $[=u_f - u_p]$
u'	fluctuating component of u_f
u'_{rms}	rms value of u'
u_τ	friction velocity $[=\sqrt{(\tau w/\rho)}]$

W	settling velocity in still water
y	distance from pipe wall
α	ratio of W(turbulent):W [eqn.4]
ϵ	diffusivity
ν	kinematic viscosity
ρ	density
τ	shear stress
ψ	scale parameter [eqn.3] [Householder (5)]

Suffices

f	fluid
p	particle
w	wall

INTRODUCTION

In pipe flows conveying solids, it is well known that where particles do not entirely follow the turbulent motions of the conveying fluid, an excess pressure drop is incurred because of the "slip" velocity difference between particle and fluid and the resulting pressure penalty caused by frictional and form drag. The drag coefficient of the particles therefore becomes important in assessing beforehand what excess pressure loss might be expected in any given pipe system layout. At least as important in the present context is the fact that where particles are large enough to result in settling, the critical velocity for the onset of such phenomena is also a function of particle drag coefficient.

In cases where settling causes a marked concentration gradient over a cross-section, efforts have been made to measure particle diffusivity ϵ_p by using the equation proposed by Batchelor (1):

$$\epsilon_p \frac{dC}{dy} + WC = 0 \qquad (1)$$

which essentially equates the rate of accumulation of particles in a certain volume due to diffusivity to the rate of loss due to gravitational settling. This equation is easily solved for C provided it can be assumed that ϵ_p is a constant over the region of interest, usually the core region of the pipe flow.

The solution is:

$$\frac{C}{C_a} = \exp\left[\frac{-W}{\epsilon_p}(y - y_a)\right] \qquad (2)$$

where C_a is the particle concentration at some reference level y_a.

Sharp (2) carried out tests to determine concentration levels at various y values in pipe flows and then used eqn.2 to calculate ϵ_p but values of the latter appeared surprisingly high.

Batchelor had suggested that W, the settling speed in a still fluid, might have to be modified by being multiplied by a pure number α to obtain a realistic settling speed in turbulent flows. Sharp's results suggested that values of α markedly less than one would be needed to give the measured concentrations.

The problem was compounded by other researchers showing results which suggested that α might even have values greater than one. Houghton (3) studied falling spheres when he vibrated an otherwise stationary fluid and found that different particles could have the same W value in still fluid but completely different values in a turbulent one. Jobson (4) quoted some cases where W was greater in a turbulent flow than in still water. This suggests that his spheres were benefitting from a lower C_D value because the combination of turbulence intensity and Reynolds number had resulted in a supercritical flow regime for each sphere. More confusing still, spheres of 123 μm diameter sometimes fell faster than those of 390 μm, suggesting at first sight a lower C_D at lower Reynolds number!

This evidence suggests that some other turbulence parameter, such as scale, might be just as important as intensity and results given by Householder and Goldschmidt (5) begin to shed some light on this. They used particles in a turbulent flow to study the ratio of particle to fluid diffusivities, ϵ_p/ϵ_f; that is, a measure of the ability of a particle to follow the motion of the local fluid in which it is suspended. Clearly, a large dense particle would normally be expected to show a very low value of ϵ_p/ϵ_f but their results suggested otherwise. They introduce a term ψ which includes the longitudinal scale of turbulence L_x. If it is justifiable to assume that L_x is proportional to pipe radius a, then their expression for ψ reduces to:

$$\psi = Re(\frac{d}{a})^2(\frac{2\rho_p}{\rho_f} + 1) \qquad (3)$$

and they plotted all their results in terms of ϵ_p/ϵ_f versus ψ. At very low values of ψ where the particle would virtually act as a tracer (very low d/a value), ϵ_p/ϵ_f approximated to unity but as ψ climbed to 10^4, corresponding to a very large particle, experimental results for ϵ_p/ϵ_f climbed towards five or six whereas logic would suggest that this ratio should tend towards zero.

Sharp (6) thereafter carried out some more tests to try to gain convincing evidence that α really was markedly less than unity. In an impressive series of experiments, he used a high speed movie camera mounted on rails to follow a flow (in the Lagrangian sense) and subsequently to evaluate a diffusion coefficient using both velocity correlations and particle displacement variances. Results for the former, normalised by division by au_T, were about 15% larger than for the latter but both were only about one fifth of the typical value of ϵ_f/au_T. Sharp's figures would therefore seem to give the conclusive result that ϵ_p/α is large not because ϵ_p is large (which he has clearly shown it is not) but because α is small. By implication, this also indicates that, in general, particle drag coefficient in a turbulent flow is higher than in a laminar one, causing a reduced settling velocity αW, so that eqn.2 should really appear as:

$$\frac{C}{C_a} = \exp\left[\frac{-\alpha W}{\epsilon_p}(y - y_a)\right] \qquad (4)$$

with α being generally less than unity.

2. THE EFFECTS OF TURBULENCE ON C_D

The usual approach to calculating drag coefficient for a spherical particle has been to use a representative Reynolds number based on slip velocity, particle diameter and fluid properties and then to pick off a value from the standard curve [e.g. (7),(8)]. Turbulence intensity I and particle roughness k/d both lower the value of critical Reynolds number at which C_D falls to much lower (supercritical) values. Otherwise a figure of about C_D= 0.47 would be assumed for Re greater than about 10^3 and higher values of C_D would be taken from the standard curve for cases in the "viscous flow" region (Re<10^3). The turbulence scale L_x/d is, however, hardly ever considered although some information is available from published literature for assessing its importance. Laufer (9) gave results for turbulence intensity in pipes but his graphs can also be used to gain an approximate idea of L_x/a values. Zenz and Othmer (10) and Soo, Ihrig and Kouh (11) also gave information on scale and, taken as a whole, the evidence suggests that L_x/a probably lies somewhere between about 0.25 and 0.5 over the vast majority of pipe radii. Since particle diameter d is normally appreciably less than pipe diameter D (=2a), the value of L_x/d is therefore always likely to be comparable with or greater than unity. Because of the level of uncertainty between the figures of 0.25 and 0.5 quoted above, the author also has attempted to measure L_x/a experimentally; results are given later in this paper.

Torobin and Gauvin (12) investigated the drag coefficient of accelerating particles in a turbulent pipe flow by a tag method and found not only that C_D can vary widely from standard curve values as turbulence intensity is increased but also that the value of critical Reynolds number drops alarmingly under the same conditions; in fact by two or three orders of magnitude. Their values of L_x/d were quoted as being generally in the range 2 to 6 and their turbulence intensities rose as high as 40% because, as they point out, the definition of I as $u'_{rms}/\bar{u}$ (where $\bar{u}$ is a time-mean velocity) involves $\bar{u}$ in this case as being the slip velocity between particle and mean flow, which is often markedly smaller than $\bar{u}$ for the mean flow itself. Their results, though spectacular, must be regarded as reliable because subsequent ones by Clamen and Gauvin (13) at higher Reynolds numbers line up with them very satisfactorily. In some cases, C_D values were obtained as high as those for flat plates perpendicular to a stream so profound flow changes are obviously involved in such situations.

Zarin (14) investigated turbulence-induced C_D changes at much lower Reynolds numbers and generally seemed to find a rise in C_D above standard curve values. His results indicate specified L_x/d values and these enable one to see a trend that increasing scale leads to: (i) higher C_D values at Reynolds numbers of a few thousands and (ii) increasingly premature breakaway of data points from the standard curve as Re is increased. Below about Re=600, turbulence seems to have little effect on C_D, presumably because viscous effects are swamping turbulence ones.

Oddly enough, refs. 12, 13, and 14 all deny any importance of turbulence scale. The first two quote only a range of L_x/d and do not attach any particular scale values to individual results but Zarin actually quotes some scale values but then discounts their importance. Meanwhile, results published by the present author (15) at relatively high Reynolds numbers seem to line up with those of Clamen and Gauvin (13) lending credence to the idea that some agency in addition to turbulence intensity is grossly affecting the level of drag coefficient. A new research project was therefore initiated to measure C_D as accurately as possible at Reynolds numbers between a few thousands and a few tens of thousands in flows where turbulence intensity and scale could be measured beforehand.

3. EXPERIMENTAL ARRANGEMENTS

In measuring the drag coefficient of spheres experimentally, there is no difference in essence between using water or air as working medium.

However, several considerations make air the favourite in practice. Turbulence can be generated using appropriate grids on the exit station of an open-section wind tunnel and the turbulence parameters can then be measured using a hot-wire anemometer technique in the region downstream of the grids. Isotropy can be ensured by not approaching the grids too closely and scale and intensity vary with distance downstream from the grids. Different characteristics can be obtained using different grids and positions.

The kinematic viscosity ν of air is about 14 times higher than for water so lower Reynolds numbers are obtained at similar speeds and sizes. Alternatively, larger spheres could be used at similar speeds to obtain similar Reynolds numbers. The rig used was therefore of the type described above and, in more detail, in ref.16., the assumption being that conclusions reached more easily in air testing would be equally applicable in water operation.

Turbulent flows were produced by an open-section wind tunnel having a contraction ratio of 8:1 and a working section measuring 405 x 240 mm. Airspeeds up to 47 m/s were attainable but the results appropriate to this paper were generally at the lower end of the speed range. Three grids of various degrees of coarseness were used to produce turbulence and the spheres, which were of diameter 37.7 mm, were supported on a traversing gear by a previously calibrated strain gauged force transducer and slim sting. Forces could be measured with an accuracy of ±0.5% at the upper end but at the lowest speeds used, accuracy was about ±10% and results at the lowest Reynolds numbers of all

should therefore be treated with care, only qualitative conclusions being justified.

Air speed was measured using a pitot-static tube and electronic digital micromanometer, able to measure pressure differences down to thousandths of a mm of water. Turbulence intensity was measured using a hot-wire anemometer with a lineariser inserted in the circuit, since intensities encountered in this project were likely to be well above the few percent figure normally used as a criterion for inclusion. Turbulence scale was measured using the same hot-wire apparatus but with a tunable bandpass filter included so that a power spectral density technique could be employed to assess scale, as detailed in E.S.D.U. Data Sheet 74031.

Six hundred and twenty six experimental runs were made but only some are relevant to the present paper; the rest are summarised in ref.16. Data were stored on a floppy disc so that experimental analysis could be easily undertaken by constructing a computer program which plotted all results within prespecified turbulence intensity and scale bands.

For measuring turbulence scale in a pipe, the same hot-wire apparatus and P.S.D. technique was employed but with the wire positioned near the exit section of a 2 inch (50.8 mm) diameter copper pipe. Air was supplied by a fan via an air box and downstream contraction to accelerate and stabilise the flow. The turbulent flow was then established over some 70 diameters of travel before meeting the hot wire.

4. DISCUSSION OF RESULTS

In trying to interpret the graphical results presented here, it is useful to bear in mind a few basic assumptions about the effects of turbulence on the flow around a particle.

It seems reasonable to assume that in general we are looking at the effects of pressure changes rather than skin friction ones. Except at low Reynolds numbers, the majority of drag on a sphere is due to pressure inequalities caused by bluffness. The boundary layer condition is therefore more important in determining the form of the wake behind the sphere than in controlling the skin friction coefficient.

At small scales of turbulence, the boundary layer transition from a laminar to turbulent condition will occur at lower Reynolds numbers than for laminar flow and therefore the well-known "C_D dip" occurs prematurely. The turbulent boundary layer separates from the sphere surface at a greater distance from the front stagnation point and gives a narrower wake and lower C_D value. At greater turbulence levels of the free stream, considerable mixing can occur between the wake and the surrounding flow leading to a reduced pressure at the rear and consequently a higher C_D value. At extreme levels of scale the sphere will simply "see" a laminar free stream of periodically varying angle of incidence and C_D levels would then be expected to return towards standard curve values.

Since the condition of the wake can be influenced not only by the state of the boundary layer but also by the state of the free stream flow, these various opposing effects are therefore not necessarily divorced from each other and wide variations in C_D can be expected.

Fig.1 shows the effects of small scale, low intensity turbulence. The expected premature C_D dip has occurred at about Re = 50 x 10^3 and if Zarin's results (14) are added and an interpolation curve is assumed to join the two sets of results it seems fair to judge that C_D has increased by about 15% at Re = 10^4 because of small scale turbulence of about I=4%. Such demonstrable drops in C_D at the right hand end of the figure could well be the cause of some authors reporting spheres settling faster in turbulent flows [e.g. Jobson (4)].

The results of increasing I to 7% can be seen in Fig.2. The critical Reynolds number has now dropped even lower and the experimental points at the left hand end of the present author's results are well above the standard curve. Zarin's results, although starting off by dipping below Achenbach's curve, soon climb as Reynolds number is increased. The I=7% curve from Clamen and Gauvin (13) has been superimposed for comparison but it should be borne in mind that this was for L_x/d values between about 2 and 6 instead of the lower values involved with Zarin and with the current project. A series of high C_D values in the same Reynolds number region is, however, indicated by their curve.

Another direct link with Zarin's results is possible when data points for L_x/d =0.2 and 0.5 and I≈10% are compared in Fig.3. This graph ought to show the effects of turbulence scale only, since all results are at roughly the same turbulence intensity. Although there is some overlap in the central regions, there are clear differences at the two ends of the Reynolds number range tested. For L_x/d=0.5, the data points leave the standard curve at about Re=850 and then maintain a rather ill-defined plateau of raised C_D values before falling to supercritical C_D values after about Re=35 x 10^3. At the smaller scale of turbulence (0.2), the data points rise above the Achenbach curve for Re>1800 before beginning a gradual drop in C_D values at Re≈ 10^4. We must therefore conclude that the smaller scale turbulence affects the sphere boundary layers more, leading to a very premature transition. At the low Reynolds number end of the scale, however, the larger scale turbulence is more successful at influencing C_D where viscous effects are beginning to predominate over inertial ones.

The interesting results of turbulence scale and sphere diameter being comparable in size are shown in Fig.4. For turbulence intensities similar to those of Fig.3, Zarin's data points for L_x/d=1 begin to leave the standard curve at Re≈600 and if the present author's L_x/d=0.8 results are added, there seems to be a sudden rise in C_D to high values prior to Re=5x10^3 with a subsequent rapid drop to the more

normal subcritical plateau values before a final critical drop seems to be setting in at about Re=40x10^4.

If scale is increased to L_x/d=1.5, there is a quite alarming drop in C_D values for Reynolds numbers around 10^4, although they subsequently rise again with Reynolds number to plateau levels before starting another drop with the L_x/d=0.8 points. The start of an S-bend shape is now evident for these points and it is interesting to note that Clamen and Gauvin's curve for an L_x/d value between 2 and 6 shows a similar shape but at different Reynolds numbers. At even higher scale values, the results from the present work are elevated to even higher C_D levels, although retaining the S-bend appearance.

For the largest turbulence scales and intensities encountered in this work, Fig.5 shows that for L_x/d=3 and I=22%, C_D values are uniformly high until a fall begins to set in at Reynolds numbers greater than about 40 x 10^3. In fact, these drag coefficients are almost as high as those for flat plates perpendicular to a flow, which is about as high as one would encounter anywhere, at these Reynolds numbers. For L_x/d=4.2, the data points suggest that C_D is starting to return to standard curve levels, presumably for the reason given earlier; the eddies are now so large that the spheres are beginning to behave as if in a laminar flow of periodically varying incidence angle.

5. APPLICATION TO PIPELINE ENGINEERING

Clearly then, if a pipeline is to convey material under conditions where a combination of slip velocity and hydraulic diameter give a Reynolds number between a few hundred and a few tens of thousands, a designer ought to take account of turbulence intensity *and* scale in assessing what C_D values are likely to be encountered in practice.

In section 2, it was pointed out that previous published results indicated values of L_x/a of between about 0.25 and 0.5, extending over the bulk of the cross-section and Fig.6 shows the results of the present tests. The power spectral density technique is not as accurate as some alternatives but was much cheaper to set up. The data points of Fig.6, though showing some scatter from these inaccuracies, give L_x/a for the pipe tested to be around 0.3, this value extending from the centreline to as near the wall as y/a =0.1. At points closer to the wall, a "log law" or "law of the wall" approach is normally required in describing the flow and concepts which are valid in the core region have to be abandoned. The idea that turbulence scale can have a value of 0.3a at a distance of only 0.1a from the wall is not the physical impossibility that it seems at first sight to be; it is merely a result of the way in which L_x is normally defined, in terms of the area under a correlation coefficient curve.

The extent of the region in Fig.6 over which L_x/a is approximately constant agrees with the energy spectrum curves given by Laufer (9). In his u' spectra, there is no substantial difference between data points for y/a=1 and for 0.691, a small difference for y/a=0.28 but not until y/a=0.074 is a marked deviation from the y/a=1 points evident. One would therefore be justified in assuming that the turbulence scale in most pipe flows has a value of L_x/a of about 0.3 over about 90% of the radius or about 80% of the cross-sectional area. The presence of solids will, of course, modify this to some extent, the turbulence being either intensified or damped out, depending on the size of particles.

If $L_x/a=0.3$ then L_x/d would be given by:

$$\frac{L_x}{d} = \frac{L_x}{a} \cdot \frac{a}{d} = \frac{L_x}{a} \cdot \frac{D}{2d} = 0.15 \frac{D}{d} \qquad (5)$$

so that if, for any given pipe flow conditions, the numerical value of 0.15(D/d) lies between about 0.5 and 6, a designer might reasonably expect to have to take turbulence scale into account in obtaining a representative value of C_D for use in pressure drop and settling velocity calculations.

6. CONCLUSIONS

This project set out to show that the behaviour of particles being transported in pipelines is likely to be affected by turbulence scale at least as much as by its intensity and it is suggested that adequate evidence has been produced to justify this suggestion.

At some Reynolds numbers, combinations of L_x/d and I can be found where artificially low values of C_D result whereas for other combinations of these parameters unusually high drag coefficients can be encountered. Each of these can account in its own way for some unusual flow phenomena reported in past literature and, in particular, a high C_D and consequent lower settling veloicty would agree with Sharp's observations (2),(6), requiring α to be less than unity.

7. ACKNOWLEDGEMENTS

Some of the research work which provided results for this paper was financed by S.E.R.C.; most of the equipment and all laboratory facilities were provided by the Mechanical Engineering Department of The CIty University, London, and equipment for measuring turbulence scale was loaned by Mr.D.M.Sykes of the Aeronautics Department. To all these, the author is extremely grateful.

REFERENCES

1. Batchelor, G.K. "The motion of small particles in turbulent flow", Procs.2nd Australasian Conf.on Hydraulics and Fluid Mechs, Auckland, N.Z., pp.19-41, (1965).

2. Sharp, B.B.and O'Neill, I.C. "A study of the behaviour of large,light particles in turbulent pipe flow", Procs.Hydrotransport 1, Warwick, U.K. Paper F4, Publ.by BHRA, Cranfield, Bedford, (1970).

3. Houghton, G. "Particle retardation in vertically oscillating fluids", Can.Jour.Chem.Eng., Vol.46, 79 (1968).

4. Jobson, H.E. "Vertical mass transfer in open channels". Ph.D.Thesis, Colorado State Univ.(1968).

5. Householder, M.K. and Goldschmidt, V.W. "Turbulent diffusion and Schmidt number of particles", Procs.A.S.C.E., Vol.95, E.M.6, 1345, (1969).

6. Sharp, B.B. and O'Neill, I.C. "Some aspects of Lagrangian studies of solid particles in turbulent pipe flow", Procs.5th Australasian Conf.on Hydraulics and Fluid Mechs., Christchurch, N.Z., pp.119-124, (1974).

7. Achenbach, E. "Experiments on the flow past spheres at very high Reynolds numbers", Jour.of Fluid Mechs., Vol.54, 565-575 (1972).

8. Achenbach, E. "The effects of surface roughness and tunnel blockage on the flow past spheres", Jour.of Fluid Mechs., Vol.65, 113-125, (1974).

9. Laufer, J. "The structure of turbulence in fully developed pipe flow" N.A.C.A.Tech.Rep.No.1174 (1954).

10. Zenz, F.A. and Othmer, D.F. "Fluidization and Fluid-Particle Systems" Reinhold Publg Corp.,New York, (1960).

11. Soo,S.L., Ihrig, H.K. and Kouh, A.F.E. "Experimental determination of the statistical properties of two-phase turbulent motion". A.S.M.E. Jour.of Basic Eng.'D', Vol.82, 609-621, (1960).

12. Torobin, L.B. and Gauvin, W.H. "The drag coefficients of single spheres moving in steady and accelerated motion in a turbulent fluid", Jour.A.I.Chem.E., Vol.7, Pt.4, 615-619 (1961).

13. Clamen, A. and Gauvin, W.H. "Effects of turbulence on the drag coefficients of spheres in a supercritical flow regime", Jour.A.I.Chem.E., Vol.15, Pt.2, 184-189 (1969).

14. Zarin, N.A. "The measurement of non-continum and turbulence effects on subsonic sphere drag", N.A.S.A. Contract Rep.No.1585 (1970).

15. Neve, R.S. and Jaafar, F.B. "The effects of turbulence and surface roughness on the drag of spheres in thin jets" The Aeronautical Journal, Vol.86, No.859, 331-336 (1982).

16. Neve, R.S. "The importance of turbulence macroscale in determining the drag coefficient of spheres", To be published in Int.Jour.of Heat and Fluid Flow, Mar.1986.

190

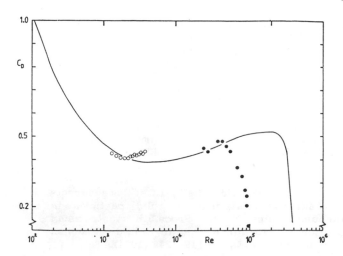

Fig. 1   Drag coefficient with small scale turbulence

```
Present work :      ●    I = 4%        Lₓ/d = 0.06
Zarin (14) :        ○    3 < I < 4.5%  Lₓ/d = 0.16
Achenbach (7):      _____
```

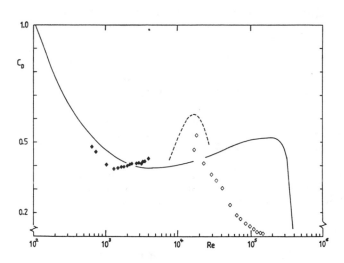

Fig. 2   Drag coefficient with small scale turbulence and I = 7%

```
Present work :      ◇    I = 7%        Lₓ/d = 0.08
Zarin (14) :        ◆    5.8 < I < 7.9%  Lₓ/d = 0.5
Achenbach (7) _____ ; Clamen and Gauvin (13) --------
```

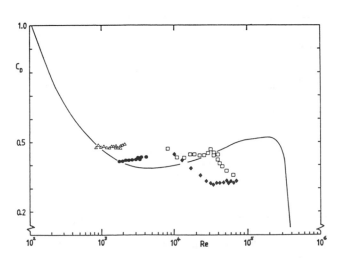

Fig. 3 Drag coefficient with small scale turbulence and I = 10%

```
Present work:       ◆    I = 10%       Lₓ/d = 0.2
                    □    I = 10%       Lₓ/d = 0.5
Zarin (14):         ●    8.8 < I < 13% Lₓ/d = 0.2
                    △    8.8 < I < 13% Lₓ/d = 0.5
Achenbach (7):      _____
```

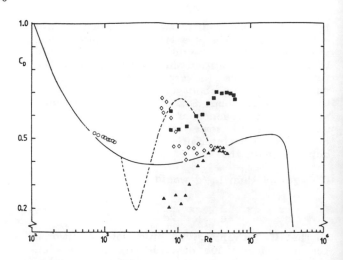

Fig. 4 Drag coefficient for sphere diameter and turbulence scale being comparable

```
Present work:       ◇    I = 10%       Lₓ/d = 0.8
                    ▲    I = 10%       Lₓ/d = 1.5
                    ■    I = 10%       Lₓ/d = 2.4
Zarin (14):         ○    I = 9%        Lₓ/d = 1.0
Achenbach (7) _____ ; Clamen and Gauvin (13) ---------
```

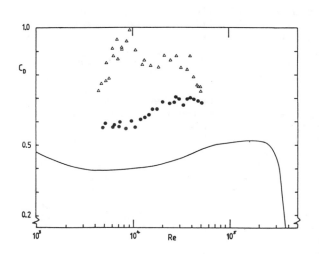

Fig. 5 Drag coefficient for turbulence scale being markedly greater than sphere diameter

```
Present work:       △    I = 22%       Lₓ/d = 3.0
                    ●    I = 22%       Lₓ/d = 4.2
Achenbach (7):      _____
```

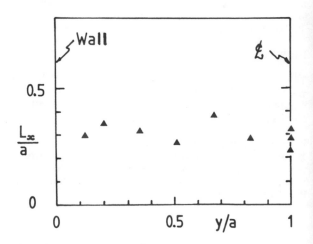

Fig. 6 Turbulence macroscale measurements in a 2 inch pipe

(Air; power spectral density technique; 15x10³ < Re < 26x10³)

10th International Conference on the
Hydraulic Transport of Solids in Pipes

HYDROTRANSPORT 10

Innsbruck, Austria: 29-31 October, 1986

PAPER F1

DETERMINATION OF SOLIDS CONCENTRATION IN SLURRIES

BY

H. NASR-EL-DIN, C. A. SHOOK AND J. COLWELL

CHEMICAL ENGINEERING DEPARTMENT
UNIVERSITY OF SASKATCHEWAN
SASKATOON, SASKATCHEWAN   S7N 0W0
CANADA

ABSTRACT

Three methods for determining solids concentration of slurry flows are compared using experimental data for a variety of slurries. Gamma ray absorption provides a mean concentration for a chord which spans the pipe and can give a reliable value of mean concentration at high concentrations or for large density differences between the phases. Sample removal can provide accurate local concentrations if thin-walled or tapered samplers are used with isokinetic sample withdrawal. However, this technique requires the local velocity to be determined.

An economical alternative to these two techniques is measurement of slurry resistivity to infer solids concentration from Maxwell's equation. Experimental measurements show that good agreement between the methods can be achieved. Procedures to accomplish this are described.

INTRODUCTION

The solids concentration in a flowing slurry is often of considerable interest: in material balance calculations or to allow use of a variable area flow meter such as a venturi or segmental orifice. Research has shown that only in special circumstances (very fine particles and/or very high velocities) is the concentration independent of position in a horizontal pipe. To avoid difficulties in the interpretation of concentration meter readings when variations occur, it is sometimes recommended that vertical flows be used. This solution becomes progressively more impractical as the pipeline diameter increases. Fairly long approach sections are required to allow the concentration variations produced at pipe bends to be destroyed by mixing processes[1] so that the flow becomes reasonably homogeneous. If appropriate approach lengths are provided, meters located on long vertical sections of pipe are difficult to maintain. The same disadvantage applies to use of vertical pressure drops as a measure of in-situ concentration. Thus vertical flow metering is probably best regarded as useful when vertical flows are essential for some other reason.

Radiation absorption is one of the most convenient industrial methods for concentration determination. With large pipelines, the beam scans a portion of the cross-section along a diametral path. Strictly speaking, interpretation of the meter signal requires a knowledge of the concentration distribution, but this complication is often ignored. Predicting the concentration distribution is difficult and few systematic investigations of this have been reported. One of the most interesting laboratory methods has employed a beam which can be traversed and rotated[2] to scan the cross section along any chord. Local concentrations can be inferred by this technique; the beam intensities allow determination of the coefficients in the functional relationship between concentration and position.

Sample removal is probably the commonest method of concentration measurement. These methods have been studied systematically in this laboratory[3], [4], [5]. The errors inherent in wall-aperture sampling and related methods have been examined experimentally and theoretically. Probe sampling using tapered or thin-walled, L-shaped samplers has been shown to give good agreement between measured collection efficiencies and those predicted theoretically. Figure 1 compares sampling performance for dilute sand slurries in terms of the particle diameter and the ratio of sampling velocity to approach velocity U_0. At the isokinetic condition, satisfactory performance can be achieved. Probes with thicker walls were shown to give systematic errors in sand-water slurries. This is a significant disadvantage since the lifetime of a thinwalled or tapered probe in a slurry could be short.

Figure 1 shows that reliable sampling requires the upstream velocity to be measured, as well as that at the sampling point. The latter can be done volumetrically, but the former requires a technique such as the probe of Brown and Shook[6].

The most convenient form of this device has a set of field electrodes and two sets of sensor electrodes[7]. The latter respond to the passage of particles which produce fluctuations in resistivity in the small spatial regions defined by the sensors (of the order of 1 mm in diameter). Since the resistivity fluctuation is produced by a concentration fluctuation, the device has the inherent possibility of being used for concentration measurement as well as velocity measurement. Evidence for this was given in reference 7.

Systematic investigations in the past two years have defined the conditions in which this resistivity method can be used. In particular, it was necessary to eliminate the effect of electrode polarization which had prevented resistance measurement from being used as a concentration sensing technique in the past. The usual effect of polarization (which had to be eliminated) is a velocity dependence in the resistivity-concentration relationship. Full details of the verification are available elsewhere[8], but it can be noted that verification of the new technique was only possible because the conditions for satisfactory performance of the reference method (isokinetic probe sampling with tapered probes) had been fully established.

Once a satisfactory method had been obtained, we decided to use it in combination with the other techniques to explore situations where the other methods might have difficulties. The first situation involved the flow of fairly coarse (d = 1.4 mm) polystyrene particles in water. At a moderate mean in-situ concentration of 34% by volume, samples could not be removed from the central portion of the pipe: the sampling probe was plugged with solids. Closer to the wall, however, sampling was possible. The present communication is an account of these and related investigations with the three techniques.

EXPERIMENTAL STUDIES

The experiments of the present study were performed in a rotatable, horizontal section of a 2 inch loop described elsewhere[6]. The probes were installed in transparent sections which allowed visual observation. The test section has an approach length of 180 pipe diameters downstream of the elbow. Previous studies have shown that this distance is adequate to eliminate flow disturbances generated by the elbow upstream.

The volume of the loop was measured by adding weighed amounts of water to fill it. The required solids concentrations were produced by adding the corresponding amounts of solids. Bulk velocities were measured by a magnetic flowmeter connected to an LSI-11 mini computer for data recording.

The solids used in this study were sands and polystyrene particles. Table 1 summarizes their properties. The continuous phase was tap water of resistivity of 2720 Ohm-cm at 25°C. The experiments of this study can be divided into two groups. In the first series, solids concentrations for polystyrene slurries were measured by isokinetic sampling and the conductivity probe. The second series involved concentration measurements for sand slurries by gamma ray absorption and the conductivity probe. In the following section, details and procedures for these methods will be given.

A. Sampling Studies

Samples were taken by an L-shaped probe constructed from 3/8 inch stainless steel tubing of 8 mm inside diameter and a probe nozzle length of seven probe radii. The flowrate of the mixture in the sampling tube was controlled by a pinch valve. Samples were collected over timed short intervals ($\pm$ 0.1s). Because the density of the particles was close to that of water, determining solids concentration by direct volume measurements was not possible. Instead, the solids were carefully collected on a fine mesh

screen and these were dried and weighed. Solids concentrations were determined from these weights and the densities of the two phases.

Local particle velocities were determined using the probe described previously. Samples were taken at radial positions 2 r/D of 0, 0.12, 0.54 and 0.81 in both horizontal and vertical planes. The horizontal plane measurements confirmed those obtained in the vertical plane and are still being interpreted theoretically so that only the latter are reported here.

During the course of the experiments, bulk velocity, loop temperature and solids concentration were kept constant. For experiments including sampling, solids concentrations were kept almost constant by adding solids equivalent to the amount removed by sampling.

The experiments in this series included studies of the effects of probe wall thickness and sampling velocity on sample concentration.

B. Conductivity

The velocity probe was also used for concentration measurements. To eliminate velocity dependence of the device, the probe was operated in a constant current mode. By maintaining a constant current between the field electrodes, polarization of these electrodes ceases to affect the potential at the sensor electrodes.

The field electrode circuit thus consists of an AC source, an ammeter and a variable ballast resistance to control the current. The sensor electrodes are connected to a voltmeter from which a time average reading can be obtained. This allows the concentration to be found.

The procedure for the conductivity probe operation was:

1. The sensor voltage was measured at various positions within the pipe.

2. These voltage measurements were corrected for the effects of temperature, position and chemical composition.

3. Concentration was obtained from a calibration curve. For spherical particles, or for broad size distributions, Maxwell's equation[9] was used. For non-spherical particles of narrow size, a calibration curve obtained with liquid-solid fluidized beds was used.

The procedure for the gamma ray measurements requires three scans. The first is done on the empty pipe in order to find the bottom of the pipe and establish the wall attenuation. The second scan is performed with the pipe filled with water. From scans one and two, one determines the path length as a function of elevation. The third scan is done for the loop running with slurry. Scans two and three then give concentration as a function of position. Although most of the experiments were performed with pipes 50 mm I.D, some tests were done in a 500 mm pipeline.

EXPERIMENTAL RESULTS

Figure 2 shows the effect of sampling velocity and probe relative wall thickness on sampling efficiency for fine (0.3 mm) polysytrene particles. Samples were taken from the pipe center at a mean concentration of 37% and a bulk velocity of 3.4 m/s with probes of wall thickness equal to 0.05, 0.5, 0.8 and 1.2 probe aperature radii. We observe that unlike the results with sand particles, shown in figure 1, the effect of sampling velocity on sampling efficiency is insignificant. This result is reasonable since polystyrene particles have a density of 1.05 kg/m^3 which is very close to water. This implies that these particles can follow fluid streamlines and consequently sampling efficiency for these particles is very close to one no matter what the sampling velocity.

An important observation is that the sampling efficiency appears to be independent of probe wall thickness at sampling velocities equal to and higher than isokinetic. This contrasts with results obtained by Nasr-El-Din and Shook[5] for sand particles. This difference can be explained as follows: In the presence of a blunt probe, the fluid streamlines deflect ahead of the probe nozzle and the deflection increases as the probe wall thickness increases. Particles of high inertia, such as coarse sand, are affected little by fluid deflection and strike the nozzle wall. Some of these bounce into the probe aperture causing higher sampling concentrations. Particles of small inertia follow fluid streamlines to a greater extent and should not strike the nozzle wall as frequently. Visual observations during the tests confirmed this tendency.

Figure 3 shows concentration profiles for 1.4 mm polysytrene particles obtained with the conductivity probe and isokinetic sampling in a vertical plane. These results were obtained at a bulk velocity of 3.4 m/s and mean in-situ solids concentrations of 21%, 27% and 34%, respectively. At 34%, the profile obtained with the conductivity probe shows that the particles are not uniformly distributed across the pipe. Instead, a strong "coring" tendency exists. We can see why sampling in this slurry was only possible at positions closer to the pipe wall, where blockage did not occur. The few samples which could be taken show good agreement with the profile obtained with the conductivity probe.

At 27%, samples could be taken from the center of the pipe and a complete profile was obtained. Again, one finds good agreement between both methods. Although the concentration profile at this concentration is similar to that obtained at 34%, there is less coring. Good agreement can be also observed at 21%. Also, there is still less coring.

Similar effects were observed for samples and probe concentrations obtained in the lateral plane through the pipe axis.

In terms of time, one scan of the conductivity probe (18 measurements per scan) takes a few minutes. Sampling measurements take much longer, especially when the local velocity is required. We conclude that conductivity has significant advantages compared to sampling.

The second series of experiments was performed with sand slurries. Here, concentration profiles were simultaneously measured with the conductivity probe and gamma ray absorption.

Figure 4 shows concentration profiles for sand particles of 0.19 mm mean diameter at a bulk velocity of 2 m/s and mean concentrations of 15% and 25%. At a mean concentration of 40%, the bulk velocity was increased to 2.3 m/s to ensure a stable flow. We observe that at 15%, the gamma ray and conductivity results are in close agreement at all positions. The local concentration is high at the bottom of the pipe and decreases continuously towards the top of the pipe. At 25%, although good agreement can be observed, especially in the upper half of the pipe, some deviation occurs near the bottom of the pipe, where the conductivity probe gives higher concentrations. At 40%, concentration profiles obtained with the two methods agree only at Y/D > 0.6, whereas at lower positions, concentrations obtained wih the conductivity probe are significantly higher than the chord-average gamma ray values.

Figures 5 and 6 show a series of measurements for the same sand, but at a bulk velocity of 3.4 m/s and mean concentrations of 15% and 44% (Figure 5) and 30% and 40% (Figure 6). Examining these profiles one observes that at a low concentration (15%), good agreement exists throughout the pipe, as the concentration increases (30%), agreement between the methods can be observed in the upper section of the pipe. Close to the bottom, the conductivity probe gives slightly higher values. As the concentration increases further (40%), agreement exists only at Y/D greater than 0.6. At lower positions, the conductivity probe gives significantly higher concentrations.

The region of disagreement increases as the concentration increases as shown in Figure 6 where at 44%, one can only find agreement at Y/D greater than 0.8. From figures 4, 5 and 6, we notice that the agreement trends with both methods are very similar at bulk velocities of 2 m/s and 3.4 m/s.

To explain this difference, one has to recall that the conductivity probe gives concentration in a small volume in space whereas gamma ray method gives concentration averaged over a pipe span. It is obvious that the concentration measurement with both methods will be equal if and only if the particles are uniformly distributed over a lateral plane. If the particles are concentrated near the center of the pipe, as a result of coring, one expects gamma ray absorption to give lower concentrations than the probe located at the pipe center. On the other hand, if the particles are concentrated near the wall, as in response to the secondary flow generated in 90 degree elbows, one expects gamma ray measurements to give higher concentrations tham the local in-situ concentration obtained by the conductivity probe.

Applying these principles to the experimental results of figures 4, 5 and 6 one can draw the following conclusions:

1. At low concentrations, good agreement between the methods is observed. This means that the solids are uniformly distributed in the lateral plane.

2. At intermediate concentrations, the conductivity probe gives slightly higher values near the bottom of the pipe, indicating that solids are uniformly distributed in all lateral planes across the pipe except near the bottom,

where solids concentration increases and particle coring occurs. At these positions, the solids are more concentrated at center of the lateral planes passing through these positions.

3. At higher concentrations, the conductivity probe gives concentrations significantly higher than the gamma ray method. We conclude that the coring tendency has become more pronounced at these high concentrations. For example, one can detect coring at positions (Y/D) up to 0.8 at a mean concentration of 44%.

Figure 7 shows a comparison between the concentration profiles obtained with the two methods for sand particles of 0.45 mm mean particle diameter at a bulk velocity of 2 m/s and concentrations of 5% and 15%. As in the case of the fine particles, one observes good agreement between the two methods at all positions at the lower concentration. At C_v = 15%, one again observes that the conductivity probe gives higher values near the bottom of the pipe. One also observes some scatter in the profile obtained with both methods near the top of the pipe where local concentrations are below 2%.

Figure 8 shows the concentration profiles for the same sand, but at a bulk velocity of 3.4 m/s. Comparing the results of figures 7 and 8, one again observes that the bulk velocity has a minor effect on the concentration profile for these particles. Only at the bottom of the pipe, where the concentration approaches 40%, is any difference detected.

All these comparisons were obtained in a two inch acrylic pipe. It was of interest to test the conductivity probe in a large pipe with a conducting wall. Figure 9 shows a comparison of the concentration profiles obtained with gamma ray absorption and the conductivity probe for sand particles of 0.3 mm mean diameter at a bulk velocity of 5.1 m/s and a mean in-situ concentration of 20%. Again good agreement can be observed at all positions, indicating that the solids are uniformly distributed in all lateral planes.

CONCLUSIONS

Isokinetic sampling is probably the most reliable technique for determining solids concentrations. At relative densities (S values) near unity, even thick-walled probes can be used and the results are insensitive to sampling velocity. Unfortunately, whenever a significant variation of concentration occurs over the cross-section of the pipe, probe wall thickness becomes important and simultaneous determination of the approach velocity becomes necessary for isokinetic sampling.

A horizontal gamma ray beam can be expected to be affected by the particle coring tendency. This is strongly dependent upon particle diameter and concentration and is relatively insensitive to velocity. Resistivity measurements appear to be at least as reliable as the other techniques of concentration measurement. Although comparatively inexpensive, this technique is still tedious because it requires the fluid resistivity (including the temperature) to be monitored. Because the effect of solids concentration on resistivity can be predicted theoretically, calibration studies with this method reduce to examining the effect of details of construction for the particular device which is used. In fact, Maxwell's equation appears to be satisfactory for slurries with broad particle size distributions so that calibration is probably unnecessary.

ACKNOWLEDGEMENTS

This research was performed with the assistance of NSERC Canada Grant A1027. A patent disclosure for the probe device has been filed.

SYMBOLS

C	Volume fraction solids (local value from probes or chord average from gamma ray beam)
C_o	Volume fraction solids in sample
C_v	Delivered mean volume fraction solids
d_{50}	Mass median particle diameter
D	Pipe diameter
U_b	Bulk or mean velocity of flow
U_o	Sample withdrawal velocity

REFERENCES

1. Nasr-el-din, H. and Shook, C.A., "Slurry Concentrations and Velocity Measurements at the Exit of 90° Elbows", submitted for publication to ASME, 1985.

2. Korbel, K., Michalik, A., Przewloski and Parzonka, W., "Determination of the Polyfractional Solids Distribution in a Pipe", Proc. 4th Int. Conf. on Hydraulic Transport of Solids, 1976, Paper A4.

3. Nasr-el-din, H., Shook, C.A. and Esmail, M.N., "Isokinetic Sampling from Slurry Pipelines", Can. J. Chem. Eng., 1984, 62, 179-185.

4. Nasr-el-din, H., Shook, C.A. and Esmail, M.N., "Wall Sampling from Slurry Pipelines", Can. J. Chem. Eng., 1985, 63, 746-753.

5. Nasr-el-din, H. and Shook, C.A., "Sampling from Slurry Pipelines: Thick-Walled and Straight Probes", J. Pipelines, 1985, 5, 113-124.

6. Brown, N.P. and Shook, C.A., "A Probe for Particle Velocities: The Effect of Particle Size", Proc. 8th Int. Conf. on Hydraulic Transport of Solids, 1982, Paper G1.

7. Gillies, R., Husband, W.H.W., Small, M. and Shook, C.A., "Some Experimental Methods for Coarse Coal Slurries", Proc. 9th Int. Conf. on Hydraulic Transport of Solids, 1984, Paper A2.

8. Nasr-el-din, H., Shook, C.A. and Colwell, J., "A Resistivity Probe for Local Concentration Measurement in Slurry Flows", submitted to Int. J. of Multiphase Flow.

9. Maxwell, J.C., "A Treatise on Electricity and Magnetism", 1881, Calendar Press, Oxford.

TABLE 1: PARTICLE PROPERTIES

PARTICLES	MEAN DIAMETER (MM)	S	SHAPE
Polystyrene	0.3	1.05	Spherical
Polystyrene	1.4	1.06	Irregular
Fine Sand	0.19	2.65	Irregular
Medium Sand	0.45	2.65	Irregular

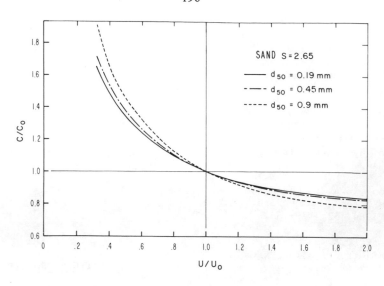

Fig. 1. Theoretical predictions of sampling efficiency.

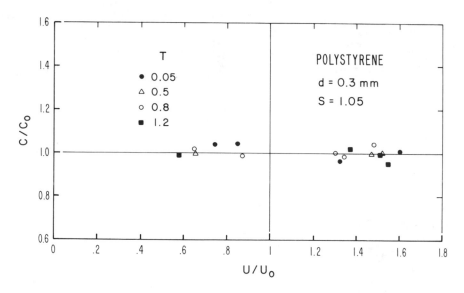

Fig. 2. Effect of sampling velocity and probe wall thickness on C/C_o for polystyrene particles.

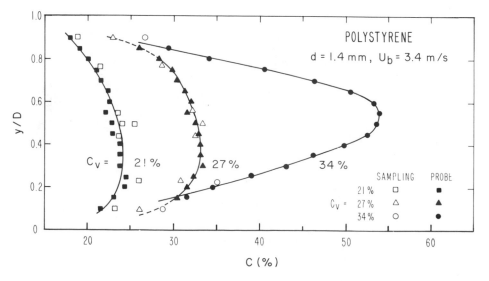

Fig. 3. Concentration profiles with isokinetic sampling and conductivity probe, vertical plane.

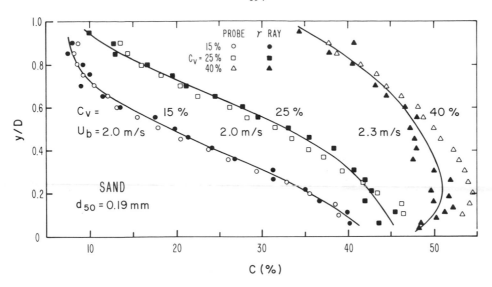

Fig. 4. Concentration profiles obtained with the fine particles, low velocity.

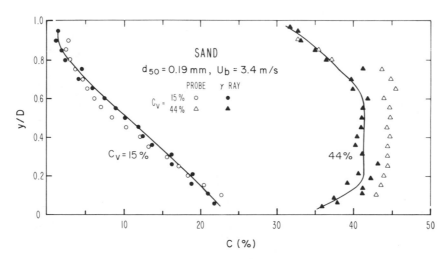

Fig. 5. Concentration profiles obtained with the fine particles, higher velocity, $C_v = 15\%$ and 44%.

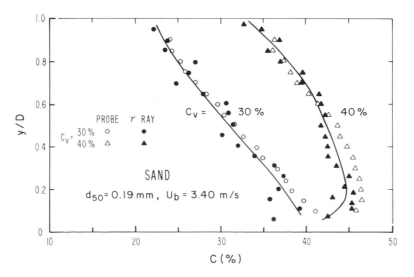

Fig. 6. Concentration profiles obtained with the fine particles, higher velocity, $C_v = 30\%$ and 40%.

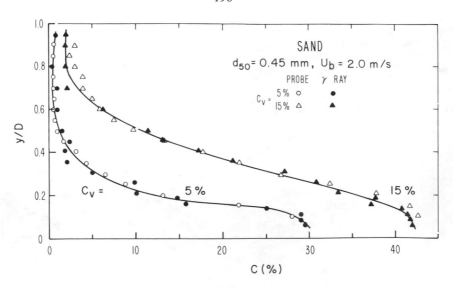

Fig. 7. *Concentration profiles obtained with the larger sand particles, lower velocity.*

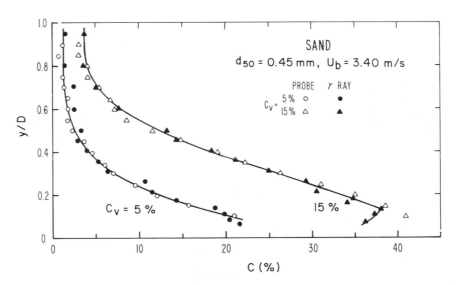

Fig. 8. *Concentration profiles obtained with the larger sand particles, higher velocity.*

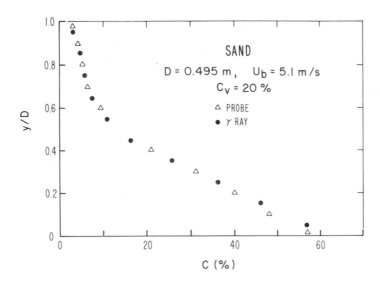

Fig. 9. *Concentration profiles obtained in a 20 inch steel pipe.*

10th International Conference on the
Hydraulic Transport of Solids in Pipes

HYDROTRANSPORT 10

Innsbruck, Austria: 29-31 October, 1986

PAPER F2

"Hydraulic Conveying of Metallic Platelets"

M. STREAT* AND A. TATSIS+

*   Department of Chemical Engineering &
    Chemical Technology,
    Imperial College,
    London.
    SW7 2BY

+   BHRA,
    Cranfield,
    Bedford.
    MK43 OAJ

Summary

One of the options for dealing with
radioactive nuclear fuel cladding waste involves
comminution as a precursor to further treatment
and thus the feasibility of handling such material
by hydraulic conveying is of interest.
Experiments have been performed with suspensions
of regular shaped metallic platelets to simulate
radioactive Magnox swarf. The terminal settling
velocity and drag coefficient of aluminium
platelets 9-12 mm long and 2 mm thick were
measured and found to be in good agreement with
theoretical predictions.

A pipe rig with 2" and 4" diameter test
sections was used to pump aluminium platelets at
volumetric concentrations up to about 10% and flow
velocities in the range 0.4-4.2 m/s. Pressure
drop data are correlated using established
empirical relationships and the mechanism of
platelet transportation is explained in terms of
an "aerofoil" theory.

SYMBOLS

		Units
A_p	particle projected area	L^2
C_D	particle drag coefficient	-
C_v	volumetric concentration of solids	-
d_s	diameter of volume equivalent sphere	L
D	internal pipe diameter	L
F_F	hydrodynamic force exerted onto the leading face of a platelet	$M\,L\,T^{-2}$
F_L	Froude number at the critical velocity for deposition	-
f_m	fanning friction factor for slurry	-
Fr	Froude number	-
F_S	hydrodynamic force exerted onto the shearing sides of a platelet	$M\,L\,T^{-2}$
F_T	hydrodynamic force exerted onto the upper surface of a platelet	$M\,L\,T^{-2}$
f_w	fanning friction factor for water	-
g	acceleration due to gravity	$L\,T^{-2}$
Ga	Galileo number	-
i	hydraulic gradient due to slurry	-
i_w	hydraulic gradient due to water	-
K	particle shape factor	-
Ke	volume coefficient of equi-dimensional particle	-
Re	Reynolds number	-
S	relative density of solids	-
S_m	mean relative density of slurry	-
t	platelet thickness	L
V_{cr}	critical velocity of solids deposition	$L\,T^{-1}$
V_m	mean slurry velocity	$L\,T^{-1}$
V_t	terminal settling velocity of a single particle	$L\,T^{-1}$
V_t	hindered terminal settling velocity	$L\,T^{-1}$
V_u	critical velocity for turbulent uplift	$L\,T^{-1}$
W_{ap}	apparent weight of a submerged platelet	M
ρ_m	mean slurry density	$M\,L^{-3}$
ρ_w	water density	$M\,L^{-3}$
ϕ	Durand head loss parameter	-
ψ	Durand velocity parameter	-

1.   INTRODUCTION

The majority of thermal nuclear reactors in
the United Kingdom use a uranium metal fuel clad
in a Magnox can. Magnox is an alloy comprising
99.3% magnesium metal and 0.7% aluminium. When
the spent fuel is decanned prior to reprocessing,
the resultant swarf is normally stored under water
in large concrete silos. Under water, magnesium
metal corrodes according to the following
reaction:

$$Mg + 2H_2O \longrightarrow Mg(OH)_2 + H_2 + Heat$$

The heat liberated by this reaction is entrapped
by the magnesium hydroxide which acts as a
convection barrier. In addition, the hydrogen gas
produced constitutes an explosion hazard if
allowed to exceed a concentration of 4% in the
surrounding air. Development work is now in
progress to empty the existing storage silos and
to evaluate alternative methodology to handle and
permanently store future arisings of Magnox
waste. One of the process options is to comminute
fresh Magnox and then transport to an
immobilisation plant. This paper considers the
possibility of hydraulic transportation of Magnox
swarf in the process cycle. Although the particle
size of comminuted Magnox swarf has not yet been
decided, it is almost certain to involve
"platelets" with a large length to thickness
ratio. Despite an abundance of literature
concerned with slurry pipelining of naturally
occurring materials, very little relates to the
unusual shapes anticipated in this work.

To avoid corrosion, aluminium platelets in the size range 9-12 mm long and 2 mm thick were used as the simulant material. Tests were performed at Imperial College with individual platelets to establish their settling characteristics and dilute suspensions of metal platelet/water mixtures were pumped in a custom designed pipe rig which was made available to us by BNFL at the UKAEA-Winfrith Laboratory.

2. THEORETICAL BACKGROUND

2.1 Head Loss in the Pumping of Settling Slurries

The main characteristic of the flow of settling slurries lies in the fact that the solid and liquid phases remain identifiable; there is no increase in the viscosity of the liquid phase due to its association with the solid particles. The primary objectives of this work are firstly to predict and secondly to optimise the head loss as a function of independent design variables. Amongst the important independent parameters for settling slurry systems are pipeline diameter, D, operating velocity V_m, mean particle size d, particle size distribution and the properties of the carrier fluid. Correlations for slurry head loss where the transporting medium is water are often given by the general relationship:

$$\frac{i - i_w}{i_w C_v} = f(Fr) \qquad \ldots (1)$$

where $Fr = \dfrac{V_m}{\sqrt{g \, D \, (S - 1)}}$

Equation 1 accounts for the frictional head loss due to the fluid and that due to the solids. The Froude number, Fr, corrected for the relative mass of solids in water relates the competing effects of inertia and gravity. Depending on the mean particle size and the superficial velocity of the mixture, transport can take place either as a suspension or in the form of a sliding bed. Where possible, the former is preferred since it offers economical advantages and minimises the probability of blockage.

Durand and Condolios (1) proposed an empirical correlation of the form:

$$\frac{i - i_w}{i_w C_v} = 150 \left[\frac{g \, D \, (S - 1)}{V_m^2 \sqrt{C_D}}\right]^{1.5} \qquad \ldots (2)$$

which for simplicity, is often quoted as $\phi = K \psi^n$.

Newitt, Richardson et al (2) pioneered a theoretical approach to the problem. By assuming no slip between the solid particles and water, they developed the following semi-empirical correlation for the fully suspended flow regime,

$$\frac{i - i_w}{i_w C_v} = 1100 \, \frac{g \, D \, V_t \, (S - 1)}{V_m^3} \qquad \ldots (3)$$

Over the years, many researchers have proposed modifications to the empirical constants in equations 2 and 3 so that the correlations will fit their own experimental data. For example, Zandi and Govatos (3) collated over 2,500 data points from various sources and compared existing correlations. They recommended three mutually exclusive correlations of their own which apply over consecutive intervals of the range for ψ.

Newitt et al (2) offered a separate expression for the case of flow with a sliding bed, i.e.

$$\frac{i - i_w}{i_w C_v} = \frac{66 \, (S - 1) \, g \, D}{V_m^2} \qquad \ldots (4)$$

Wilson et al (3) used a bed slip model approach which is based on an analysis of the hydrostatic type forces acting on a bed of particles within a conduit. They showed that the total pressure drop with a sliding bed is given by

$$i = 2f_s \, (S - 1) \, C_b \left[\frac{\sin\Theta - \Theta \cos\Theta}{\pi}\right]$$

$$+ \, 2f_b \frac{\rho_{mb}}{\rho_w} \frac{\Theta}{\pi} \frac{V_b^2}{Dg} + 2f_t \frac{\rho_{mt}}{\rho_w}$$

$$\frac{(\pi - \Theta)}{\pi} \frac{V_t^2}{Dg} + \frac{2f_s \, f_b}{\tan\phi} \frac{\Theta}{\pi} \frac{(V_t - V_b)^2}{Dg} \qquad \ldots (5)$$

where the symbols are defined in the appropriate reference.

Wilson (4) developed this concept further into a unified analysis for pipeline flow by considering separately the effects due to suspended and contact load respectively. Clift et al (5) have used this method for scaling mixed regime slurries in the suspended and bed load regimes. They distinguished between the head loss gradient for the suspended component and the contact load component and showed that the overall head loss is given by

$$\frac{i - i_w}{S_m - 1} = K \, B \, V_m^{-m} + A \, i_w \, (1 - K \, V_m^{-m}) \qquad \ldots (6)$$

where the parameters A, B and m are properties of the fluid and have to be determined experimentally.

2.2 Single Particle Hydrodynamics

Hydraulic conveying may be examined analytically by considering the hydrodynamic forces acting upon the individual particles. Although this approach is simplistic, it is nevertheless useful in providing an insight to the transport mechanisms. At low flow rates, particle transport takes place in a sliding bed. The particles move in "unison" so that particle drag is irrelevant and solid friction at the pipe wall accounts for most of the pressure drop incurred. At higher flow rates, however, the particles travel in suspension and whilst particle drag becomes important, the net result is a reduction in the frictional head loss. The mean slurry velocity which marks the onset of settling in a given pipe, commonly referred to as the "critical velocity of deposition", is often used as a criterion in deciding which mathematical model is most applicable to the type of flow. The type of flow attained in a given pipe diameter is largely determined by the particle size and shape and hence by the particle drag. The drag coefficient for a sphere falling freely at its terminal velocity through an unbound liquid is given by

$$C_D = \frac{4 \, g \, d \, (S - 1)}{3 \, V_{ts}^2} \qquad \ldots (7)$$

Similarly, for a flat prismatic object of cross-sectional area A_p and thickness t ($\sqrt{A_p} \gg$ t) settling with its plane normal to the direction of motion it can be shown that

$$C_D = \frac{2 \, g \, t \, (S - 1)}{V_{tp}^2} \, . \qquad \ldots (8)$$

The calculation of drag for a non-spherical particle becomes increasingly difficult as the particle shape deviates from spherical. This is partly due to inadequate methods of defining shape but more importantly due to the complex trajectory described by the falling particle. It is generally accepted, however, that when a non-spherical particle is settling in the turbulent flow regime, then it orientates itself into the position of greatest drag (this provides greatest stability) and then follows a sinuous path. The complete analysis of drag variation with particle shape is beyond the scope of this paper and reference is made to the work carried out by Heywood (6). The particle terminal velocity V_t required in equation 7 is derived from the following fluid-particle expression:

$$Ga = 18 \, Re_{ts}$$

$$\text{if } Ga < 3.6 \qquad \ldots (9)$$

or

$$Ga = 18 \, Re_{ts} + 2.7 \, Re_{ts}^{1.687}$$

$$\text{if } 3.6 < Ga < 10^5 \qquad \ldots (10)$$

or

$$Ga = 1/3 \, Re_{ts}^2$$

$$\text{if } Ga > 10^5 \qquad \ldots (11)$$

$$\text{where } Re_{ts} = \frac{V_t \, \rho \, d_s}{\mu} \qquad \ldots (12)$$

For a non spherical particle the settling velocity is adjusted for the effect of shape by multiplying with an appropriate shape factor K, i.e.

$$V_{tp} = K \, V_{ts} \qquad \ldots (13)$$

Heywood (6) has shown that a reliable estimate of K may be obtained using the definition of an equivalent sphere based on the same projected area as the particle, i.e.

$$K = \frac{Ke \, t}{\sqrt{A_p}} \qquad \ldots (14)$$

where

$$Ke = \frac{\text{Volume of particle}}{(\text{equivalent diameter})^3} \qquad \ldots (15)$$

Finally, Heywood (6, 7) gave empirical correlations which may be used to obtain a better estimate of K.

Any additional effects which might arise from hindered settling may be accounted for by using the Richardson and Zaki (8) correlation, i.e.

$$V_t' = V_t \, (1 - C_v)^a \qquad \ldots (16)$$

where a is a function of the particle Reynolds number.

Two of the most commonly used expressions which relate the critical settling velocity of deposition to the terminal settling velocity are given below

$$V_{cr} = F_L \sqrt{(2 \, g \, D \, (S - 1))} \qquad \ldots (17)$$

Durand (1) and

$$V_u = 0.6 \, V_t \sqrt{2/f_w} \, e^{45 \, d/D} \qquad \ldots (18)$$

Wilson and Watt (9).

3. EXPERIMENTAL

3.1 Experimental Equipment and Procedure

The terminal settling velocity of individual aluminium platelets was determined using a glass column as shown in Fig. 1. A closed loop test rig, shown in Fig. 2, was used to pump dilute suspensions of platelet/water mixtures. Although the rig was built with 4" I.D. piping, provision was made to replace the test sections with different pipe diameters - 2" and 4" configurations were used in this work.

Flow rate was measured by an electromagnetic flowmeter and the mixture density was adjusted by controlling the rate of platelets' entrainment in the suction end of the pump. Both these quantities could be checked by diverting the slurry for a given time into a weigh tank. Slurry velocity was controlled by using either a variable speed drive on the centrifugal solids-handling pump or a bypass.

The frictional pressure drop across the test section of horizontal pipeline was measured using two strain gauge type transducers. The pressure difference was reflected as a voltage signal which was filtered for noise and relayed to a multi-channel UV recorder. Prior to taking a pressure drop reading the flow would be established in the circuit and the rig would be allowed to run for a few minutes to attain a uniform concentration throughout the loop.

A detailed description of the apparatus is given in Ref. 10.

3.2 Experimental Material

It was estimated that about 250 kg of light-metal "platelet material" was required to reach an upper volumetric concentration of 30% using the

Winfrith pipeline. The maximum particle dimensions should be approximately 1/10 of the pipe diameter.

To avoid the manufacturing cost of producing the material out of commercial aluminium sheeting, the possibility of using the waste products of standard operations was explored. Stampings produced in the making of Dexion angle offered an excellent alternative and 410 kg of mixed Dexion stampings were supplied by the Dexion Company.

Fig. 3 shows an isometric view of a Dexion angle length comprising one cycle of perforations. The pattern consists of nine holes (or stampings) which come in four different sizes, "A, B, C" and "D" with maximum particle size ranging from about 58 to 10 mm respectively. Apart from Item D, which has an ordinary disc shape, each stamping is made up of three sections; two semicircles separated by a rectangular piece in the middle. The length to breadth ratio varies from unity for "D" to about 5.7 for "A". The simplicity of geometry facilitates the calculation of the projected area. Following some commissioning tests in the 4" pipeline it became clear that platelets "A" were unsuitable for pumping due to the large d/D ratio (= 0.6).

Platelets "A" were then discarded by sieving and Table 1 shows the size classification of the remaining items (about 260 kg), before and after pumping in the two pipe diameters. The reduction in mean particle size is attributed mainly to wear within the pump volute despite the use of a recessed impeller type pump. Plate 1 is a close-up photograph of two used "B" items confirming that apart from some curling of the material occurring at the distant edges, the overall shape has been retained. Finally Table 2 provides equivalent particle dimensions which are representative of all platelets used in the tests.

3.3 Hydrodynamic Properties of Dexion Stampings

The prediction of terminal settling velocity for a non-spherical particle falling freely through water requires the knowledge of a suitable "shape factor". In the case of Dexion stampings, appropriate shape factors were found using the method outlined under single particle hydrodynamics (see Table 3). Alternatively, the terminal settling velocity of individual platelets was determined experimentally by means of a 4" water column. Initially, ballotini spheres were used in the size range 3-12 mm in order to establish the effect of the column-walls on axially descending particles. The agreement obtained between predicted and experimentally determined values of the terminal settling velocity was to within 5%, suggesting that for particles up to 12 mm in diameter the effects due to the pipe-wall are negligible and can be safely ignored.

When Dexion stampings were tested, their mode of descent was observed carefully and the various effects are summarised in Table 4. It is interesting to note that the small deformation of used platelets resulted generally in a reduction of the number and intensity of secondary motions.

Table 5 shows the experimental results obtained for the various items and draws a comparison with the predicted values of terminal settling velocity. The best agreement, to within 1.5%, was obtained using an approximate volumetric factor Ke = 0.560 which was proposed by Heywood [7] for rounded isometric irregular shapes.

4. RESULTS AND DISCUSSION

4.1 Correlations Using the Durand Parameters

One hundred and twelve useful data points were obtained from pressure drop tests in the 4" pipe diameter. When the results were correlated using the dimensionless groups proposed by Durand and Condolios [1] the following expression was obtained

$$\phi = 265 \, \psi^{1.38} \qquad \qquad \ldots (19)$$

where $\phi = \dfrac{i - i_w}{i_w \, C_v}$ and $\psi = \dfrac{g \, D \, (S - 1)}{V_m^2 \, \sqrt{C_D}}$

Each data point was evaluated using i, V_m and C_v as the input parameters whilst C_D was calculated using equation 8 and i_w was obtained from separate water tests. Fig. 4a shows some experimental scatter at extreme values of the ψ parameter. At low flow rates (i.e. large ψ) the scatter can be easily explained in terms of operating instabilities introduced by the use of a bypass. Deviations from the straight line at high mean velocities (i.e. low ψ), were also observed by Newitt et al [2]. They pointed out that in this flow regime the quantities i and i_w become large and of similar magnitude. Consequently the dimensionless group ϕ becomes inaccurate since it is calculated by taking the small difference between two large values. The Durand and Condolios correlation evaluated with their own coefficients has also been included for comparison. Fig. 4b shows the results obtained from the 2" diameter tests. The data may be correlated by the equation

$$\phi = 188 \, \psi^{1.44} \qquad \qquad \ldots (20)$$

The general comments made on the 4" diameter results apply here also.

The differences in the values of the exponents in equations (19) and (20) from the value of 1.5 proposed by Durand are not considered to be significant and are well within the scatter of the experimental results. The differences in the constants, however, are significant and suggest that the inclusion of the drag coefficient to account for the effect of particle shape is not entirely satisfactory and that there must be a fundamental difference in the conveyance of platelets as opposed to rounded particles.

In order to establish the validity of the Durand and Condolios approach for scale-up calculations the results from each pipe diameter were plotted collectively in Fig. 5, giving the following equation:

$$\phi = 238 \, \psi^{1.41} \qquad \qquad \ldots (21)$$

The comparison suggests that the results obtained from the two pipe diameters may be grouped together without significant loss of accuracy — the use of such empirical correlations for modest scale-up is thus justified. A comparison between equations 19 and 20 suggests, however, that the scale-up correlation is affected by the pipe diameter. Specifically, it might be expected that the ratio of pipe diameter to mean particle size D/d would affect the $\phi - \psi$ correlation and additional experimental work is needed with metallic platelets of various sizes in several pipe diaemters to establish this effect.

4.2 Interpretation of the Results

The fact that the results correlated well using the Durand and Condolios approach suggests that the platelets were suspended over a substantial part of the velocity range. In the absence of independent experimental evidence, this hypothesis was tested by adopting the Clift et al (5) approach: a plot of $(f_m - f_w)/(S_m - 1)$ as a function of V_m may be used to show whether values of V_m are sufficiently high to sustain the solids in suspension. The asymptotic part of the curves shown in Figs. 6a and 6b confirms that fully suspended flow was actually achieved above mixture velocities of about 1.7 and 2 m/s for the 2" and 4" pipe diameters respectively. Hence, rewriting equation 6 in terms of friction factors, and using it in its asymptotic form at high velocities, i.e.

$$\frac{f_m - f_w}{(S_m - 1)} = A\, f_w \qquad \ldots (22)$$

the constant A was determined to be 8.6 and 2.4 for the 2" and 4" pipelines respectively. If it is assumed that there is no slip between the particles and the liquid, then the solid/liquid suspension behaves as a fluid of a higher density ρ_m with the result that A = 1. The fact that in the present study A is much larger, suggests that platelets have an effect in addition to that of merely increasing the effective density of the slurry. It appears that this arises from the finite relative motion between the platelets and the fluid which results in the generation of eddies. The latter is associated with increased frictional dissipation which results in additional pressure drop. This conclusion is corroborated by the findings of Pouska and Link (11) who made measurements of the frictional pressure drop in the pumping of dilute suspensions of oil-shale in water. Their results yield values of the constant A which are greater than unity and are approximately equal in magnitude to those obtained in the present study.

An independent estimate of the threshold velocity for turbulent uplift can be obtained using equation 18. This predicts the minimum velocities required for turbulent suspension as 7.4 and 4.6 m/s respectively. These values are considerably greater than those deduced from Figs. 6a and 6b. There may be a number of reasons why equation 18 is not suitable for use with platelet material. Two possible reasons are discussed below:

(i)   Although the value of V_t employed in equation 18 is experimentally determined it is still necessary to estimate an appropriate value for d; of the various alternatives, a sphere settling at the same rate as the platelets was selected. The diameter of such a sphere was estimated to be 1.5 mm as opposed to the maximum platelet dimension of about 12 mm. Use of this equivalent sphere diameter makes equation 18

self consistent, but unfortunately, it does not simulate an aluminium platelet in physical terms.

(ii)   Secondly, the predicted velocity is highly sensitive to the value of the constant in the exponential factor. Wilson (9) determined this constant by correlating a large amount of data for which d $\ll$ D. In the present study, however, d is of the same order as D and since the particle size is a significant fraction of the macro-scale of turbulence, the predetermined value of K = 45 cannot be expected to apply rigorously.

Using the Durand and Condolios correlation, given by equation (17), critical settling velocities of 1.7 and 2.5 m/s were predicted for the 2" and 4" pipe diameters respectively. These predictions agree closely with the velocities anticipated from Figs. 6a and 6b but there are several reasons why in general the Durand correlation may be invalid for use with platelets.

Durand and Condolios based their correlation on tests using granular material with a top particle size of about 3 mm; extension of their results to aluminium platelets with maximum dimension about 10 mm has been made on the assumption that the Froude number remains constant. Furthermore, there is no experimental evidence that platelets travel as a sliding bed.

Therefore, in demonstrating that both the Wilson and Durand correlations are inadequate for the case of flat objects, the need for a fresh, physically based approach which takes into account the actual shape of a platelet becomes evident.

4.3 The Aerofoil Approach

Consider a single aluminium platelet resting on the invert of a horizontal pipe. The vertical component of the reaction at the interface counterbalances the apparent weight of the submerged platelet whilst the horizontal component provides a direct measure of the frictional resistance at the solid boundary. Under these conditions the platelet remains stationary and in contact with the pipe wall. Under steady flow conditions, however, a dynamic equilibrium exists due to the hydrodynamic forces which are exerted on the platelet. These forces are generated mainly due to the dynamic pressure acting on the leading face and the shearing stresses associated with fluid drag on the free sides of the platelet, see Fig. 7c. To facilitate an estimate of these forces, a number of simplifying assumptions were made:

(i)   It is assumed that the radius of curvature of the containing pipe is large so that the platelet lies flat and no fluid flow occurs at the interface.

(ii)   The x-y dimensions of the platelets used are approximately equal so that the maximum total force may be assumed independent of particle orientation with respect to the direction of flow. However, an order of magnitude analysis shows that the force exerted on the leading face is by far the greatest which suggests that the magnitude of the force itself will be sensitive to the profile of the leading edge. Fig. 8 shows the definition of a typical platelet profile (type B) which, if projected to the flow, is expected to develop the lowest head compared with other orientations and platelet shapes. For the preservation of shape continuity between adjoining surfaces a slight refinement has been introduced by using the equation of a semi-variable ellipse to describe the profile of the leading face. The resulting contour varies between a semi-circle at $\Theta = 0°$ and a straight line at $\Theta = 90°$.

(iii) It is further assumed that for reasons of maximum stability the platelet will tend to orientate itself so that the direction of flow is parallel with one of the long axes.

(iv) Figs 7a and 7b illustrate the main simplifications made regarding the flow around each platelet. Basically, it is assumed that the fluid comes to rest isentropically against the leading face, that new turbulent boundary layers develop along the shearing faces and that there is no flow around the trailing edge.

Assumption (iv) implies frictionless flow. Hence, using the Bernoulli equation it can be shown that total force due to the stagnation pressure acting on the leading face is given by

$$F_f = 2 \left[\frac{1}{2} \int_o^{\pi/2} \int_o^t \rho_w V_h^2 \frac{y}{2} \cos \Theta \cos \Omega \, dh \, d\Theta \right] \qquad \ldots (23)$$

where the local velocity V_h may be estimated using the appropriate Von-Karman equation for a logarithmic velocity profile. Similarly, it can be shown that the shearing force acting on the elevated sides is given by

$$F_s = 2 \left[\frac{1}{2} \int_o^t \rho_w V_h^2 C_D (x - y) \, dh \right] \qquad \ldots (24)$$

where C_D can be expressed as a function of the Reynolds number.

Finally, the shearing force acting on the top surface of the platelet, is given by

$$F_T = \frac{1}{2} \rho_w V_h^2 C_D \left((x - y) + \frac{\pi y^2}{4} \right) \qquad \ldots (25)$$

Thus, by using an estimated initial value for the mean slurry velocity, the sum total of all the hydrodynamic forces may be calculated by integrating over the dynamically active area of the platelet. The result is compared with the horizontal component of the reaction at the wall and the value of the mean velocity is iterated until

$$F_F + F_S + F_T = W_{ap} \tan(\phi) \qquad \ldots (26)$$

i.e. the instant the platelet begins to slide. The sliding angle (ϕ) was determined experimentally by noting the inclination to horizontal at which several platelets just began

to slide down a submerged steel platform. This was found to be approximately $22°$ compared with about $25°$ under dry conditions.

Besides causing the platelet to slide, the hydrodynamic forces provide up-turning moments about the bottom edge of the trailing face which compete with the stabilising effect of the weight. Thence, it is possible to obtain a new threshold slurry velocity at which the moments balance-out and the platelet becomes "buoyant" by levering against the pivoting edge (Fig. 9). Once the platelet is inclined, a lift force will be generated due to differential pressure created by fluid flowing faster over the upper than lower surfaces respectively. At this instant, the platelet behaves like an aerofoil, it lifts and becomes "fluid borne". Thereafter it is expected to follow a trajectory similar to that shown in Fig. 9; i.e. the platelet accelerates to the local velocity of the fluid by projecting its plane to the direction of flow. At this position the driving force is removed, the platelet "stalls", it then settles under gravity and the process repeats itself. As the pipe velocity is increased, excess lift will be available even at small platelet inclinations so that relatively long periods of suspension may be obtained.

Using the method set out above, critical velocities for the onset of sliding were calculated and were found to be 0.53 and 0.61 m/s for the 2" and 4" pipe diameters respectively. These result are in good agreement with the lowest velocities at which it was possible to pump platelet/water mixtures, see Figs. 6a and 6b. Similarly, the minimum velocities required for platelet uplift were estimated at 1.74 and 2.14 m/s in each case. These predictions are in excellent agreement with the velocities at which asymptotic values are reached in Figs. 6a and 6b.

Although these results were based on the behaviour of a single particle, it is expected that the theory applies to dilute suspensions e.g. less than about 10% by volume. The theory was also tested by plotting the results according to equation 3 proposed by Newitt (2) for heterogeneous suspensions – a reasonable correlation was obtained. The theory was further tested by means of the bed slip model given by equation 5. A comparison of the experimental results with predictions provided by equation 5 confirmed that above a pipe velocity of about 2 m/s, platelet transport takes place in the fully suspended flow regime.

4.4 The Effect of Particle Shape on Head Loss

The effect of particle shape on head loss can be predicted using the Durand equation with drag coefficients of 1.360 for platelets and 0.444 for volume equivalent spheres. In this case, the calculated ratio of ϕ for spheres and platelets is independent of velocity at a value of 2.31.

Fig. 10 shows the actual head loss for platelets obtained experimentally and given by equation 27 compared with the predicted head loss for spheres. In this case, the ratio of ϕ for spheres and platelets is a function of mean velocity:

$$\frac{\phi_{spheres}}{\phi_{platelets}} = \frac{1.505}{V_m^{0.18}} \qquad \ldots (27)$$

Fig. 10 predicts that head loss in conveying metallic platelets is less than for volume equivalent spheres. For example, the ratio of

$$(\phi_{spheres}/\phi_{platelets})$$

falls from 1.71 at 0.5 m/s to 1.17 at 4 m/s. In general, it can be concluded that particle shape is important, though further experimentation is required to rigorously quantify the effect.

5. CONCLUSIONS

A large sample of aluminium platelets has been classified before and after use in hydraulic conveying tests. The hydrodynamic properties of individual platelets have been analysed by carrying out a series of terminal settling velocity measurements. A typical aluminium platelet was observed to descend with its plane perpendicular to the direction of motion, the rate of free settling being about 200 mm/s. Using a modified shape factor based on the particle thickness and projected area, i.e.

$$C_D = \frac{2 \, g \, (S - 1) \, t}{V_t^2} \qquad \ldots (28)$$

it was possible to predict the experimental result to within 1.5%.

A pipe-rig fitted with 2" and 4" diameter pipes was used to pump dilute suspensions of metallic platelets at flow rates ranging from 0.5 to 4 m/s and concentrations up to about 10% by volume. The results were found to correlate well using the Durand type equations, and the following expressions were obtained:

$$\phi = 188 \, \psi^{1.44} \qquad \ldots (29)$$

for the 2" pipeline,

$$\phi = 265 \, \psi^{1.38} \qquad \ldots (30)$$

for the 4" pipeline,

$$\text{or} \quad \phi = 238 \, \psi^{1.41} \qquad \ldots (31)$$

using all the results collectively. The latter provides evidence in favour of using the Durand equation for limited scale-up. A plot of the non-dimensionless parameter for head loss $(f_m - f_w)/(S_m - 1)$ versus mean slurry velocity was used to assess the transition velocities. The results were interpreted by analysing the transport mechanisms associated with the hydraulic conveying of platelets. The use of existing theoretical or semi-empirical equations for the prediction of the critical velocity of deposition was unreliable due to lack of physical rigour. For this reason an alternative analysis was developed by considering the platelet as a thick aerofoil. Thus, by making a number of assumptions regarding the geometry and fluid flow around the platelet it was possible to estimate mean slurry velocities at which platelets either begin to slide or become fully suspended, i.e.

Pipeline Diameter	$V_{sliding}$ m/s	$V_{suspension}$ m/s
2"	0.53	1.74
4"	0.61	2.14

These velocities are in excellent agreement with the experimental results and it is on this basis that the aerofoil approach is proposed as a reliable method for the prediction of threshold velocities in the handling of platelet material at low concentrations.

Finally, the effect of particle shape on head loss has been considered by comparing a notional head loss calculated for volume equivalent spheres with the actual experimental head loss obtained for platelets. For the range of velocities considered, in this study the head loss ratio for the two shapes takes the form

$$\frac{\phi_{spheres}}{\phi_{platelets}} = \frac{1.505}{V_m^{0.18}} \qquad \ldots (32)$$

$$0.5 < V_m < 4 \text{ m/s}$$

Equation 34 confirms that the pumping of platelets is less energy intensive than volume equivalent spheres, but further work is required to establish the extent to which head loss is affected by shape.

6. ACKNOWLEDGEMENTS

The authors are grateful to B.N.F.L. for permission to publish this work and for the loan of their pipe test facility and to the personnel at the Technological Division of UKAEA-Winfrith who gave assistance in operating the rig.

The financial support of the Science and Engineering Research Council is gratefully appreciated.

Thanks are also due to the Council and Chief Executive of BHRA for their permission to publish this paper.

7. REFERENCES

1. Durand, R. and Condolios, G.
    The Hydraulic Transport of Coal and Solids in Pipes, Colloquium on Hydraulic Transport, National Coal Board, London, 1952.

2. Newitt, D.M., Richardson, J.F., Abbott, M. and Turtle, R.B.
    Hydraulic Conveying of Solids in Horizontal Pipes, Trans. Inst. Chem. Engrs., Vol. 32 (II), pp.93-113, 1955.

3. Wilson, K.C. Streat, M. and Bantin, R.A.
    Slip-Model Correlation of Dense Two-Phase Flow. Hydrotransport 2 (BHRA), Warwick (U.K.), Paper C4, pp.1-10, Spetember, 1972.

4. Wilson, K.C.
    A Unified Physically Based Analysis of Solid-Liquid Pipeline Flow, Hydrotransport 4 (BHRA), Alberta, Canada, Paper A1, pp.1-16, May, 1976.

5. Clift, R., Wilson, K.C., Addie, G.R. and Carstens, M.R.
    A Mechanistically Based Method for Scaling Pipeline Tests for Settling Slurries. Hydrotransport 8 (BHRA), Johannesburg (S.A.), Paper B1, pp.91-101, August, 1982.

6. Heywood, H.
   Symp. Interaction Fluids and Parts. Inst.
   Chem. Eng., pp.1-8, London, 1962.

7. Heywood, H.
   Powder Metallurgy, Vol. 7, pp.1-28, 1961.

8. Richardson, J.F. and Zaki, W.N.
   Sedimentation and Fluidisation. Trans. Inst.
   Chem. Engrs., Vol. 32 (I), pp. 35-53, 1954.

9. Wilson, K.C. and Watt, W.E.
   Influence of Particle Diameter on the
   Turbulent Support of Solids in Pipeline
   Flow. Hydrotranpsort 3 (BHRA) Colorado
   (U.S.A.), Paper Dl, pp.1-9, May, 1974.

10. Tatsis, A.
    Hydraulic Conveying of Metallic Platelets.
    PhD Thesis, London University, 1985.

11. Pouska, G.A. and Link, J.M.
    Investigation of head Loss in Coarse Oil Shale
    Slurries. Hydrotransport 5, Paper H2, May,
    1978.

TABLE 1 : GROUP % CONCENTRATION OF ITEMS A, B, C AND D IN THE OVERALL SAMPLE OF DEXION STAMPINGS

ITEM	No. OF PLATELETS	% CONCENTRATION OF PLATELETS IN	
REFERENCE	HAND SORTED	OVERALL SAMPLE	MATERIAL CYCLE
A	141	6.5	11.11
B	1,086	50.21	44.44
C	474	21.91	22.22
D	462	21.36	22.22
Total = 2,163 platelets			

TABLE 2 : PRINCIPAL PARTICLE DIMENSIONS

A : DIMENSIONS OF A TYPICAL PLATELET USED IN THE 4" PIPE TESTS

Mean Length	$\bar{x} = 12.4664$ mm
Mean Breadth	$\bar{y} = 9.6149$ mm
Mean XS-Area	$\bar{A}_p = 99.5464$ mm^2
Mean Height	$\bar{t} = 1.9193$ mm
Mean Volume	$\bar{V} = 191.0594$ mm
Mean Weight	$\bar{W} = 0.5023$ g
Diameter of volume equivalent sphere:	$d_v = 7.1459$ mm
Diameter of XS-Area equivalent sphere:	$d_a = 11.2580$ mm

B : DIMENSIONS OF A TYPICAL PLATELET USED IN THE 2" PIPE TESTS

Mean Length	$\bar{x} = 11.5486$ mm
Mean Breadth	$\bar{y} = 9.5768$ mm
Mean XS-Area	$\bar{A}_p = 90.6288$ mm^2
Mean Height	$\bar{t} = 2.0142$ mm
Mean Volume	$\bar{V} = 182.5445$ mm^3
Mean Weight	$\bar{W} = 0.4799$ g
Diameter of volume equivalent sphere:	$d_v = 7.0381$ mm
Diameter of XS-Area equivalent sphere:	$d_a = 10.7420$ mm

207

TABLE 3 : STEPWISE CALCULATION OF THE TERMINAL SETTLING VELOCITY FOR PUMPED PLATELETS USING THREE ALTERNATIVE VALUES OF Ke

PLATELET DIMENSIONS & OTHER PARAMETERS	4" PIPE TESTS			2" PIPE TESTS		
l (mm)	12.466			11.5486		
b (mm)	9.615			9.577		
t (mm)	1.919			2.014		
d_a (mm)	11.258			10.742		
Ga_{da}	2.269×10^7			1.971×10^7		
Re_{da}	8,250.06			7,689.41		
V_{ts} (mm s^{-1})	736.345			719.272		
ISOMETRIC SHAPE	CYLINDRICAL	CUBIC	ROUNDED	CYLINDRICAL	CUBIC	ROUNDED
Ke	0.785	0.696	0.560	0.785	0.696	0.560
$A_p = \dfrac{\pi (d_a)^2}{4}$ (mm^2)	99.543			90.629		
$K = \dfrac{Ke\, t}{(A_p)^{0.5}}$	0.1509	0.1338	0.1076	0.1661	0.1473	0.1185
K_A	0.3546	0.3290	0.2896	0.3762	0.3480	0.3047
$V_{tp} = V_{ts} K_A$ (mm s^{-1})	261.14	242.28	213.25	270.56	250.29	219.16

TABLE 4 : DIAGRAMATIC REPRESENTATION OF THE MODE OF DESCENT FOR ALUMINIUM PLATELETS

PLATELET TYPE (d_a)/mm	Re_{da} of DESCENT	VERTICAL	SPIN	TILTING	SWINGING	SLIPPAGE
THE TRUE MOTION IS OBTAINED BY SUPERIMPOSING THE APPROPRIATE CONSTITUENT SECONDARY MOTIONS						
NEW (D) d_a = 10.444 mm	2,086	**	*	*	*	
NEW (C) d_a = 11.397 mm	2,287	**	*	*	**	*
NEW (B) d_a = 12.411 mm	2,464		*	**	***	**
USED (D) d_a = 10.355 mm	2,176	*	*	**	**	
USED (C) d_a = 10.657 mm	2,249	*	*	*	*	
USED (B) d_a = 10.947 mm	2,478	**	*	*	*	
c.f. GLASS SPHERE d_a = 11.827 mm	8,435	***				

(*): Number of (*) denote the relative intensity of each effect (approximate).

(H)* Denotes horizontal position.

TABLE 5 : COMPARISON BETWEEN EXPERIMENTAL AND PREDICTED TERMINAL SETTLING VELOCITY (T.S.V.) FOR ALUMINIUM PLATELETS

	EXPERIMENTAL				PREDICTED	
GROUP	NEW PLATELETS V_t (mm s^{-1})	USED PLATELETS V_t (mm s^{-1})	CHANGE ON V_t %	MEAN (T.S.V.) $\overline{V_t}$ (mm s^{-1})	PLATELETS USED IN 4" PIPE TESTS	PLATELETS USED IN 2" PIPE TESTS
					Ke = 0.560	
					K_A ...	K_A ...
(D)	200.6 ± 7.7	211.5 ± 6.0	+5.4	206.0 ± 6.9		
(C)	201.6 ± 5.2	212.0 ± 6.2	+5.1	206.8 ± 5.7	(see Table 3)	
(B)	199.4 ± 5.9	227.4 ± 11.2	+14.0	213.4 ± 8.9		
OVERALL MEAN (T.S.V.)	220.2 ± 9.2			210.2 ± 7.8	213.3	219.2

1.4% OVERESTIMATED

0.5% UNDERESTIMATED

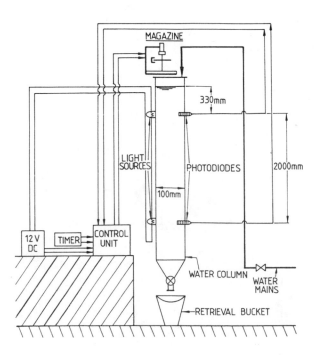

Fig. 1. A schematic presentation of the platelet characterisation apparatus.

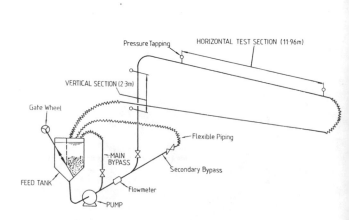

Fig. 2. Schematic presentation of 4" hydraulic conveying rig.

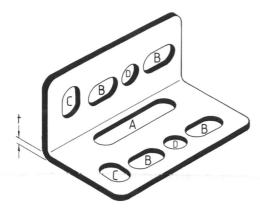

(a) Isometric view of Dexion Angle (one cycle).

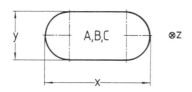

(b) Plan view of typical Dexion Stamping.

Fig. 3.

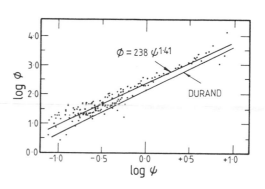

Fig. 5. Log ϕ vs log ψ data collected for 2" and 4" horizontal sections.

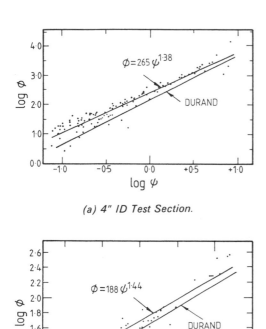

(a) 4" ID Test Section.

(b) 2" ID Test Section.

Fig. 4. Log ϕ vs log ψ data for individual horizontal sections.

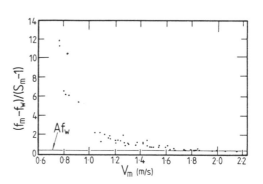

(a) 2" ID Test Section.

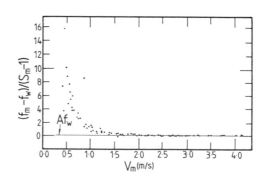

(b) 4" ID Test Section.

Fig. 6. $(f_m - f_w)/(S_m - 1)$ vs V_m data—horizontal.

(a) Qualitative streamline pattern.

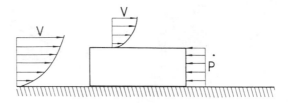

(b) Assumed velocity and pressure profiles.

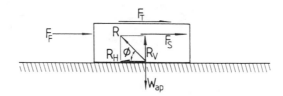

(c) Main forces acting on the platelet.

Fig. 7.

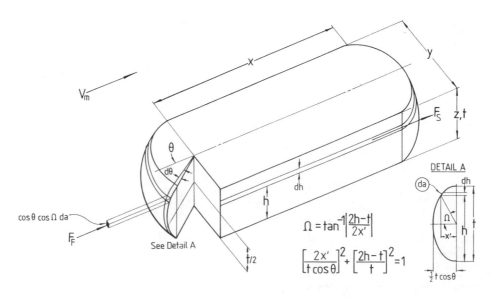

Fig. 8.   Definition of a typical platelet profile.

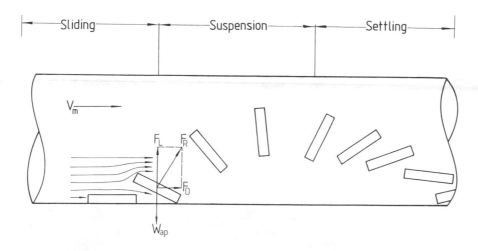

Fig. 9.   *Likely trajectory of individual platelet in fully developed pipe flow.*

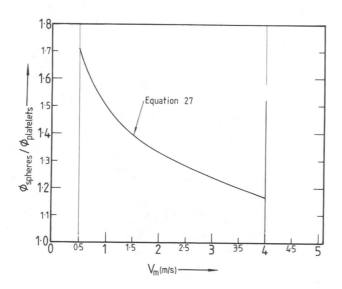

Fig. 10.   *The influence of particle shape on head loss.*

212

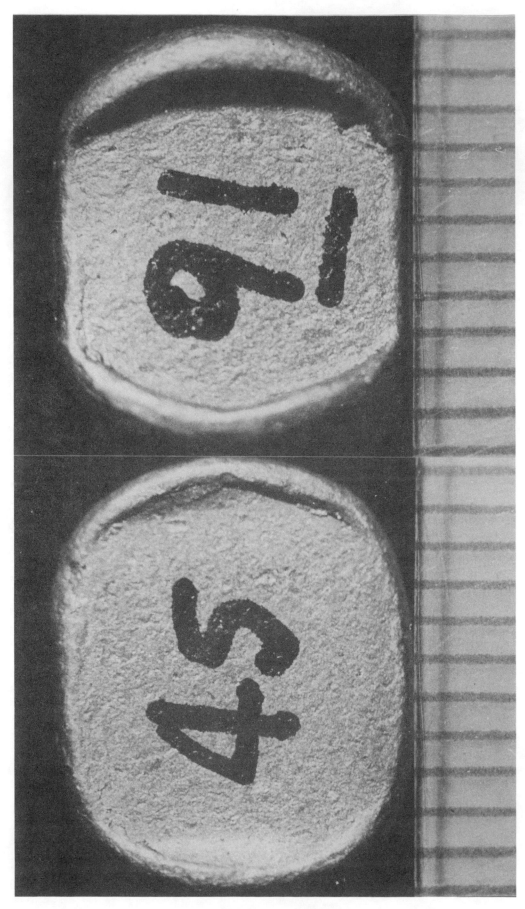

*Plate 1.   A close-up view of two used platelets showing the curling
at the edges. Group B, Scale: mm.*

10th International Conference on the
Hydraulic Transport of Solids in Pipes

HYDROTRANSPORT 10

Innsbruck, Austria: 29-31 October, 1986

PAPER F3

HYDRAULIC BEHAVIOUR OF ICE PARTICLES IN WATER
Some preliminary results related to the handling
of ice from ice-based heat pumps.

A. Sellgren

Department of Water Resources Engineering (WREL)
Luleå University of Technology, Sweden.

SYNOPSIS

Heat pumps based on the phase change from ice to
water require large quantities of ice being
transported hydraulically. Some preliminary
pipeline transport parameters have been evaluated
from pilot-scale experiments. Furthermore, the
dynamics of a large spherical ice particle was
observed and compared to the governing
hydrodynamic equation.

NOMENCLATURE

A   = projected area in the direction of flow

D   = diameter of pipe

U   = velocity of fluid

V   = volume of particle

C_A = added mass coefficient

C_D = drag coefficient

d   = diameter of sphere

d   = differential operator

f   = Darcy-Weisbach friction factor

g   = acceleration due to gravity

t   = time

u   = velocity of particle

d_{50} = average particle size

u_0 = initial velocity

Ø   = denotes "function of"

α   = coefficient

μ_0 = viscosity of water

ρ   = density of fluid

ρ_s = density of solids

Re = particle Reynolds number

Re_D = pipe Reynolds number

INTRODUCTION

Sweden has no domestic resources of oil, gas or
coal. Electric energy is obtained from hydropower
and nuclear power. Since the oil crises in the
1970s, new energy sources and technologies such as
natural and waste heat, solar energy, heat storage
systems and heat pumps have been developed and
introduced.

Tens of thousands of heat pumps have been
installed in houses in recent years. Heat
extracted from municipal waste water and natural
surface water is usually used in large district
heating units.

Recently, heat pumps of over 100 MW have been
installed at a municipal waste water plant in the
city of Stockholm. Furthermore, an installation
of over 10 MW has been taken into use for
extraction of the heat from water courses
adjoining the city. The heat generated will save
large quantities of oil in district heating
systems.

Most heat pumps are driven by electric motors and
generally the pumps consume electric energy
corresponding to about one third of the heat
energy generated.

The government energy policy, which includes the
closing down of all nuclear power plants in the
period up to 2010 and the present transfer from
oil to electricity in the Swedish energy system
means that it will be extremly important to use
electric energy effectively in the future.
Therefore, there is a need for development of heat
pumps which will be driven by other fuels than
electricity or oil.

The water in Swedish lakes, rivers and estuaries
contains enormous amounts of energy. The use of
this seasonal resource is partly limited by the
climatic conditions. When the water temperature
falls to the freezing point, the efficiency of
conventional heat pumps decreases. In order to
utilize the seasonal resources more effectively
storage systems have been developed, for example
in the ground (boreholes in rock, tubes in clay
and peat etc.).

Alternatively, development of heat pumps based on
the heat released at the phase change from water
to ice can be used in wintertime as a complement
to the conventional heat pumps which are used
during the summer.

Ice-based heat pumps

The energy released at the phase change from water
to ice is a source of great potential,
corresponding to an temperature increase of about
80 Centigrades for one m^3 of water.

An ice-based heat pump of 140 kW was installed and
tested at the Swedish State Power Board's
Alvkarleby Laboratory in central Sweden (Lindström
1982). Water at a temperature of about 0 Centi-

grades from a nearby river was spread over a plate heat exchanger where it froze. The ice layer was removed by a temporary increase of the temperature of the plates. The ice was then crushed and mixed with water and pumped back into the river. The energy-efficiency was demonstrated, but due to mechanical problems, the system was too unreliable to be used permanently in the heating system of the Laboratory.

A modified version of this ice-based heat pump has been installed for heat production to a group of houses in Sälen. The installation was based on a feasibility study (Persson 1981) and testing started in February 1986.

Svensson (1983) developed and tested an arrangement where drops of water rise in a liquid medium which has a lower freezing-point than water (Figure 1). During their passage through the medium (freon), the water drops form ice-crystals and heat is transferred to the liquid. This heat is then extracted in a conventional heat pump.

In the system in Figure 1 the liquid is circulated in a closed loop where the ice particles are continuously separated and pumped with surplus water as an ice-slurry to a recipient.

Svensson's experimental results confirm the principle, and drops of sizes of about 1-2 mm were obtained with a freon temperature of about -3 Centigrades. However, in these preliminary tests the losses of freon with the ice-water slurry were too large to be economically and environmentally acceptable. Therefore, further development and testing of the exchanger is needed.

An ice-based heat pump produces large quantities of ice, for example a 1 MW-unit generates about 200 tonnes of ice per day. The discharge system must be carefully designed in order to avoid formation of ice jams in the transport system or in the recipient (Figure 2).

Svensson (1983) suggested that the ice-slurry should be pumped and extrained in the recipient at a suitable depth from which the ice particles rise, melt and form water. In river stretches with no ice cover frazil ice is formed during cold periods. Discharge of large quantities of ice-slurry under these conditions promotes formation of flocks of slush which freeze together and form ice floes. Floating slush and ice floes can form a fragmented ice cover. An ice jam is created if the floating ice sweeps under the edge of the ice cover and is packed underneath it.

The handling of ice from ice-based heat pumps includes many problems which can be related to hydraulic transport and to the hydrodynamic behaviour of ice particles.

OBJECTIVES

The aim of this study has been to evaluate some hydraulic transport parameters through pilot-scale experiments in a open loop test facility. Another objective was to study the dynamics of ice particles theoretically and experimentally.

EXPERIMENTAL STUDY

The hydraulic characteristics of ice-slurry transport in pipelines were studied in a loop with a pipe diameter of 0.105 m and a total pipeline length of 25 m (Figure 3). The experiments were carried out with a crushed ice product with the particle size distribution shown in Figure 4.

The temperature of the water and the whole facility were held at approximately 0 Centigrades and the ice was succesively introduced into the pump through a separate pipe with a direct connection to a cone just under the opening of the return pipe. The flow rate was measured using a magnetic flow meter and by diverting the flow to a sampling tank. The ice concentration was estimated by control of the volumes of particles and water in circulation. The ice content was also determined from samples where the ice was screened off.

The lowest velocity observed was 0.7 m/s and this represents conditions where all the ice was in the upper half of the pipe section. The temperature of the pumped mixture was 0 to -1 Centigrades during the test period.

Dynamics of ice spheres

The dynamics of spherical ice particles were investigated experimentally when dropped from a height of about 0.8 m into still or upward flowing water in a transparent pipe of a diameter of about 0.1 m, Figure 5.

Ice spheres of diameters of about 38 mm were produced by controlled freezing of water-filled table tennis balls. The density was about 915 kg/m^3.

The motion of the ice sphere was recorded by a film camera operating at 18 frames/s. A ruler was placed at the side of the transparent pipe for measurements. The particle path was evaluated by subsequent frame-by-frame analysis of the film.

The upward flow of water in the pipe was fixed at 0 and 0.1 m/s. The motion was only evaluated in the inner part (0.75D) of the pipe where the local water velocity exceeded the average value. If the particle reached the wall region, a slower motion was observed and the test was cancelled.

With an entry velocity of 3.8 m/s the ice particle decelerated and reached a maximum depth of about 0.3 m in still fluid and 0.25 m in upgoing water flow at a velocity of 0.1 m/s. The maximum depth was reached after 0.4-0.6 s after which the sphere gradually accelerated upward until the particle reached a steady rising velocity. The ratio of the cross-section of the sphere to the cross- section of the pipe was not negligible, which influenced the relative velocity in the pipe. The steady rising velocities were reduced considerable.

DISCUSSION OF RESULTS

Hydraulic transport

The pipeline energy loss data were evaluated in terms of the Darcy-Weisbach parameters where the Reynolds number for the mixture was represented by the clear water value (Figure 6). The divergence of the slurry curve from the clear water values in Figure 6 can be related to increased separational tendencies of the components for lower velocities with ice concentrated in the upper half of the pipe section.

Dynamics of ice particles

The experimental data for the transient downward motion of a single ice sphere below a free water surface in a vertical pipe (Figure 6) were analysed theoretically by the classical one-dimensional hydrodynamic model where forces due to gravity, acceleration, and drag are balanced. The influence of the last two terms was related to by added mass and drag-coefficients denoted C_A and C_D. The equation can be expressed as follows with the direction of motions shown in Figure 7.

$$(\varrho_s + C_A \varrho) \, V \frac{du}{dt} = (\varrho_s - \varrho) V g - \varrho A \frac{1}{2} C_D |u-U| \, (u-U) \quad (1)$$

where

V = volume of particle

ϱ_s = density of solids

C_A = added mass coefficient

ϱ = density of fluid

u = velocity of particle

g = acceleration due to gravity

A = projected area in the direction of flow

C_D = drag coefficient

U = velocity of fluid

Inserting geometric factors

$$V = \frac{\pi d^3}{6}$$

$$A = \frac{\pi d^2}{4}$$

equation (8) reads:

$$(\varrho_s + C_A \varrho) \frac{du}{dt} = (\varrho_s - \varrho) g - \frac{3 \varrho C_D |u-U| \, (u-U)}{4d} \quad (2)$$

The last term in Equation (2) expresses the drag force or flow resistance, which includes pressure or form drag and friction or surface drag. The resistance is proportional to the square of the relative motion, and the drag coefficient is expressed as a function of the particle Reynolds number here defined as follows:

$$Re = \frac{|U-u| \, d\varrho}{\mu} \quad (3)$$

The drag coefficient C_D is a functional relationship of Re

$$C_D = \emptyset \, (Re) \quad (4)$$

With values of Re less than 10^5, the relationship between C_D and Re can be expressed empirically in formulas of the form:

$$C_D = A \, Re^\alpha \quad (5)$$

where A and α are constants, which can be empirically fitted to the standard curve for spheres. For low values of Re (<1) Stokes' law is applicable and A = 24 and α = 1.

In the model discussed here (Equation 1) the influence of acceleration history has been neglected. A further discussion of this subject is given by Sellgren (1983).

The axial movement of a particle in a circular tube is reduced because of the flow being restricted with a wall. The resulting increase in the drag coefficient was accounted for in the model, which means that other values of A and α were applied.

Clauss (1978) experimentally and theoretically investigated the settling velocity of ceramic spheres (density = 2500 kg/m^3) in an upward pipe flow of water (pipe diameter = 0.1 m). He related the increase in velocity in the annulus between sphere and pipe wall and the corresponding change in pressure to a relationship which expresses an increase in the standard drag coefficient. Clauss' results were applied to the motion of ice spheres studied here and it was found that the C_D-coefficient should theoretically be about 1.0. The small density difference between ice and water means that tendencies of lateral motion are easily initiated. Therefore, the large values of C_D observed here (1.25) may primarily be related to three-dimensional motion and particle rotation.

The classical added mass coefficient, C_A = 0.5, is theoretically applicable to irrotational motion only, a condition hardly fullfilled by the experimental results obtained here. Odar et al (1964) theoretically and experimentally investigated the motion of spheres which oscillated harmonically in an otherwise still fluid. Their experiments covered values of Re up to about 60. They found that the value of C_A varied with velocity and acceleration. If velocity was small compared to acceleration, the relationship was reduced to the classical value of 0.5. If their results are applied to the conditions investigated here, a nearly constant value of 1 for C_A is obtained.

Based on the previous discussion, the governing one-dimensional equation (1) was applied with C_A and C_D = 1. The results are compared to the experimental data in Figure 8.

It follows from the comparisions in Figure 8 that the dynamic behaviour of the ice particles was simulated with resonable accuracy. However, the spheres were strongly decelerated after the entrainment. During the descent of the sphere through the water, large cavities of air were sometimes entrained and tendencies toward lateral particle movement were observed. Therefore, it cannot be precluded that the agreement between theoretical and observed behaviour during the initial stage is partly explained by errors which are compensated for rather than relevant physical modelling.

CONCLUSIONS

The energy released at the phase change from water to ice is a source of great potential for the second generation heat pumps in cold regions. Ice handling includes many problems which can be related to hydraulic transportation and to the hydrodynamic behaviour of ice particles.

Preliminary results with pumping of an ice-slurry (d_{50} = 12 mm, D = 0.1 m) indicated that the energy losses were up to 50 % larger than the clear water losses for velocities of 1-1.5 m/s with ice concentrations of 13-15 % by volume. With larger velocities the losses agreed with the clear water values.

The dynamics of a spherical ice particle (d=38 mm) when dropped into still or upward flowing water in a vertical pipe was predicted theoretically by the governing one-dimensional hydrodynamic relationship where forces of gravity, acceleration and drag are balanced.

In the comparision with experimental results it was found that theoretically relevant coefficients of added mass (C_A = 1) and drag (C_D = 1) adequately simulated the transient motion of the ice particles. However, further investigations are needed in order to clarify the general usefulness of the model. The agreement between results with the model and observed behaviour may partly be explained by errors which are compensated for rather than relevant physical modelling.

ACKNOWLEDGEMENTS

This investigation was partly financially supported by The Swedish Council For Building Research.

REFERENCES

Clauss, G. (1978): Hydraulic Lifting in Deep-Sea Mining, Marine Mining, Vol. 1, No 3.

Lindström, H.O. (1982): Ice-Based Heat Pump at the Alvkarleby Laboratory, internal report. The Alvkarleby Laboratory (in Swedish).

Odar, F. and Hamilton, W.S. (1964): Forces on a Sphere Accelerating in a Viscous Fluid, J Fluid Mech. Vol 18.

Persson, S-E. (1981): Ice-Based Heat Pump For 43 Houses in Sälen, Report, R 49:1981, The Swedish Council For Building Research (in Swedish).

Sellgren, A. (1983): Unsteady Motion of Particles in Water Dynamic Modelling Based on Experiments With Ice and Silica Spheres, Research Report, Series A No 119, Dept. of Water Res. Eng. Luleå Univ. of Tech.

Svensson, T. (1983): A System for Heat Production With a Heat Pump Based on the Phase-Change from Water to Ice, Report, Div. of Hydraulics, Chalmers University of Technology (in Swedish).

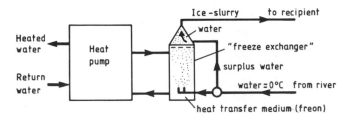

Fig. 1. *Conventional heat pump supplemented with 'freeze exchanger'. After Svensson (1983)*

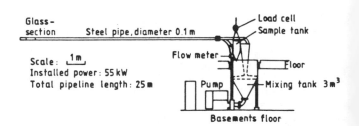

Fig. 3. *Open loop test facility—schematic layout.*

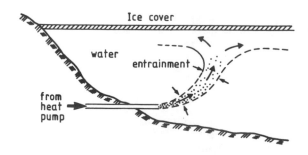

Fig. 2. *Schematic representation of the discharge of ice-slurry in a recipient.*

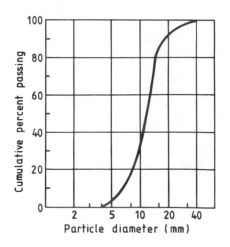

Fig. 4. *Particle size distribution of the crushed ice used in the pipeline test.*

217

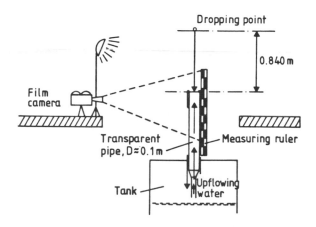

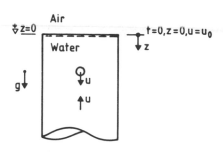

Fig. 7. Downward motion of a sphere in a fluid flowing upward in a pipe.

Fig. 5. Experimental set-up for observation of the transient motion of an ice sphere in a vertical pipe.

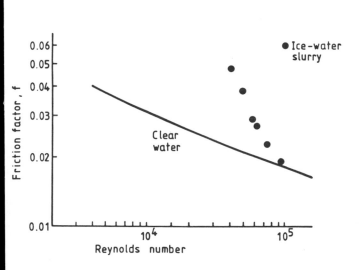

Fig. 6. Representation of experimental results with ice-slurry expressed in term of Darcy–Weisbach parameters. The ice concentration by volume was 13–15%.

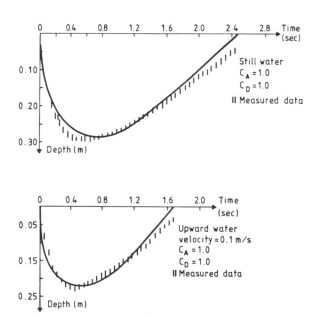

Fig. 8. Comparison of experimentally determined distance–time data for an ice sphere to theoretical simulations.

10th International Conference on the
Hydraulic Transport of Solids in Pipes

HYDROTRANSPORT 10

Innsbruck, Austria: 29-31 October, 1986

PAPER F4

EXPERIMENTS WITH COARSE PARTICLES IN A 250 mm PIPELINE

C.A. Shook, University of Saskatchewan, L. Geller, Energy Mines and Resources Canada, R.G. Gillies, W.H.W. Husband and M. Small, Saskatchewan Research Council.

SYNOPSIS

To broaden the data base for coarse-particle slurry flows, experiments were conducted with sands of mean diameter 0.29, 0.33, 0.55 and 2.4 mm at volumetric concentrations below 35%. Some variation of fluid viscosity was provided in the experiments.

The data were interpreted with a two-layer force balance model. By including the results of other experiments in this laboratory, an expression for the contact load fraction was inferred. This showed pipe diameter to be of greater importance than previous work has suggested.

The experiments suggest that the viscosity of the carrier fluid is of importance for such coarse, narrow distribution materials flowing at moderate concentrations. Viscosity modification with clay addition produced very complex behaviour.

INTRODUCTION

Prediction of pipeline headlosses for slurry flows continues to attract research effort because of its importance and its intractability. One of the most useful approaches has been that of Wilson and coworkers (1,2) whose original treatment of coarse- particle, high concentration flows has been extended to other heterogeneous flows. The simplifying assumption of their model: that the cross-section of the pipe could be visualized as two layers, each of uniform concentration, allows a rational analysis of the problem in terms of known physical laws.

It is worth noting that the form in which the model is to be used does have an effect on the way it is presented. Recognizing that designers (and many experimenters) are interested in the headloss-velocity-delivered concentration relationship, Wilson's algebraic solutions have allowed a simple and elegant graphical presentation of such predictions. These predictions use assumed value of parameters which unfortunately were rarely measured in experimental work.

A laboratory conducting experiments with indust-

rial slurries in large pipelines has many of the practical problems which affect the operator of a commercial pipeline. One of these difficulties is the fact that it is the in-situ concentration and not the delivered concentration which can be measured quickly and reliably. The latter is particularly difficult to obtain if the slurry contains coarse particles. In addition to being more easily measured, the in-situ concentration (and particularly its variation with height in the pipe) is of direct application in the model. Thus the approach taken in this laboratory differs in detail from that used by Wilson. Although the immediate goal has been to interpret experimental results, our version of the model can also be used to predict headlosses. At the expense of some algebraic elegance, we retain the in-situ concentration as an independent variable and treat the delivered concentration as a predicted quantity. A consequence of our approach is that headlosses reduce to solving a set of simultaneous equations which are easily programmed on a microcomputer. The simplifying assumptions and the empirical equations in the model can of course be altered systematically in searching for improvements.

Previous studies in this laboratory using coarse coal had established[3] that the model could be used for coals with broad particle size distributions. Particle degradation is always a problem in coarse coal experiments and Canadian coking coal is particularly friable so that it became necessary to use a non-degradable material (sand) in further work. Using a 263 mm I.D. pipeline, our first study of this type[4] established that the effect of the fines in a mixture was, _inter alia_, to increase the bouyant force on the coarse particles.

In planning the next phase of the investigation the existence of a number of coarse-particle investigations in other laboratories was recognized. Since there appeared to be a gap in the data base for the transition category between coarse-particle and fine-particle flows, this was the region which was selected. The experiments were planned to contribute to the development of the model and accordingly, a number of additional parameters had to be determined. These included the concentration of a settled particle bed, the kinematic coefficient of particle-wall friction using the method of Eyler et. al.[5], the pipe wall roughness, particle drag coefficients and the viscosity of the carrier fluid. The latter was to be varied in the experiments since it had been found to be important in previous (proprietary) studies with commercial slurries. The experiments were funded by the Canada Centre for Mineral and Energy Technology, Energy, Mines and Resources Canada. A complete record of the data is available[6].

THEORETICAL CONSIDERATIONS

Appendix A contains a complete statement of the model which we used and an explanation of how the equations are solved. For the hypothetical two-layer mixture shown in Figure 1, the model contains steady flow equations specifying:

   a) mass balances for fluid and solids and

b) forces balances for each layer.
The latter requires the stresses acting at S_1, S_2 and S_{12} to be evaluated. These depend upon the density of the (fluid + suspended solids) mixture and its effective viscosity. In the upper layer, this density is easily obtained from the concentration of suspended particles. In the lower layer the density (equation A11) reflects the fact that the sum of these three volume fractions (fluid + suspended solids + "contact load" solids) must be unity.

For particles of the size which were used in this investigation, viscometric determination of the viscosity-concentration relationship is not practical. Thus three empirical relationships had to be tested, and perhaps modified in the light of the experiments. These were:

1. the fraction of the total solids content (C_r, the in-situ mean concentration) which is transported as contact load (C_c);
2. the interfacial friction factor used in calculating the interfacial shear stress;
3. the viscosity-concentration relationship for the (fluid + suspended solids) mixture.This affects the boundary stress at the perimeter S_1.

Previous work had suggested initial values for these quantities but because somewhat different equations had been used, it was necessary to reconsider the original data. This feature of a multi-equation model should be emphasized: changing one of the equations will affect the inferences which are drawn from experimental results. The version of the model in Appendix A is the simplest one likely to be of practical value. Because other versions have been used in publications from this laboratory, we emphasize its distinctive features:

i) a constant concentration (C_2) in the lower layer;
ii) no slip between fluid and solids in the lower layer;
iii) the interfacial shear stress at S_{12} makes no direct contribution to the particle-wall stress at S_2. The stress at S_{12} is therefore "fluid-like" rather than "solids-like".

For the conditions of these experiments ii) is a reasonable representation of physical reality. The other two features represent a convenient simplification (i) and a reasonable hypothesis (ii).

EXPERIMENTAL PROCEDURE

The pipelining tests were carried out in a 250 mm diameter (nominal) pipeline loop which is shown schematically in Figure 2. A differential pressure transducer was used to measure headlosses over a 29 m long horizontal test section. The inside diameter of the test section was 263 mm. The test section was originally coated internally with epoxy resin to provide a constant (and low) pipe roughness factor. However, the epoxy coating had worn from the lower third of the test section exposing the carbon steel pipe. Frequent clear water headloss versus velocity measurements were necessary during the test program to obtain accurate estimates of the pipe roughness during each slurry test. These remained low, of the order of 5 μm.

A 190 kW electric motor was used to power the pump. A variable speed fluid drive provided flow control and the slurry flowrate was measured using a magnetic flowmeter.

A chilled ethylene glycol-water mixture was circulated through the annulus of a pipe-over-pipe heat exchanger to maintain a constant slurry temperature. Temperature was controlled within 1°C.

The slurry flow regime was determined by visual inspection through a transparent acrylic section of pipe and with a Snamprogetti detection probe[7]. The probe was flush-mounted on the pipe wall. The probe detects fluctuations in slurry electric resistance which result from the "macroturbulent" particle concentration fluctuations in the flow. These fluctuations increase in amplitude as the slurry velocity decreases and then disappear entirely if the probe becomes covered with a stationary deposit of particles.

A Hewlett-Packard 9825A computer and a NEFF 620 multiplexer system measured and recorded all sensor readings. Instrument calibrations were performed prior to the sand slurry tests. Pressure transducers were calibrated using U-tube manometers. The magnetic flowmeter was verified in comparison with a standard orifice plate using water.

A slurry test consisted of the following steps:

1. The pipeline loop was filled with water.
2. A pre-determined amount of sand was dropped into the pipeline through the feed tank. The pipeline was operated at 3 to 4 m/s during the addition of solids.
3. Headloss versus flowrate measurements were made starting at the highest flowrate. The flowrate was reduced step-wise until a stationary deposit was observed.
4. Samples for particle size analysis were removed at the pump discharge as soon as the headloss determinations were completed.
5. The concentration profiles were determined at two velocities using a gamma ray gauge. At the same time, particle velocities were determined[5] using the probe described previously[5].

Sampling

Samples were withdrawn isokinetically from the pump discharge section of the test loop to determine the particle size distribution of the solids in the delivered mixture. It was assumed that the solids are uniformly distributed over the cross-section of the pipe near the discharge of the pump. A 40 mm diameter sampler tube was used in the isokinetic sampling. The pipeline was operated at 4 m/s during sampling.

"Head" samples were obtained by removing portions of the solids as they were added to the loop. These were combined and analyzed to obtain the initial pipeline in-situ solids particle size distribution.

Concentration Measurement

Since the pipeline test loop was horizontal and recirculating, the in-situ solids concentration was independent of slurry velocity. A traversing gamma ray gauge was used to measure in-situ solids volume fractions in the 263 mm diameter test section.

Chord-average values of in-situ solids concentration were determined using a Cs-137 source and a

Ortec detection system. The beam was collimated to a height of 6 mm. Absorption coefficients were determined in separate calibration tests. The average in-situ solids fraction was determined by integrating the local solids fractions over the pipe cross-section.

$$C_r = \sum_{j=1}^{\eta} c_j L_j \bigg/ \sum_{j=1}^{\eta} L_j$$

where C_r and c_j are the average and local volume fractions and L_j is the chord length at vertical position j. The measurments were made at ten equally spaced vertical positions.

Delivered solids concentrations were obtained by integrating the product of local concentration (gamma ray gauge results) and local particle velocity. The latter were obtained with a probe located in a pipe section with an inside diameter of 254.5 mm. The pipe section could be rotated to allow sampling at any position. Figure 3 shows the locations of the sampling points used in the present study. By assuming that the flow is symmetrical about the vertical axis of the pipe, the average velocity could also be determined.

Solids Property Determinations

Experiments were conducted with washed sands of SiO_2 content greater than 99%. The clay added to the pipeline was sodium bentonite.

Particle size distributions were determined by sieving. Samples were wet screened using a sieve with 200 mesh (0.074 mm) openings to remove the fines portion. The material passing through the 200 mesh sieve was collected and dried to determine the weight of fines. The +200 mesh portion was dried and screened and using a series of Tyler sieves and a Ro-tap shaker.

Pycnometers were used to determine the density of the sand samples. Particle drag coefficients were determined from measurements of the terminal settling velocity, V_t of individual solids particles in a column of water.

Viscosity Determinations

The carrier viscosity was monitored using a Brookfield concentric cylinder viscometer. The minus 0.074 mm fraction of the slurry was considered to be the carrier for the viscosity tests. The slurry sample was withdrawn from the pipeline through the centre stream sampler and the plus 0.074 mm fraction was removed by screening. The water and the minus 0.074 mm fines were transferred to the viscometer and tested as soon as possible to avoid significant settling prior to the test. It was planned to study the carrier viscosity effects systematically by adding bentonite clay to the slurries.

The properties of dilute bentonite-water mixtures were established in preliminary tests. Additions were planned to produce Newtonian slurries of viscosity in the range 2-3 centipoise.

However, on addition to the pipeline the clay produced complex rhelogical behaviour. Carrier samples exhibited much higher apparent viscosities and shear history dependent properties. To avoid complicating the model analysis unnecessarily, a more straightforward method of varying the carrier viscosity was required. As an alternative to

adding clay, the carrier viscosity was changed by increasing the temperature of the slurry from the normal 15°C to 40°C.

Runs were also performed with clay added. Although the rheological properties of the clay-contained carrier were not quantified and the results were not used in the model development, the observations were very interesting.

RESULTS

Figures 4a and 4b show the initial solids size distributions for each series of slurry tests. The plotted distributions represent the initial in-situ particle size since head samples were used. The solids mean size did not change significantly during the tests. However, the pump discharge sample size distributions showed that re-circulation in the test loop generated some fines. These fines comprised less than 1% of the solids for the 0.29 and 0.55 mm sands but ranged between 1.8 and 6.1% for the 0.33 mm material. Unfortunately, pump discharge samples were obtained incorrectly for the 2.4 mm gravel tests. As a result, the amount of fines could not be quantified.

Figures 5 to 9 show slurry headloss as a function of solids concentration and slurry velocity. A stationary bed exists at the lowest velocity point for each slurry concentration. Generally, the measured headloss decreases with decreasing velocity and then increases at the stationary deposit point. The reason for this increased headloss is not entirely clear. With this comparatively large pipe, the thickness of the stationary deposit is so small that any increased flow resistance would be negligible. The possibility of an experimental artifact must be considered since flows in the vicinity of deposition are not steady and deposits may not form uniformly throughout the whole pipe. A deposit near the downstream pressure tapping would produce an apparent increase of headloss. Conversely, one would find the reverse if the first deposit formed upstream.

Deposition occurred between the two lowest points plotted on these diagrams. These velocities were generally in reasonable agreement with Wilson's nomogram[8]. For the coarser particles the presence of long, slowly moving dunes near deposition would probably require pipelines to operate somewhat above the deposition condition.

Figures 5 and 6 show the strong sensitivity of headloss to mean particle diameter in the range between 0.29 and 0.55 mm. The relevant variable to characterize a mixture which is not too broad (and which does not have a gap in its size distribution) appears to be its d_{50} value. This can be seen by comparing Figures 6 and 7. The latter mixture contained far more coarse particles but its headlosses were substantially lower than the narrow distribution whose headlosses are shown in Figure 6. Instead, headlosses were close to those measured for the 0.29 mm narrow distribution material. The latter contained virtually nothing closer than 1 mm while the former contained 20% of such material.

With a further increase in particle size, headlosses continued to increase, as shown in Figure 8. These headlosses are very high. In fact, they are so high that lower headlosses can be achieved by mixing a given quantity of this high headloss material with 0.2 mm sand. This is shown in Figure 9 and is further evidence of the buoyancy increase produced by the finer particles.

The effect of a fluid viscosity decrease, produced by an increase in temperature, was substantial at a d_{50} of 0.55 mm (Figure 10). It was small but significant (Figure 11) for the finer 0.29 mm sand. Figure 12 shows that it was insignificant for the broad distribution, which contained coarse solids, but whose d_{50} was close to that of Figure 11. Corresponding (but much larger) changes in behaviour were produced by adding very small quantities of clay. The 0.55 mm sand showed a drastic reduction in headloss but little change in deposition velocity (Figure 10). The finer sand showed a substantial reduction in deposition velocity (Figure 11) which ultimately produced lower headlosses. The broad size distribution mixture showed only small reductions in both headloss and deposition velocity. These effects of clay addition are consistent with the two layer model. For the 0.55 mm sand the contact load is normally high but it is drastically reduced by the clay. One cannot tell whether this is a yield stress effect or not. For the other sands contact loads are much lower and there is less scope for headloss reduction. Clay addition produced a more uniform concentration distribution in all cases. The fact that this altered the deposit velocity for the finest particles was interesting.

For the coarsest particles the effects of temperature change and clay addition were quite different from those observed with the other sands. When the temperature was raised, a significant increase in fines content occurred (Figure 13) indicating that this material had contained a number of aggregates which could survive transport at 15°C but which decomposed at 40°C. These fines were evidently very effective in reducing contact loads because the headlosses dropped substantially, in contrast with the earlier results. There was also a synergistic effect with the clay because, on addition of the latter, a further headloss reduction occurred. From a mechanical point of view, the causes of these effects must be subtle because the gamma ray absorption measurements showed only small differences in the concentration distributions in these three cases. Further investigations of the phenomenon are planned.

A practical conclusion from these experiments is that even materials such as sand and gravel can be affected profoundly by factors which have often not been quantified in experimental studies: operating temperatures and the presence of small quantities of clay and other fines. This increases the difficulty of generalizing experimental measurements, especially for large pipelines. It is by no means certain that the viscosity of the medium will provide a simple explanation of these effects. Samples of the fines-containing carrier fluid were time-dependent when tested viscometrically, but no such effect was observed in the loop tests.

MODELLING

To generalize these measurements, the two-layer model of Appendix A was employed. Because of the complications associated with clay addition, these results were not used. Similarly, the mixture containing two different particle sizes was considered to be of limited practical interest. The remaining data were interpreted by selecting three experimental points per run.

In considering the model, it was first discovered that for the conditions of these experiments (C_r less than 0.35 and comparatively large particles) no slurry viscosity effect could be detected. This rather surprising result is consistent with observations in another recent study.[9] However, the constraints on this observation are important: water slurries of sand particles with d_{50} ranging between 0.18 and 2.5 mm and C_r less than 0.35. For finer particles and/or higher concentrations, an effect similar in its consequences to an increase slurry of viscosity can be anticipated.

Using an iterative procedure, values of the ratio (C_c/C_r) were inferred from the experimental results. In this determination, the interfacial shear stress was calculated using the (Fanning) friction factor f_{12} defined by:

$$f_{12} = 2/[3.36 + 4 \log_{10} (D/d)]^2 \qquad (1)$$

The coefficient of kinematic friction in the sand and gravel runs was 0.50 and C_2 was taken as 0.6 on the basis of the concentration measurements at shutdown.

An empirical correlation which generalizes the contact load values was obtained from these experiments and other recent measurements in this laboratory. Because the effect of relative density is important, recent data obtained with coarse coal slurries were included. For the coal slurries solids finer than 0.074 mm were considered to be associated with the fluid in all cases, producing a small increase in viscosity (this increase had been measured) and density.

The correlation can be expressed in a variety of ways, depending upon the choice of dimensionless groups. Its form was chosen to reflect the fact that C_c/C_r should approach zero when the suspending power of the flow is high and unity at the other limit. In terms of the Archimedes number for particles in the carrier fluid, the correlation is:

$$(C_c/C_r) = \exp [-0.124 \ Ar^{-0.061} \ Frp^{-0.028}$$
$$(D/d)^{0.431} \ (S_s-1)^{-0.272}] \qquad (2)$$

Although subject to modification in the light of subsequent experiments, this expression does give a reasonable fit for data in a wide range of experimental conditions. This can be seen in the comparisons of Figures 14 and 15 where values of (C_c/C_r) from the experiments are compared with equation (2). This maximum value of Ar was 3×10^5.

The correlation shows a strong pipe diameter dependence which was not reflected in earlier expressions. On the other hand, fluid viscosity and mixture velocity (in Ar and Fr) seem to be less important than previously indicated. However, there is evidently a residual velocity effect not taken into account by the correlation since the data for any one set of experiments (when all other factors are fixed) show considerable spread in some cases. Thus a better correlation could perhaps be obtained from this data with a different function.

Delivered concentrations from the model were in reasonable agreement with those calculated from the particle velocities and concentrations, except for the coarsest particles. To improve the agreement, the interfacial friction factor f_{12} in the model should be increased. Since there was little data on which to base a modification, the change which we recommend should be regarded as tentative. We suggest that f_{12} from equation (1) should be multiplied by the factor $(1 + Y)$ where:

$$Y = 0 \qquad (d/D) < 0.0015 \qquad (3)$$
$$Y = 4 + 1.42 \log_{10} (d/D); \; 0.0015 < (d/D) < 0.15$$

Further experiments are planned to improve these methods for predicting C_c and f_{12}. At present, it is recommended that equations (2) and (3) be restricted to $Ar < 3 \times 10^5$.

It is worth emphasizing that in principle, the model can be used for all "settling" slurries within which some heterogeneity or concentration variation occurs. Slurries of this type cannot be tested viscometrically. If viscometry is possible, the well known methods for homogeneous slurries should be used.

CONCLUSIONS

1. A simple two-layer model describes flows of particles of intermediate size. The effect of pipe diameter seems to be much greater than previous correlations have indicated.

2. For solids with narrow size distributions, Wilson's nomogram gives reasonable estimates of deposit velocities in a 263 mm pipeline.

3. The effect of clay and other fines on pipeline performance can be profound, even at low concentrations.

4. Fluid viscosity appears to be the appropriate correlating parameter for coarse particles ($d_{50} > 0.18$ mm and above) at in-situ concentrations less than 35% by volume.

5. Further research to improve the model is still required.

ACKNOWLEDGEMENTS

This research was undertaken with the technical assistance of Messrs. J. Lychak, D. McEwen, P. Schergevitch and J. Wright.

REFERENCES

1. Wilson, K.C., Streat, M. and Bantin, R.A., "Slip Model Correlation of Dense Two Phase Flow", Proc. 2nd Int. Conf. on Hydr. Transport of Solids in Pipes (Hydrotransport 2), BHRA Fluid Engineering Cranfield, U.K., September 1972, Paper B1.

2. Wilson, K.C., "A Unified Physically Based Analysis of Solid Liquid Pipeline Flows", Hydrotransport 4, 1976, Paper A1.

3. Shook, C.A., Gillies, R., Husband, W.H.W. and Small, M., "Pipeline Flow of Coarse Coal Slurries", J. Pipelines, 1981, 1, 83-92.

4. Shook, C.A., Gillies, R., Husband, W.H.W. and Small, M., "Flow of Coarse and Fine Sand Slurries in Pipelines", J. Pipelines, 1982, 3, 13-21.

5. Gillies, R.G., Husband, W.H.W., Small, M. and Shook, C.A., "Some Experimental Methods for Coarse Coal Slurries", Hydrotransport 9, 1984, Paper A2.

6. Gillies, R.G., Husband, W.H.W., Small, M., "A Study of Conditions Arising in Horizontal Coarse Slurry Short Distance Pipelining Practice", Saskatchewan Research Council Report R-833-2-C-85.

7. Ercolani, D., Ferrini, F. and Arrigoni, V., "Electric and Thermic Probes for Measuring the Limit Deposit Velocity", Hydrotransport 6, 1979, Paper A3.

8. Wilson, K.C., "Deposition Limit Nomograms for Particles of Various Densities in Pipeline Flow", Hydrotransport 6, 1979, Paper A1.

9. Shook, C.A., "Experiments with Concentrated Slurries of Particle with Densities Near That of Carrier Fluid", Can. J. Chem. Eng. 1985, 63, 861-869.

10. Churchill, S.W., "Friction Factor Equation Spans All Fluid Flow Regimes", Chemical Engineering, 1977, 84(24), 1977, 91.

APPENDIX A

THE TWO-LAYER MODEL

The model employs mass balance and force balance relationships for the flow depicted schematically in Figure 1. The hypothetical horizontal interface separates the contact particle free upper layer from the lower layer of concentration C_2. The mass balance for the total slurry is:

$$A V = A_1 V_1 + A_2 V_2 \qquad A(1)$$

Force balances can be written for each layer and these are coupled through the stress on the interface. Assuming study flow:

$$i\rho_f g = (\tau_1 S_1 + \tau_{12} S_{12})/A_1 \qquad A(2)$$
$$i\rho_f g = (\tau_2 S_2 - \tau_{12} S_{12})/A_2 \qquad A(3)$$

The first step in using the model to calculate headlosses is to select a value for the lower layer concentration C_2. In this study, C_2 was set equal to 0.6 which is close to the experimentally determined value for concentration in a stationary deposit.

The next step is to obtain a value for C_c, the total contact load. The correlation given in equation (2) may be used. Once a value is determined for C_c, the suspended load C_1 can be calculated:

$$C_1 = C_r - C_c \qquad A(4)$$

Once C_1 and C_c are fixed, the area A_2 can be calculated:

$$A_2 = A C_c/(C_2 - C_1)$$

Knowing A_2, the angle β is determined iteratively:

$$A_2 = D^2 (\beta - \sin\beta \cos\beta)/4$$

From geometry:

$$A_1 = A - A_2$$
$$S_1 = D (\pi - \beta)$$

$$S_2 = D\beta$$

$$S_{12} = D\sin\beta$$

Equations 1, 2 and 3 are solved iteratively for the unknown variables V_1, V_2 and i. First, a value is selected for V_1. Selection of the correct value for V_1 will result in a headloss, i from equation 2 matching the headloss calculated by using equation 3. As a starting point for the iteration, let V_1 = V. V_2 is then calculated from equation 1:

$$V_2 = (A V - A_1 V_1)/A_2$$

The stresses are calculated as follows:

$$\tau_1 = f_1 V_1 | V_1 | \rho_1/2 \qquad \text{A(5)}$$

$$f_1 = f_1 [Re_1, \varepsilon/D] \qquad \text{A(6)}$$

Churchill's equation (10) may be used to calculate f_1:

$$f_1 = 2[(8/Re_1)^{12} + (a+b)^{-3/2}]^{1/12}$$

$$a = \left(2.457 \ln \frac{1}{\left[\frac{7}{Re_1}\right]^{0.9} + \frac{0.27}{D}}\right)^{16}$$

$$b = (37530/Re_1)^{16}$$

Also,

$$Re_1 = \rho_1 DV/\mu_1 \qquad \text{A(7)}$$

$$\rho_1 = \rho_s C_1 + \rho_f(1-C_1) \qquad \text{A(8)}$$

In principle, $\mu_1 = \mu_1[\mu_f, C_1]$ but it was found that $\mu_1 = \mu_f$ gave satisfactory results in this study. However, a concentration dependent viscosity function may be required for slurries with a very high solids fractions.

Stress τ_{12} depends on the particle diameter and the difference in velocity between the layers:

$$\tau_{12} = f_{12} (V_1 - V_2) | V_1 - V_2 | \rho_1/2 \qquad \text{A(9)}$$

The interfacial friction factor is given by equation 3. In mixtures, the particle diameter, d_m was chosen such that the weight fraction of the particles with diameter less than d_m would be C_1/C_r. For narrow distributions, d_m is the median particle diameter.

Stress τ_2 includes two effects: fluid-wall friction in the lower layer and particle-wall sliding friction. Neglecting transmission of stress τ_{12} through the lower layer and considering the effects of buoyancy produced by suspended particles in the lower layer:

$$\tau_2 S_2 = (f_2 V_2 |V_2| \rho_1 S_2)/2 + \qquad \text{A(10)}$$
$$\{(\rho_s-\rho_2) g D^2 \eta_s (C_2 - C_1)(\sin\beta - \beta\cos\beta)\}/2$$

The density of the suspended mixture in the lower layer is:

$$\rho_2 = [C_1\rho_s + (1-C_2)\rho_f]/(1 - C_2+C_1) \qquad \text{A(11)}$$

The friction factor, f_2 is approximated using f_1. The experimentally determined coefficients of kinematic friction were:

$$\eta_s = 0.4 \text{ for coal}$$

$$\eta_s = 0.5 \text{ for sand}$$

SYMBOLS

A	=	cross-sectional area of pipe
Ar	=	Archimedes number $4gd^3(\rho_s-\rho_f)\rho_f/3\mu_f^2$
c	=	local solids volume fraction
C_c	=	contact solids volume fraction
C_D	=	particle drag coefficient
C_r	=	average in-situ solids volume fraction
C_{vd}	=	solids volume fraction in delivered mixture
d	=	particle diameter
D	=	pipe internal diameter
f	=	fanning friction factor
F_{rp}	=	Froude number V^2/gd
g	=	gravitational acceleration
i	=	headloss gradient due to flow (m of water/m of pipe)
L	=	pipe chord length for gamma ray gauge scan
Re	=	Reynolds number
S	=	perimeter
S_s	=	density ratio (particle/fluid)
v	=	local particle velocity
V	=	average slurry velocity
V_p	=	average particle velocity
V_t	=	terminal settling velocity of solids
Y	=	correction factor for f_{12}
β	=	angular position in pipe defined in Figure 3
τ	=	pipe roughness
η	=	coefficient of kinematic friction
μ	=	dynamic viscosity
ρ	=	density
τ	=	shear stress

Subscripts

1	=	upper layer
2	=	lower layer
12	=	interface value
d	=	delivered

f = fluid

j = position in pipe cross-section

s = solid

r = in-situ
or
t

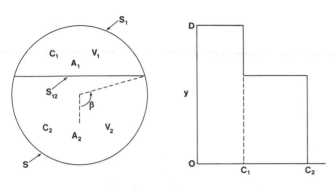

Fig. 1. Hypothetical two-layer model.

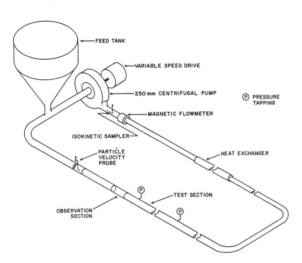

Fig. 2. Experimental system.

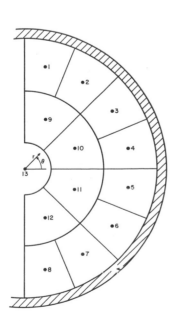

Fig. 3. Sampling positions for velocity measurements.

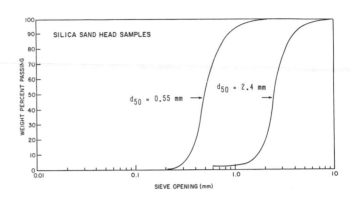

Fig. 4. Sand size distributions.

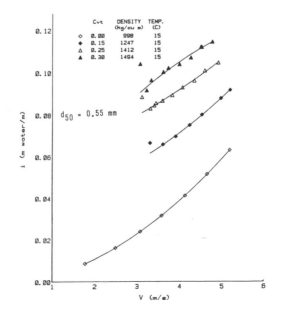

Fig. 5. Headlosses for 0.55 mm sand.

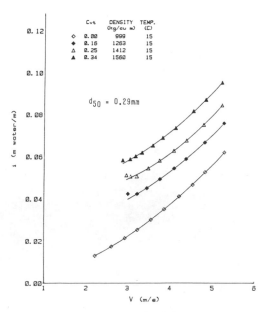

Fig. 6. Headlosses for 0.29 mm sand.

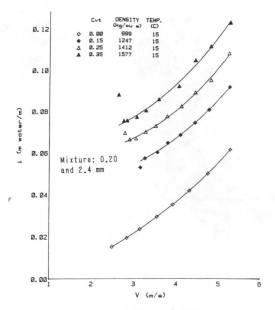

Fig. 9. Headlosses for 50:50 mixtures of 0.2 mm and 2.4 mm sand.

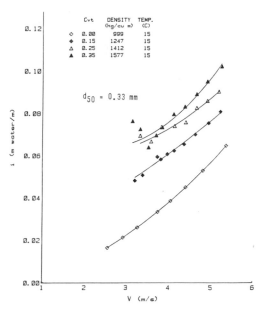

Fig. 7. Headlosses for 0.33 mm sand.

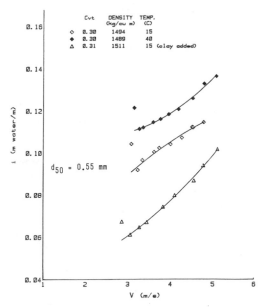

Fig. 10. Effect of temperature and clay addition for 0.55 mm sand.

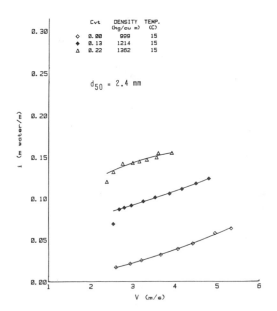

Fig. 8. Headlosses for 2.4 mm sand.

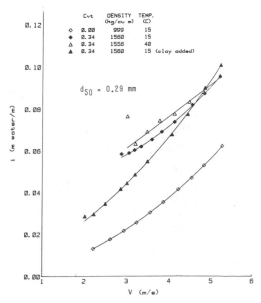

Fig. 11. Effect of temperature and clay addition for 0.29 mm sand.

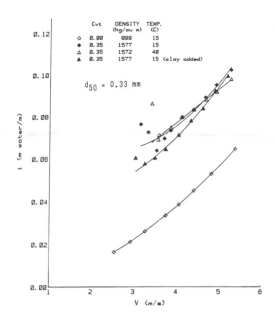

Fig. 12. *Effect of temperature and clay addition for 0.33 mm sand.*

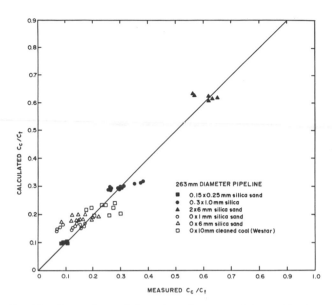

Fig. 14. *Comparison of C_c values from correlation with those inferred from headlosses.*

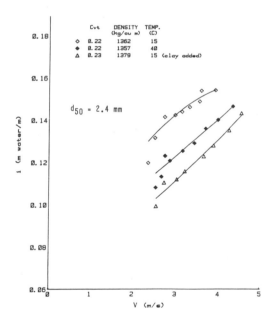

Fig. 13. *Effect of temperature and clay addition for 2.4 mm sand.*

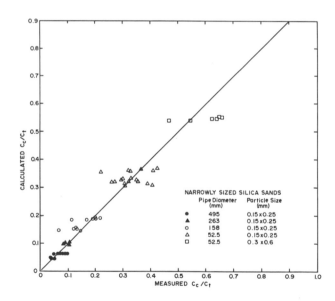

Fig. 15. *Comparison of C_c values from correlation with experimental results in various pipelines.*

10th International Conference on the
Hydraulic Transport of Solids in Pipes

HYDROTRANSPORT 10

Innsbruck, Austria: 29-31 October, 1986

PAPER G1

SOME SLURRY HANDLING PROBLEMS
IN THE CHEMICAL INDUSTRY

A. W. ETCHELLS
ENGINEERING SERVICE DIVISION
ENGINEERING DEPARTMENT
E. I. DU PONT DE NEMOURS CO.
WILMINGTON, DELAWARE U.S.A.

SUMMARY

The current status of slurry handling in the
chemical industry is discussed. Problems of
interest to this industry such as slip,
centrifugal pumping and the like are discussed.

PURPOSE

The purpose of this paper is to discuss the
current status of slurry handling in the chemical
industry from the viewpoint of an internal
consultant in a large chemical company in the
United States. By discussing what can be done
and what are problem areas it is hoped to direct
the attention of workers in the field to problems
of general usefulness whose solutions would have
wide applicblility.

BACKGROUND

The chemical industry's slurries include ores,
coal, pigments, minerals, metallic catalysts,
radioactive materials, organic polymers, and
intermediates, heavy and light organic solids,
continuous phases both water and organic, low and
high densities, solids that sink and solids that
float, thin water like slurries transporting lead
compounds and thick slurries that can be written
upon.

INDUSTRIAL GOAL

In process design the chief interest is in
transporting or storing a slurry. Often this
slurry is an intermediate in the process.
Formulating and detailed characterization of a
final product slurry is left to others. The
object is to transport and hold without plugging
or settling. The goal is trouble-free operation.
It is preferred that this be done at low cost.
However, with many of the high value-in-use
materials that the industry handles as slurries,
the cost of pumping is small compared to the
cost of not pumping due to pluggage. This is not
true of some other industries, but it is true in
many cases and influences design and the
requirements of design equations.

The things to be determined are, then, the
conditions, typically flow rate, that give
pluggage-free transport for a given
configuration of equipment and the cost in terms
of pressure drop to be overcome by the pumping
device.

These conditions must also be determined with the
minimum amount of information or basic data, such
as physical properties and only simple
observations or measurements. Pilot-scale tests
are expensive and often impractical because of
safety requirements.

SLURRY CHARACTERIZATION

Slurries can usually be divided into two groups,
fast and slow settling. A slow-settling slurry
is one that takes more than fifteen minutes to
separate. This may seem simplistic, but it
immediately determines a path to be followed for
further tests. Note that this test requires
direct observation of the slurry. With the
variety of slurries occurring in industry direct
observation of the slurry is required.
Concentration, particle size and composition are
just not enough.

Fast-settling slurries are associated with low
concentrations of big particles (10 microns or
larger). Slow settling slurries have higher
concentrations, fine particles and are almost
invariably non-Newtonian. The most common appear
to be yield stress materials, or very shear
thinning.

Basic data for fast-settling slurries are the
densities of liquid, usually correct, and solid,
often incorrect, but easily determined. The
particle size is useful, but emphasis must be on
the largest particle size. Elaborate size
distributions are not warranted. Viscosity of
the fluid is also important.

For non-Newtonian slurries, rheology is always
required to get sufficiently accurate estimates
of pressure drop. Unfortunately, the quick,
simple devices available at most plant sites are
totally inadequate. They tend to be low shear
rate devices which give falsely high apparent
viscosities which often mislead. Samples should
be sent to a laboratory where Couette rheology
can be run. It would be preferable for scale-up
purposes to run capillary or pipeline rheology.
However, such tests take many times longer than
the Couette tests and cannot show effects like
settling, shear thickening or hysteresis.

For slow-settling slurries a battery of tests can
be run in the lab to check for concentration
dependence, usually very strong, air entrainment,
hysteresis (thixotropy), dilatancy at high shear
rates and anything else peculiar. This is not
done for all fluids; often a visual inspection

can tell if a material is a troublemaker.

Many of our fluids can be characterized as Bingham Plastics with a yield stress and infinite shear viscosity. In some cases this is a simplification but is used it because it is conservative, fast, and easy to explain to others. Generally the infinite shear viscosity is low and the yield stress moderate (100-500 dynes/cm2). For discussion purposes the description in Table I has been useful.

DESIGN PROCEDURES

The type of slurry usually classifies the flow regime, fast-settling slurries always in turbulent flow to avoid settling and slow settling slurries in laminar flow, usually non-Newtonian. If a slow-settling slurry is handled at conditions that result in turbulence, it is then treated as a fast-settling slurry.

The following basic questions must be answered:

o   For a given pipe size, what is the minimum transport velocity and, therefore, flow?

o   What is the pressure drop at the required flow to size a pump?

o   Is this the optimum pipe size?

o   What are the design criteria for laying out the pipe?

MINIMUM TRANSPORT VELOCITY

For fast-settling slurries the minimum transport velocity in a horizontal pipe is a critical piece of information. For industrial size particles, under 1 cm , transport is usually in suspended but not uniform flow. Sliding beds are not used intentionally. To predict the minimum transport velocity for this condition equations of the form suggested by Durand (Ref.1) are used.

$$VM = F*(2*G*(S-1)D) .5$$

Where VM - minimum transport velocity

F - empirical Durand factor

G - gravitational acceleration

S - density ratio - solid to liquid

D - pipe diameter

The Durand factor depends on concentration and particle size. Charles (Ref. 2) showed the variation in F for a wide variety of materials. Typical range is .5 to 1.5. For conservative design 1.5 is often used. Where more accuracy is required, such as rating an existing system, then the method of Davies (Ref. 3) for estimating F is used. The effects of concentration and particle diameter are weak as proved by experiment and confirmed by Davies' and others' theories.

For slow-settling slurries the need for a minimum velocity is uncertain; typically a value of .3 meter/sec is used as a lower bound.

PRESSURE DROP

The purpose of pressure drop calculations are to size pumps or other transport devices. High accuracy is usually not required. Most of the correlations in the literature are very empirical and occasionally based on false premises. Many of them attempt to correct water pressure drops for the effect of solids (Ref. 4,5). A method based on a pseudo-homogeneous pressure drop seems more theoretically sound. An average density of the slurry and the viscosity of the continuous phase are used. This method has been used by Turian (Ref. 6), and at velocities higher than the transport velocity, the effect of solids over pseudo- homogeneous pressure drop is very small.

For non-Newtonian slurries the equations of motion apply exactly, given correct rheology (Ref. 7). Using conservative estimating techniques of a Bingham Plastic fluid based on Couette rheology, it is possible to successfully size pumps and piping systems. The basis of this method is not always theoretically sound.

The transition from laminar to turbulent flow is important. Experience shows that this occurs at velocities near 2 to 3 meters/sec and, therefore, can occur under normal pumping conditions.

SOME PROBLEMS OF INDUSTRIAL INTEREST

There are a number of areas of industrial interest that are problem areas. Some of the more interesting ones are listed.

o   Pressure drop equations for fast settling turbulent flow based on theory and experiment. The equations should incorporate slip and critical velocity concepts within the form.

o   Confirmation of the use of Couette rheometry to predict pipe pressure drop for non-Newtonian slurries. Couette rheometry is fast and reproducible and able to measure time effects. The easier to scale up capillary measurement is time-consuming and can be thrown into error by time effects such as thixotropy and settling.

o   Predictive equations for critical velocity and pressure drop for mixed solids sizes. Should the critical velocity be based on the largest particle? How should averages be made?

o   Predictive equations for large particle transport in non-Newtonian slurries. How does the rheology of a moving non-Newtonian fluid affect the settling rate of large particles? Is it possible with a material with a yield stress to carry unmoving in the core large chunks of heavier material? Will these solids settle out when the flow approaches turbulent?

o   Predictive equations for effect of slurries on centrifugal pumps, for both types of slurries. What is the effective viscosity for a non-Newtonian fluid in a pump turning at 1800 rpm? Centrifugal pumps in our experience are very forgiving of yield stress materials. Do heavy solids reduce the efficiency of pumps? Is the head generated by a slurry equivalent to that for a heavy single phase fluid? Is this theoretically sound?

These are only a few of the many topics of interest. Several have been picked for further discussion because they are important and are probably susceptible to theoretical attack and therefore of interest to academics. A few others will be expanded upon.

SLIP

It is well known that in horizontal and vertical fast settling slurry flow, the solids travel at a different in situ velocity then the liquid. In vertical flow the slip velocity is assumed to be the settling velocity. (Ref. 8) In horizontal flow the slip velocity is considerably greater than the settling velocity and therefore more important. If the slip velocity is large, then the ratio of in situ volume fraction to input is large. At high in situ volume fractions the line may plug. A factor of two for this ratio is not uncommon with large heavy solids. In addition most in-line concentration instruments measure in situ not input, concentration causing a significant error. The slip velocity is somehow related to the critical velocity but we are not familiar with a method for its prediction that is totally satisfying for horizontal flow. (Ref. 9,10)

PARTICLE DAMAGE

Often in the transport of slurries the energy for transport causes size reduction or damage of the particles. This is particularly true of organic crystals or flocculated materials. Often in transferring material from a hold tank to a centrifuge or filter, the material is so badly damaged that significant reductions of separation rates take place, often a factor of two or four. The chief culprits seem to be centrifugal pumps and valves. However, positive displacement pumps can cause significant damage under certain circumstances. What is needed is a way of evaluating on the lab scale the strength of the slurry and then characterize on the operating scale the various components that could cause damage. This involves estimating turbulence parameters or shear rates in various unusual geometries.

CENTRIFUGAL PUMPS

Such pumps are popular for handling slurries both fast and slow settling. Correlations exist for correcting pump performance for viscosity, and for high viscosity these corrections are significant. Are these relevant for highly shear thinning slurries? To calculate an effect the shear rate in the pump must be known. What is it? The insides of a pump may be highly turbulent, what then is the effective viscosity? If low shear viscosity is used, the pump manufacturers say the pump will not work but there is evidence to suggest that thick slurries behave as thin liquids in centrifugal pumps.

SUMMARY

There is enough information available to design relatively trouble free systems for the transport of solids in the chemical industry, for both fast and slow settling slurries. Several areas require more work to improve existing correlations and to develop new ones. Such correlations should be developed based on basic fluid dynamic principles combined with carefully taken experimental data to confirm the theory. This combination has produced design correlations that have been useful in the past. Purely theoretical relations are suspicious because the situation is often too idealized. Purely empirical correlations are suspicious because invariably they will be used outside the range of data that forms their basis. Design procedures that require a large body of knowledge of slurry esoteric properties have little help to the average process designer.

Only design equations based on theory confirmed by data can be extrapolated.

REFERENCES:

1. Durand, R. Minn. Int. Hydraulics Conv. Proc. p.89, Int. Assoc. for Hydraulic Research, 1953.

2. Charles, M. E., W. Parzonka, J. M. Kenchington Canadian J. Chem. Engr. 59, 291-296 (1981)

3. Davies, J. Univ. of Birmingham U.K. personal communication, to be published.

4. Govier, G. W. and Aziz, K. "The Flow of Complex Mixtures in Pipes" Van Nostrand N.Y. 1972 p.667 - 690

5. Turian, R. and A. R. Oroskar AIChE J 26, 550-558 (1980)

6. Turian, R. and V. D. Sirrakos submitted for publication.

7. Govier, op. cit. Chapter 5

8. Govier, op. cit. p.469 - 475

9. Viswanathan K. and P. Mani AIChE J 30, p.682-684 (1984)

10. Govier, op. cit. p.665 - 667

TABLE OF YIELD STRESSES FOR NON-NEWTONIAN SLURRIES

INFINITE SHEAR VISCOSITY ABOUT 100 MILLI-PASCAL SEC. OR LESS

DENSITY ABOUT 1000 TO 2000 KGM/M3

YIELD STRESS	DESCRIPTION (PASCAL-SEC OR DYNES/CM2/10)
less than 10	easy to pour, milky
10 to 20	thick, pours easy, thin milkshake, conventional liquid designs will work
30 to 40	thick, hard to pour, forms peaks, can write name on surface, difficult flow to pump suction
40 to 100	flows poorly, may need to shove into suction, will hang up in moderate line sizes under gravity
greater than 100	can build with it, must be moved by positive devices, will stay in inverted jar

10th International Conference on the
Hydraulic Transport of Solids in Pipes

HYDROTRANSPORT 10

Innsbruck, Austria: 29-31 October, 1986

PAPER G2

MATON ROCK PHOSPHATE CONCENTRATE SLURRY PIPELINE

S. Anand, S.K. Ghosh, S. Govindan and D.B. Nayar

Engineers India Limited, New Delhi, India.

SYNOPSIS

On behalf of Hindustan Zinc Limited, Engineers India Limited conducted a feasibility study in 1978 for transportation of 400 tonnes per day of rock phosphate concentrate from Maton mines to Debari, by pipeline and road modes. The pipeline mode was found to be substantially cheaper. Following this study, the pipeline system was implemented by Engineers India Limited in early 1983. This paper presents the salient aspects of this pipeline system.

INTRODUCTION

Hindustan Zinc Limited (HZL) own and operate a phosphoric acid plant (PAP) at Debari zinc smelter, located 15 km east of Udaipur in the State of Rajasthan. The PAP receives rock phosphate concentrate from nearby Maton mines, also owned and operated by HZL.

In 1977, Engineers India Limited (EIL) was assigned to conduct a techno-economic feasibility study for transportation of 400 tonnes per day (TPD) of rock phosphate concentrate from Maton mines to Debari by pipeline and road modes. Options to feed the concentrate to PAP in dry or wet form were also considered. The pipeline mode with wet feed to PAP was found to be substantially cheaper. Following this study, the pipeline system was successfully designed, engineered and commissioned by EIL in early 1983. Salient aspects of this system are described below:

EXPERIMENTAL WORK AND ANALYSIS

Designs for the feasibility study were based on bench and pilot scale tests conducted in 1977. These tests were carried out on a dry sample as was being produced at Maton mines. The bench scale programme included physical characterisation and settling behaviour of both solids and slurry (prepared using tap water) and determination of limiting gradient for pipeline. Pilot scale runs were conducted on 54 and 105 mm diameter loops to study pressure drop and suspension behaviour at various velocities and concentrations up to 57 % solids by weight (%Cw), corrosion-erosion characteristics and shutdown and restart behaviour.

Test work for finalisation of designs for implementation was carried out in 1980 on two samples: i)Sample A - dried concentrate as was being produced at Maton mines; and ii)Sample B- wet concentrate (at 55-65 %Cw, containing process water and reagents) from underflow of thickener at Maton mines as would be available for transportation by pipeline.

Test programme for Sample A included tests as carried out for preparing the study report plus rheological studies on Weissenberg rheogoniometer. Owing to limited availability of Sample B, only bench scale tests were conducted. This programme included physical characterisation and settling behaviour of both solids and slurry and rheological studies on Weissenberg rheogoniometer. Rheological data on Sample B is given in Figures 1 and 2.

Bench scale data on Sample B was analysed to select hydraulic parameters for pipeline design. Bench and pilot scale data on Sample A was used to validate the correlation models used for analysis of bench scale data on Sample B. Erosion-corrosion allowance and limiting pipeline gradient were based on data on Sample A.

DESIGN PARAMETERS

Based on the test work mentioned, the following design parameters were selected: i) particle size distribution : 100 % passing 52 BS mesh with 75-90 wt% below 300 BS mesh (Figure 3); ii) slurry concentration : 58-68 %Cw; iii)minimum operating velocity: 1.25 - 1.55 metre/second depending on slurry concentration; iv) erosion - corrosion allowance: 0.3 mm/year; and v) limiting pipeline gradient: 8 %.

A 73 mm outside diameter pipe with 7.14 mm wall thickness (including 4.5 mm erosion-corrosion allowance for 15 years) was selected after optimisation (Figure 4). The optimum slurry concentration for transportation through the selected pipe diameter was 65 %Cw.

SYSTEM DESCRIPTION

The rock phosphate concentrate slurry pipeline system (Figure 5) consists of a despatch terminal at Maton mines, the main pipeline from Maton to Debari and the receiving terminal at Debari. The system is briefly described below:

Despatch Terminal

Slurry from underflow of the existing thickener at Maton mines is transferred by centrifugal pumps(1+1) to a 360 cubic metre capacity slurry storage tank. A sampling probe downstream of the pump facilitates check on quality of feed. The slurry enters the tank through a safety screen and is kept in suspension by a rubberlined axial flow turbine type agitator. The agitator is driven by a 45 kW motor through a gearbox and a fluid coupling. Two suction nozzles have been provided for withdrawal of slurry from the tank. A third nozzle has also been provided at agitator impleller level to drain/flush the tank during emergencies. Low and high level alarms facilitate monitoring of level of slurry in the tank.

Slurry from the tank is withdrawn by centrifugal booster pumps(1+1) and discharged to the mainline pumps(1+1). The section between the booster pumps and the mainline pumps is instrumented to monitor slurry quality, suspension, pressure, flow, density and pipeline corrosion. A density control loop has also been provided for dilution of slurry in case its concentration exceeds 65 %Cw.

In case the slurry is off specification, provision has been made to recirculate it to the tank or to the thickener. This recirculating line - tapped downstream of the flow and density instruments - also serves as the test section. To facilitate testing, a differential pressure manometer and an observation chamber have been installed in the horizontal section of the recirculating line.

Mainline pumps are triple acting vertical plunger type. These are driven by 153 kW electric motors and develop a maximum discharge pressure of 14 MPa. Each pump has been provided with a variable speed fluid drive to cater to start up requirement, variation in throughput desired during operation and gradual changes in flow rate anticipated due to pipeline wear. A speed control loop linked to the flow and density instruments through a throughput controller has also been provided to maintain a constant solids throughput.

Shutdown interlocks have been provided for abnormal flow, pressure and density conditions. In addition, shear relief valves have been provided on the discharge of the mainline pumps for over pressure protection.

For slurry piping at terminals, 45⁰ reinforced laterals, 3 D bends and hardened lubricated taper plug valves have been used.

Inhibitor dosing is recommended during line flushing only. An agitated inhibitor tank and a dosing pump have been provided for the purpose.

Main Pipeline

The pipeline is 10.5 km in length with an outer diameter of 73 mm. It generally has a flat profile. Elevation of Debari terminal is about 25 m lower than of Maton mines.

Linepipe conforms to API 5L Gr. B and is of seamless construction. The pipeline has been designed to ANSI B 31.4 by taking into account the maximum operating pressure, hydraulic surge and set point of the shear relief valve of the mainline pump. A shear relief valve has also been provided upstream of the block valve at Debari terminal to protect the pipeline against accidental closure of the block valve.

No slack flow is involved during pipeline operation.

Pig launching and receiving traps have been provided at the terminals for periodic cleaning of the pipeline.

Coated with coal tar enamel, the pipeline is buried underground and is cathodically protected.

Receiving Terminal

Slurry from the mainline is received in a 360 cubic metre capacity agitated slurry storage tank, similar to that at Maton mines, through a pipe header provided with station block valves, corrosion probe and spools, densitymeter, flowmeter and a pressure transmitter. Alarms are provided for abnormal conditions and corrective actions are taken after consultation with the despatch station on the VHF communication system. Off specification slurry is discharged into a dump pond of 100 cubic metre capacity.

Centrifugal pumps(1+1) transfer the slurry from the above tank to the feeding facilities of the phosphoric acid plant.

Communication

Communication between the terminals is maintained through portable, battery powered, VHF trans-receiver sets, one at each terminal. A third set is mounted on a patrol vehicle used by the maintenance crew.

Operation and Control

System operation is primarily manual. Suitable instruments have been provided at either terminal to monitor the system performance. Abnormal conditions are indicated by alarms at the respective control panels. Shutdown interlocks have been provided at Maton for density, flow and pressure variations beyond acceptable limits. In addition two loops, one each for density and throughput control have been provided at Maton.

SYSTEM PERFORMANCE

The mainline pumps had some start up difficulties due to the presence of oversized particles in slurry. Remedial action was taken by installing strainers in the suction lines to the booster pumps.

The system was successfully commissioned in March 1983 and has since been operating satisfactorily, close to the design values specified.

ACKNOWLEDGEMENT

The authors are grateful to Managements of HZL and of EIL for according their approval to present this paper.

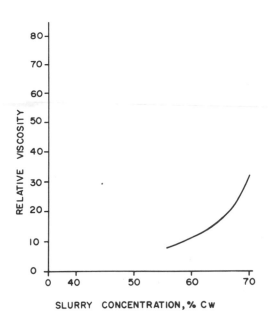

Fig. 1. Relative viscosity of slurry (Sample B).

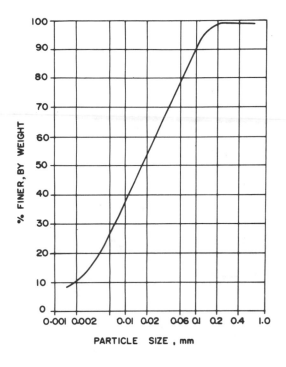

Fig. 3. Particle size distribution of solids.

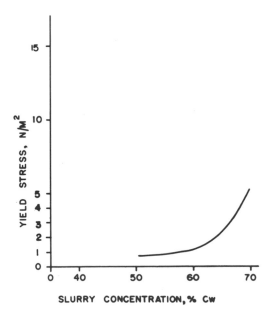

Fig. 2. Yield stress of slurry (Sample B).

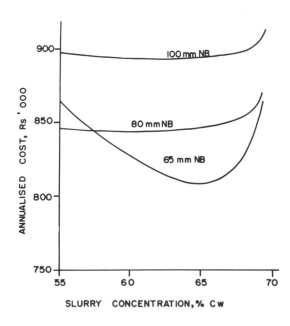

Fig. 4. Optimisation for selection of pipeline diameter and slurry concentration.

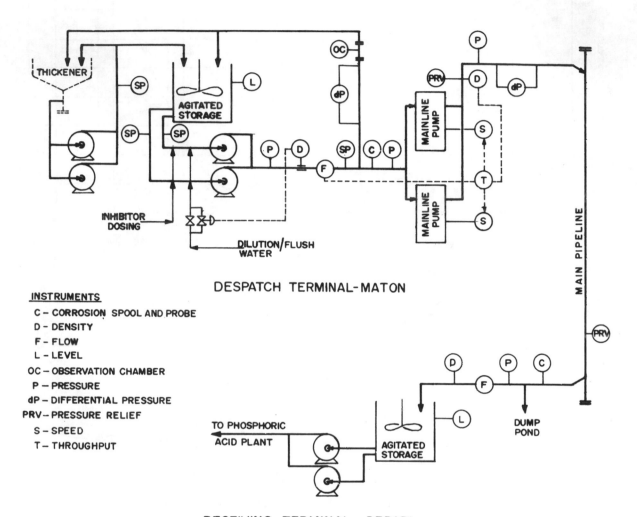

INSTRUMENTS

C – CORROSION SPOOL AND PROBE
D – DENSITY
F – FLOW
L – LEVEL
OC – OBSERVATION CHAMBER
P – PRESSURE
dP – DIFFERENTIAL PRESSURE
PRV – PRESSURE RELIEF
S – SPEED
T – THROUGHPUT

DESPATCH TERMINAL–MATON

RECEIVING TERMINAL – DEBARI

Fig. 5. System flow diagram.

10th International Conference on the
Hydraulic Transport of Solids in Pipes

HYDROTRANSPORT 10

Innsbruck, Austria: 29-31 October, 1986

PAPER G3

THE NEW ZEALAND STEEL IRONSAND SLURRY PIPELINE

P B Venton

Principal Engineer, Slurry Systems Pty Limited,
Sydney Australia

N T Cowper

Managing Director, Slurry Systems Pty Limited,
Sydney Australia

SYNOPSIS

The New Zealand Steel Ironsand Slurry pipeline
system was commissioned in January 1986. The
project is the first high pressure pipeline in
the world to use positive displacement pumps and
a buried, polyurethane lined pipeline to
transport abrasive granular solids.

The design, development, and technical aspects of
the slurry pipeline operation of the pipeline are
discussed.

INTRODUCTION

New Zealand Steel (NZS) Limited manufactures
steel from titano-magnetite ironsand ore, mined
and concentrated at its Waikato North Head open
cut mine, approximately 80 km south of
Auckland, New Zealand. Concentrate is transported
20 km north for processing into semi-finished and
finished steel at the Glenbrook (South Auckland)
Plant.

Primary ironsand concentrate is 55-60% Fe, and is
processed to iron by direct reduction with coal
in rotary kilns, followed by melting in electric
furnaces. Further processes convert the iron to
steel, and recover byproduct vanadium from the
slag. The process was developed for ground and
pelletised concentrate, but experience during
initial operation showed that the process
operated more efficiently using granular
ironsand.

An expansion of the Steel Mill was proposed in
the late 1970's and a separate Company, NZS
Development (NZSD) Limited, was formed to
investigate and implement the expansion.

In October 1981, approval was given for NZS to
enter into a five fold expansion program to
increase production of semi-finished steel from
150 000 to 770 000 tonnes per year. This required
NZS Mining Division increase its annual
production of ironsand concentrate to 1.5 M
tonnes from 10 M tonnes of ironsand deposit mined
annually.

Prior to the expansion, ironsand (240 000 t/a)
was transported by road. As part of NZSD's pre-
commitment investigations, a detailed study of
transport options was made in 1979. Options
studied included:-

Aerial Ropeway
Barge
Conveyors
Pneumatic Capsule
Railway
Road
Slurry Pipeline

The conclusion from this study was to proceed
with the project using rail transport, which
offered significant cost advantages to NZS, and
was a fully proven technology. A buried pipeline
was considered the most environmentally
acceptable option, but the conventional
centrifugal pumping technology and the forecast
replacement life of the pipeline made this option
unacceptable (the estimated pipeline cost was 1.7
times the rail cost).

The land between the Mine and the Steel Mill is
intensively used for horticultural and dairy
production, and significant opposition to the
rail construction developed from landowners
throughout the region.

In July 1982, the authors' company was visited by
a representative of the Concerned Landowners
Association to search for a slurry pipeline
option which could satisfy the technical concerns
of NZS and NZSD, and compete with the rail
transport cost. From this meeting the proposal
which subsequently won the international tender
for a slurry pipeline system developed.

In October 1983, a contract for the turnkey
project was awarded, subject to completion of
sufficient evaluation, testing and demonstration
to satisfy NZS that the design and the equipment
selected, complied with the design parameters,
and offered the best long term solution.

DESIGN PARAMETERS

NZSD required the slurry pipeline system meet the
following design parameters:-

Throughput (annual)	1.5M t/a (5% moist basis)
Throughput (weekly)	30 000 t
Catch-up Capability	30 000 t/month
Weekly Operation	120 h (incl. maintenance)
Design Life	25 year
Noise Emission (plant)	80 dBA at 3 m

In addition, the design must:-

Minimise environmental impact
Be technically Proven
Offer competitive transport costs

IRONSAND PROPERTIES

Ironsand concentrate has the following typical properties:-

Specific Gravity   4.76
Mean Particle Size   120 micron (94% between 212 and 75 micron)

During transport, up to 1% of the ironsand solid is released as clay slimes.

PROJECT DEVELOPMENT

Earlier experience with the Waipipi ironsand shiploading projects convinced the designers that positive displacement pumps and lined pipe were essential to satisfy the design requirements.

Positive displacement have three advantages over centrifugal pumps:-

- High head capability (to 14000 kPa) (minimising the number of pump stations required)

- Pump discharge is independent of discharge head (ensuring that transport velocity is maintained independent of operating conditions)

- Pump efficiency is high (minimising installed and operating power)

Lined pipe offers the potential to provide a pipeline capable of satisfying the design life without maintenance or replacement, allowing the pipeline to be buried as a conventional high pressure slurry pipeline.

Two system designs were considered:-

- A single pump station/250 NPS system operating at 30% solids concentration and a maximum pressure of 14 000 kPa

- A two pump station/200 NPS system operating at 50% solids and a maximum pressure of 10 000 kPa

After award, a study group comprising NZSD, NZS, the Contractor and the Design Engineer considered these designs and all other possible slurry pipeline, pipe material and pump options, for the 25 year project life. This extensive study concluded the two positive displacement pump station/200 NPS/50% concentration option, using polyurethane lined pipe offered the most economic solution. However three technical concerns remained with NZSD:-

- Were P D pumps capable of reliably pumping high specific gravity coarse granular material, without excessive maintenance or Operator attention

- Was polyurethane lined pipe capable of withstanding 25 years of operation at high and cyclic operating pressures

- NZSD preferred the security of a welded pipeline joint, which required development for polyurethane lined pipes

Pump Testing

Tests were conducted by potential pump suppliers in Holland and Texas (USA) to determine whether PD pumps could pump ironsand. The concerns were:-

- Would the rapid settling ironsand (25 mm/s mean settling velocity) created an operational or a wear problem when superimposed on the velocity cycle of a PD pump

- Could a pump be restarted after shutdown with ironsand slurry at normal concentration

- Would suction lines plug

- Would standard slurry valves perform reliably at the service pressure

A model of each proposed pump constructed with perspex viewing covers was tested. Figure 1 shows the fluid end used in the diaphragm and the flushed plunger pump, and indicates the areas of concern through the stroke. Each test showed there was sufficient turbulence during the stroke prevent deposition, although settlement was evident during the low velocity portions of the cycle.

Further tests showed there was no problem restarting a pump after shutdown with slurry at normal concentrations. The pump cylinder on the suction stroke progressively fluidised the settled ironsand, while ironsand in the cylinder on the discharge stroke was displaced sufficiently for the stroke, without evidence of jamming. The diaphragm pump was assisted by the flexibility of the diaphragm, while the flush water used in the plunger pump partially fluidised the settled solids ahead of the plunger.

Model pumps were operated through a range of speeds to assess whether suction and discharge pipes could be plugged. These tests showed that when the flowing velocity drops below the deposit velocity, ironsand deposits until the superficial velocity equals the deposit velocity (Figure 2). Immediately the velocity is increased the settled bed erodes to maintain the superficial velocity at the deposit velocity until full suspension is obtained. The pipes could not be plugged provided the pressure drop over the settled bed was lower than the available pressure.

Detailed study (and correlation with other experience) confirmed the maintenance requirements offered by suppliers were realistic.

Polyurethane Lined Pipe

The pipe problem was more difficult. Although polyurethane lining has been used in slurry pipelines since the 1970's, there was little published history, and certainly no history of service in long term high pressure ironsand pipelines. The designers were aware of negative opinions in some mining operations which were traced to costly failures resulting from poor application, rather than polyurethane failure in service.

Data obtained from rubber lined pipe in similar ironsand service for 9 years showed only minor wear. This experience, correlated with the limited information available for polyurethane pipe linings in the UK, USA and Australia provided confidence in polyurethane lining. In addition:-

-   4-6 mm of polyurethane was considered necessary for the 25 year service life (9.5 mm thickness was used).

-   The polyurethane must be a polyether. Wear data and hydrolytic stability data for toluene di-isocyanate (TDI) and methylene di-isocyanate (MDI) pre-polymers showed no clear advantage either type in the service conditions. The MDI pre-polymers offer an advantage in production application because the mix ratio is less critical than with TDI materials and they are processed at lower temperatures, minimising the potential for degradation on storage.

-   The highest level of applicator competence and quality control was essential if the potential for polyurethane failure was to be avoided.

Accelerated tests were performed by the DuPont polyurethane laboratory to evaluate the performance of the lining and the bond line under cyclic pressures from 10 000 kPa to full vacuum. These tests showed there was no deterioration evident after 7500 cycles (the pipeline will go through 1500-2000 pressure cycles throughout its 25 year life), supporting other data that the lining selected was capable of achieving the design life.

Pipe Jointing

There was a strong preference for a welded pipeline over a mechanically jointed (flange or high pressure victaulic) jointed line, because of the potential for failure in any of the 1500 joints required.

Unlined steel pipe could be welded, but the expected wear rate of 1 mm/a (erosion and corrosion, based on data from NZS Taharoa shiploading pipeline) required a pipeline 25-30 mm thick to provide the design life. World experience with high pressure polyurethane lined pipe was at that time limited, and welded jointing was not practical because of damage to the lining.

The Contractor developed (and has a provisional patent on) a coupling which allows welded pipe joints rated to ANSI Class 600 (Figure 3). Extensive testing was performed in New Zealand and the USA to prove the design could be welded without damage to the polyurethane, and without disbondment in a production welding environment, using stovepipe welding techniques. Continuous cooling is required from the time the root weld is completed until the final weld pass is completed to maintain the temperature at the

polyurethane/steel bondline below 100 deg C (the temperature limit for disbondment). The weld quality was assured by radiographic examination, and the bond quality by a continuous temperature record made at two points on each weld.

In July 1984 NZSD authorised the Contractor to proceed with the project.

PIPELINE SYSTEM DESIGN

Route

The pipeline route shown in Figure 4 is across undulating farmland and along country roads. The pipeline crosses approximately 6 km of reclaimed swamp where the water table lies to within 0.6 m of ground level. This region is crossed with numerous drainage ditches which are machine cleaned annually. The minimum pipeline cover in this region is 0.75 m with 1.0 m cover provided below the invert of drains.

The pipeline travels across undulating country to pump station 2 (kmP 9.2). At chainage 12.0 km the pipeline enters the road easement and is constructed approximately 1 m from the tar seal at a minimum depth of 1 m.

The pipeline is constructed at a maximum slope of 15% uphill in the direction of flow, and 12% downhill, to limit accumulation at low points on shutdown.

Pipeline Hydraulics

The particle size analysis and specific gravity of the ironsand dictate a high energy heterogeneous slurry pipeline.

Care was required to select an operating velocity sufficiently above the deposit velocity to provide safe operation with minimum hold-up, while not penalising the operating cost.

These parameters were restricted by the available pipe sizes, the weekly operating cycle, and system operation as the pipe lining wears. The final design is summarised in Table 1.

The projected operating hydraulic gradient was calculated from the known solids properties.

Full advantage was taken of the smooth finish provided by centrifugally cast polyurethane lined pipe in determining the design hydraulic gradient.

TABLE 1.

Pipeline System Design

Minimum Operating Velocity	3.6	m/s
Design Operating Velocity	3.9	m/s
Concentration	48.5	wt % solids (dry basis)
Pipeline Inside Diameter	187.4	mm
Design Hydraulic Gradient	1024	Pa/m
Pipeline Length Section 1	9200	m
Section 2	8800	m
Maximum Operating Pressure	ANSI Class 600	

SYSTEM DESIGN

A schematic of the Slurry pipeline system is presented in Figure 5.

The slurry pipeline system accepts ironsand either direct from the Concentration Plant product cyclones, or reclaimed from the plant stockpile (or both), and delivers the ironsand to a belt filter dewatering plant or an emergency cyclone at the pipeline Terminal.

Slurry Concentration Control

One of the benefits of positive displacement pumps is the guarantee of constant volumetric throughput. This allows the slurry concentration entering the pipeline to be controlled by controlling the ironsand mass flow into the pipeline feed tank. Ironsand is delivered at a controlled rate over a safety screen to the constant level pipeline feed tank. The ironsand settles to the tank outlet, and is delivered to the pump at the correct concentration.

The pipeline inlet density meter provides a safety backup to the mass flow control, and will over-ride the mass flow control if pipeline density increases beyond a safe value.

This control system has proven simple and reliable. It also enables the pipeline operation to be switched between "feed" to "flush" by simply starting and stopping the ironsand feed. No valves or flush pumps are required.

Pump Station Design

The pump stations are designed for maximum simplicity. The Terminal elevation is approximately 10 m above pump station 1, allowing the system to be operated without block valves. Rising stem gate valves fitted with replaceable liners are provided at each pump station to allow isolation for maintenance only. Similarly flush and drain valves are provided to clean the pump fluid ends for maintenance.

Each pump set is fitted with a scoop controlled fluid coupling and sufficient cooling capacity to dissipate 90% of the drive power to provide capacity to restart the pipeline should it become overloaded through poor operation or instrument failure. Pump Station No 1 is driven at 100% speed and Pump Station No 2 is driven at controlled speed (through the fluid coupling) to maintain an adequate suction pressure.

The two pumps at each station are driven through a common drive train. This provides a capital cost saving, but more importantly, allows the pump crankshafts to be aligned for minimum pulsation, simplifying the problem of pulsation control.

Terminal

Slurry is delivered to a Dewatering facility at the Steel Mill. Normal deliveries are to a belt filter dewatering plant provided by NZSD, and emergency discharge through a cyclone is provided to allow the pipeline to continue operation should the belt filter plant fail. The pipeline discharges into a feed tank and slurry is pumped to primary dewatering cyclones above the belt filter. These are operated at 35% concentration

and are used to separate ironsand and the clay slimes liberated during pumping (effectively upgrading the ironsand product).

Overflow water is clarified to a high standard prior to disposal and the reclaimed slimes are further dewatered prior to disposal with other refuse from the steel mill.

Control System

One of the attractions of slurry pipeline transportation is its suitability for automation, and integration with other processes controlled from a central control room.

The pipeline system is fully automated, and operated by the Mine Site Concentration Plant Operator. Pump Station No 2 is unattended, except for an inspection each shift by an attendant.

A complete programmable logic controller (PLC) based control system was provided with the pipeline. A master PLC located at Pump Station No 1 controls the operation of the pipeline and ironsand reclaim systems. Pump Station No 2 has an independent PLC which is programmed to safely control the operations of the pump station irrespective of the state of the communication link. The Dewatering Plant also has an independent PLC, and is controlled from the Iron Plant Control Room at the Steel Mill by the duty Operator.

The PLC's communicate through leased telephone lines. Each PLC is responsible for operation of equipment under its direct control, and will, if necessary for safe operation, trip its plant. It immediately signals the Master PLC at Pump Station No 1 which initiates the necessary control action for the remainder of the system, and informs the Operator.

The Pipeline Control System is interfaced with facilities installed by NZS at the respective control rooms. At the Minesite, the plant is operated from a Modicon "Modvue" touch screen interface, and the process is controlled through a Taylor process control computer. This equipment is linked to the pipeline master PLC (and to the mine PLC's) through communications network provided with the PLC's. At the Terminal, the Dewatering Plant is operated from a console, or from the Iron Plant Taylor Process Control Computer.

The Pipeline operating display contains two touch buttons, one to start and stop the pipeline system, and one to start and stop the ironsand feed. These are the only buttons necessary to operate the pipeline.

Alarm conditions associated with the system operation are displayed on the Modvue monitor.

The Pipeline Process monitor displays each analog signal in the process (Pump Station 1, 2 and the Terminal) and allows the Operator to select the ironsand feed rate and the Pump Station No 1 speed. Other control loops and limits to the process (such as maximum pipeline density, high discharge pressure control, and pump acceleration control) are contained in the PLC program and are not Operator adjustable.

Alarm conditions associated with the process are

displayed on this monitor.

PIPELINE SYSTEM OPERATION

The slurry pipeline system was commissioned in January 1986. Owing to the reduced demand for ironsand while the expanded steel mill is commissioned, only 90 000 t of ironsand (300 h) of operating experience is available at the time this paper was prepared.

The pipeline system was commissioned with water to evaluate and test the control system performance in a safe condition. Ironsand commissioning required approximately two weeks, after which time a 60 000 t (200 hour) performance test was undertaken to determine the unit energy consumption and the throughput rate for contract guarantees.

Initial system performance details are:-

Ironsand Transport

The ironsand is transported through the pipeline in the heterogeneous flow regime. Material is transported by drag forces exerted by the vehicle fluid, implying a velocity gradient between the fluid and the particles being transported. Little data is available to allow accurate prediction of the extent of this phenomena and its effect on the pressure gradient in the pipeline. Because this is a long pipeline operating at a constant velocity, it provides an excellent opportunity to develop a better understanding of the phenomena.

The theoretical pipeline transit time is 76 minutes. It requires 93 minutes for the ironsand to arrive at the terminal, and approximately 98 minutes for ironsand to be flushed from the pipeline.

At the normal operating concentration, the solids travel at a velocity of approximately 3.3 m/s, when the theoretical velocity is 4.0 m/s. For this to occur, the insitu concentration must increase from a theoretical of 48.5% to an actual concentration of 54%.

Measurements from the Terminal density meter confirm the calculated concentration.

Figure 6 illustrates the tail-out which occurs when the pipeline is flushed. There is no clear slurry/water interface, rather a gradual reduction in concentration as solids are flushed from the line.

Pipeline Pressure Gradient

Figure 7 presents design predictions for the pipeline pressure gradient, together with various data points measured during early operation. Data is for normal velocity operation.

The measured pressure gradient is within 3 % of the design value.

Energy Consumption

One of the requirements of a turnkey project is a performance guarantee. This project included a guarantee of throughput, and unit energy consumption. After Commissioning, the pipeline

system was operated to transport 60 000 t during a two week performance test. This demonstrated:-

- The pipeline system operates at the design throughput, at the design pressure gradient.

- The unit energy consumption is 7.82 kWh/t. This is 10 % below the guarantee.

- The pipeline system was designed with a 15% margin on pressure. The pipeline can be safely operated at 106-109% of design throughput, using some of the pressure margin.

Restart

Restart of the pipeline system when filled with ironsand at 50% concentration is normally accomplished by the control system using the normal start sequence to accelerate the flow to full speed in 3 minutes. Acceleration is smooth, and the discharge pressure increases with flow, without significant overpressure.

During restart the pump station discharge pressure is controlled. It will over-ride the acceleration ramp if necessary to maintain safe discharge pressures.

Ironsand is a granular material with no fines to bind the settled solids on shutdown. Test work early in the project demonstrated that the settled ironsand is readily resuspended once the flowrate is increased beyond the deposit velocity. Physical measurements on the pipeline have not been possible at this time, however the restart performance appears to confirm the initial observations.

No provision was made to flush the Geho pumps on pipeline restart (based on initial testing), and operation has confirmed this capability.

Concentration Control

Ironsand slurrying and concentration control loops have proven successful in providing satisfactory control for pipeline operation. There is a lag of approximately six minutes from the reclaim system to the pipeline inlet density meter. Most of the lag is provided by the feed tank, and this serves to smooth any bumps in the feed rate arising from the the reclaim system.

Control System

The control system has demonstrated that it is practical to fully automate a major slurry pipeline. The only decision required from the Operator is when to start and stop the pipeline, and when to introduce ironsand into the system. Pipeline operation is shared with Concentration plant operating duties by one shift operator and one plant attendant.

Pipeline Surge

Surge was considered a potential problem because the system is a high velocity, high energy operation. Surge pressures are limited to the effect of pump station failure because pump station isolating valves are very slow closing. Operation has shown surge is within the limits predicted, and the control system is successfully used to operate the pump stations to control surge.

Geho Pumps

This is the second pipeline system in the western world to use Geho piston diaphragm pumps. The equipment was selected primarily for the maintenance savings expected in the pump fluid end.

Initial observations include:-

- The pumps perform well

- The pumps are capable of restart without flushing

- The automatic diaphragm travel control system operates effectively except when the suction pressures exceed the operating pressure of the propelling oil circuit. This creates a problem at the second pump station when the discharge pressure at station 1 is lower than at pump station 2 (during pipeline flush, or when ironsand feed is lost for a period during operation). This allows the pumps at station 1 operate at a slightly higher speed than the pumps at station 2 and causes the suction pressure at station 2 to increase.

Geho control logic latches an alarm if the diaphragm travel is outside limits for more than 3 minutes, creating a problem for a remote, unattended pump station. The problem was solved by increasing the operating pressure of the propelling fluid. A better solution for multiple pump station systems using these pumps is to specify the capacity of the downstream pumps approximately 3% higher than the initial station, allowing the suction pressure control to maintain the normal set point.

- Heat exchangers are recommended for the pump lubricating oil whenever the pump is expected to operate at high loads. The pumps were supplied without heat exchangers, because the manufacturers considered radiative losses were adequate. At ambient temperatures of 23-17 deg C (day-night), the oil temperature stabilised at approx 60 deg C, and the pressure dropped to 50% of the recommended limit.

While the equipment was still being adequately lubricated (lubricant flow, not pressure is the important factor), the performance was not adequate for system control. This problem was resolved by installing larger capacity pumps.

CONCLUSION

The New Zealand Steel ironsand slurry pipeline has demonstrated that high pressure long distance pipeline technology can be economically used for granular, abrasive, high specific gravity solids. Significant technological advances were made in the application of polyurethane linings, positive displacement pumps, and automatic control of slurry pipeline systems in the design of the pipeline which will be of benefit to the future of slurry pipeline technology.

REFERENCES

LYE M, IRONSAND SLURRY PIPELINE - SELECTION AND DESIGN CONSIDERATIONS, IPRNZ Auckland Branch Meeting 17/10/84

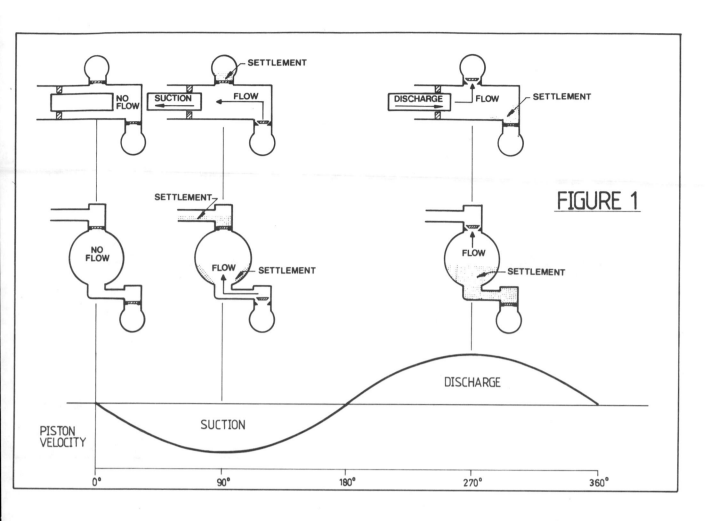

FIGURE 1

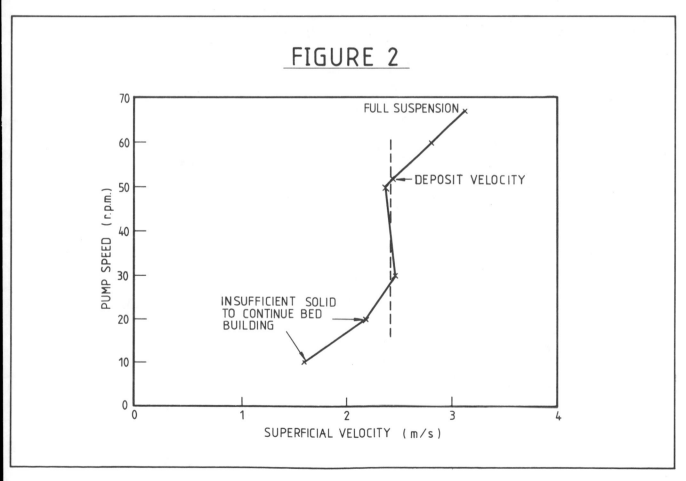

FIGURE 2

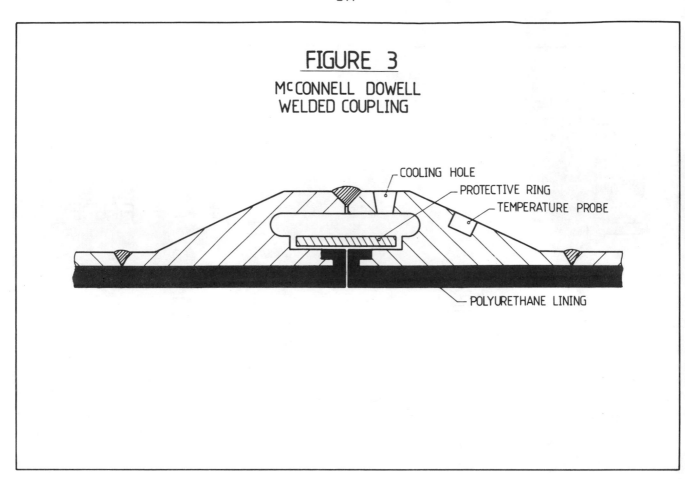

FIGURE 3
McCONNELL DOWELL
WELDED COUPLING

COOLING HOLE

PROTECTIVE RING

TEMPERATURE PROBE

POLYURETHANE LINING

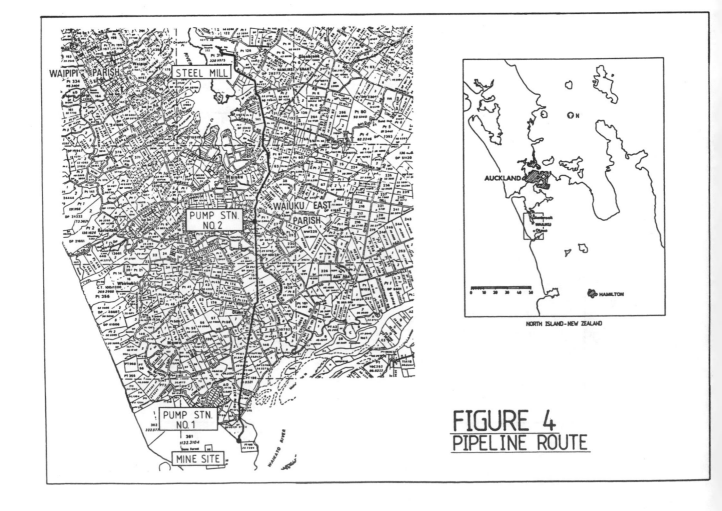

FIGURE 4
PIPELINE ROUTE

WAIPIPI PARISH

STEEL MILL

PUMP STN. NO.2

WAIUKU EAST PARISH

PUMP STN. NO.1

MINE SITE

AUCKLAND

HAMILTON

NORTH ISLAND - NEW ZEALAND

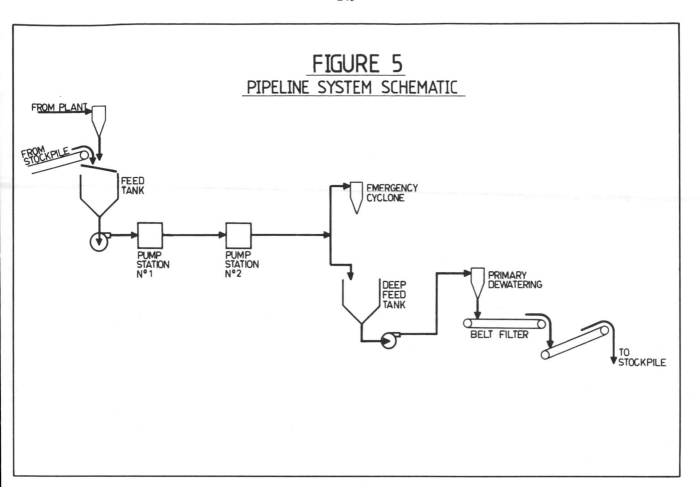

FIGURE 5
PIPELINE SYSTEM SCHEMATIC

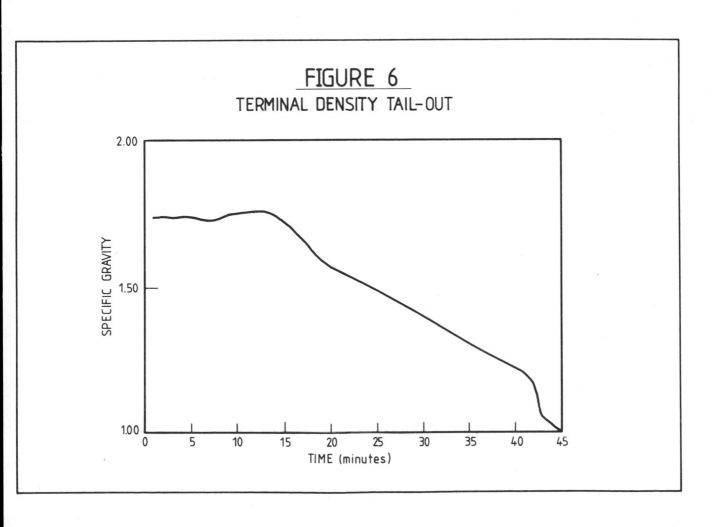

FIGURE 6
TERMINAL DENSITY TAIL-OUT

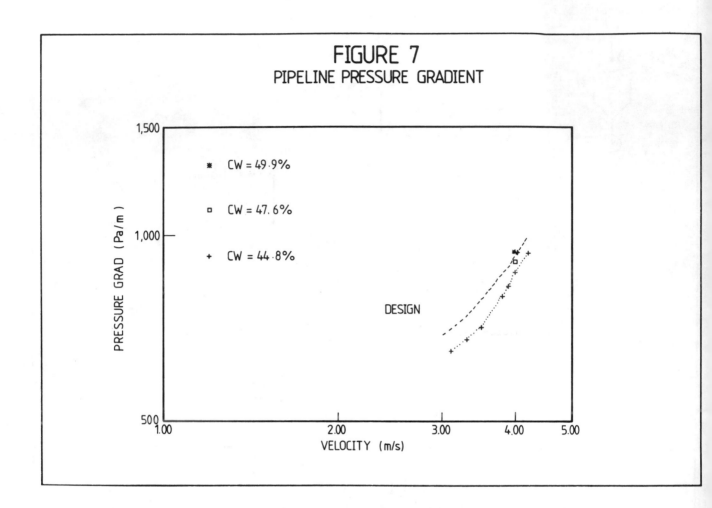

FIGURE 7
PIPELINE PRESSURE GRADIENT

10th International Conference on the
Hydraulic Transport of Solids in Pipes

HYDROTRANSPORT 10

Innsbruck, Austria: 29-31 October, 1986

PAPER G4

DEVELOPMENT AND WEAR TEST OF NEW POLYURETHANE-
LINED LINEPIPE FOR SLURRY TRANSPORTATION

Y. Ikeda, A. Misawa, S. Minamiya, S. Owada,
I. Kondo

Kawasaki Steel Corporation, Japan

T. Kawashima

Tohoku University, Japan

SYNOPSIS

Among anti-abrasion pipe materials for slurry
transport, polyurethane is known for its superior
performance against hard particles, such as
silica sand, and is used in hydro-transport of
deposit sand at the Sakuma Dam in Japan (reported
at Hydrotransport 7, 1980, Sendai, Japan).
Conventional polyurethane linings, however, posed
the following problems for practical use.
In manufacturing process, a lot of energy was
required for thermosetting of polyurethane.
Welded joints could not be lined on-site.
Manufacturing and instruction costs are still
expensive. To solve them, we developed a new
type polyurethane lining. Abrasion and other
tests were performed to confirm the quality of
the lining. The newly developed pipe
demonstrates excellent results with various types
of slurries.

NOMENCLATURE

R.T.V. : Room temperature vulcanization

T D I   : Toluene diisocyanate

PTMG   : Polytetramethylene ether glycol

MOCA   : Methylene bis ortho-chloroaniline
          (Trade name of Dupont Company)

"NCO"   : Terminated difunctional raw material
          used to make prepolymers

"OH"   : Terminated difunctional resin used
          to make prepolymers and cure
          prepolymers

"Amine" : Terminated difunctional material
          used to cure T D I based prepolymer

R T P   : Newly developed prepolymer

R T C   : Newly developed curing agent

INTRODUCTION

We are promoting a total engineering program for
research and development in the area of slurry
transportation. Earlier, results were published
at the 9th Hydraulic Transport of Solids
Conference (1984, Rome) in regard to analysis of
hydrotransport of high concentration coal and ore
slurries (Ref. 1). The internal polyurethane
lined steel pipe has become of much interest from
the view point of extending the service life of
abrasive slurry transportation pipeline.
Polyurethane lining, however, has not widely been
applied up to now. The reasons are high
manufacturing cost, expensive joint, and less
practical reliability. To solve these problems,
we undertook development of a new type
polyurethane linepipe which can be performed
easily at room temperature, (R.T.V. lining).
This paper describes the economical manufacturing
process and the characteristics of new lining in
various types of slurries. Furthermore,
utilizing features of the lining, we have
developed a new field-lining technology for
uncoated portions of welded joints of pipe lines.

MANUFACTURING PROCESS

R.T.V. lining process is shown in Fig. 1. R.T.V.
resin must be poured into the pipe and leveled
quickly to keep efficiency within the pot life.
Taking too long in pouring and leveling of R.T.V.
resin produces bad effects in the uniformity of
lining thickness and surface quality.

R I M (Reaction Injection Moulding) machine that
can supply a large amount of R T V resin in a
short time is applied to the new manufacturing
process. :

1. Improvement of Reaction Characteristic of
   Polyurethane

The composition of R T V is basically similar
to the conventional thermosetting polyurethane
which is composed of T D I diisocyanate and
PTMG polyether-polyol and curing agent which is
composed of MOCA diamine. In R T V system,
however, both prepolymer (RTP) and curing agent
(RTC) were modified to improve the chemical
cross-linking reactivity at room temperature.

2. Development of Primer System

In order to apply R.T.V. to a polyurethane lined
pipe, and optimum primer system which offers
strong adhesion between steel and polyurethane
elastomer without heating had to be developed.
The double layer primer system was newly applied
to R.T.V. lined pipe. Fig. 2 shows the structure
and the function of the new primer system.
The first primer is applied between the steel and
the second primer layer. It offers strong
adhesion between them and prevents the steel
surface from corroding. The second primer is
applied between the first primer layer and the
polyurethane elastomer. It adheres well to both
the first primer layer and the polyurethane
elastomer. It offers water-permeation resistance
too. The special feature of the second primer is
to form a moisture curing polyurethane resin.

This makes it possible to finish the bonding reaction at room temperature.

3. Molding Condition

Typical molding conditions of conventional thermosetting polyurethane and R T V Elaster is shown in Table 1. These factors have a major effect on both productivity and quality of polyurethane lined pipe.

4. Centrifugal Casting

The R T V lined pipe is formed by centrifugal casting method immediately after the pouring process. It provides a smooth, glossy surface and a void-free lining.

5. Curing of R.T.V.

The R.T.V. system needs no curing treatment with heating. It is only recommended that R.T.V. lined pipe is cured and aged at room temperature for 1 ~ 2 weeks before shipping.

6. Available Size   Fig. 3 shows the lining plant.

The available size of R T V lined pipe is as follows.:

Pipe Diameter	(O.D)	(mm)	89.1 ~ 406.4
Thickness of Pipe		(mm)	3.2 ~ 16.0
Pipe Length		(mm)	max. 6,000
Thickness of R T V elastomer	(mm)		3 ~ 10($\pm$ 1)

Manufacturing of 12 meter long pipe is ready to apply.

DURABILITY OF R.T.V.

1. Wear resistance

The authors performed the slurry abrasion test to clarify the wear resistance of R T V compared to that of other materials. Fig. 4 shows a chart of relative wear resistance on coarse and fine silica stone slurry obtained by rotary abrasion test. This test simulates the impingement and sliding attack encountered in slurry pipeline. Abrasion test was carried out by using a drum tester shown in Fig. 5. Specimens were applied to the inside surface of a drum where the slurry concentration stayed stable. From these results, it is predicted that R T V has excellent wear resistance for both impingement attack and sliding attack by abrasive particles in slurry pipeline relative to other materials including conventional thermosetting polyurethane.

2. Effects of physical properties of R.T.V. on wear resistance

Relation between wear resistance and physical properties of R T V and other materials are shown in Fig. 6. These figures show relative resistance to wear by coarse silica stone, compared to mild steel in rotary wear test. It is found that as tensile strength, tensile elongation and tear strength increases, conversely hardness and Taber abrasion loss decreases, wear resistance of polyurethane elastomers increases. These results confirm the report by Pitman (Ref. 2). We consider that the fracture energy of material follows integral equation (1).

$$E_f = \int_0^{\varepsilon_b} \sigma \, d\varepsilon \quad (1)$$

where : E_f = fracture energy
   σ = stress
   ε = strain
   ε_b = strain at break

Equation (1) simply means that a tough material needs to have a large tensile strength and a large tensile elongation at the same time. Fig. 6(e) shows the relation between wear resistance and a product of tensile strength and elongation. It is easily understood that polyurethane elastomer has an excellent wear resistance relative to other materials. Polyurethane elastomer has the largest product of tensile strength and elongation. In other words, R T V is the toughest material.

3. Hydrolytic resistance

The hydrolytic resistance of polyurethane is greatly affected by the composition of diisocyanate system. Pentz, et al, pointed out that PTMG polyether polyurethane shows greater hydrolytic resistance than polyester polyurethane and other polyether polyurethane (Ref. 3). Meier, et al, indicated that the long term hydrolytic stability lasts more than 20 ~ 30 years when TDI-MOCA polyurethane in usual slurry service environment is used (Ref. 4).

4. Corrosion resistance

The corrosion resistance of lining materials is the principal factor in severe corrosive slurry environment. R T V elastomer is considerably resistant to the mild conditions of common acids, bases and organic reagents encountered in usual slurry environment as same as the conventional polyurethane.

DURABILITY OF R.T.V. LINED PIPE

The wear resistance and the water-permeation of R T V lined pipe were evaluated by means of the loop test and the service test in field line.

1. Loop test

(1)Experimental Equipment

We have three loop test facilities : large scale loop (150mm and 200mm in dia.), small scale loop (89mm and 114mm in dia.), and another loop (89mm in dia.). Photo 1 and Fig. 7 show the view of loop lines and schematic drawing of small scale loop used in the present experiments respectively. The total length of the small scale loop is 70m. Flow rate and slurry temperature are automatically monitored. The pump, control valves and measuring instruments are automatically operated and controlled.

(2)Experiment

(a)Sample

Specimens and materials are shown in Table 2. Steel pipes lined with R T V polyurethane and conventional thermosetting polyurethane lining were used as the samples. As the reference, mild steel pipes were mainly used, and high-carbon steel pipe and 1%Cr steel pipe were used partially. Table 3 shows the chemical composition and mechanical properties of sample steels. The samples were straight pipes of 89mm O.D. x 500mmL or 114mm O.D. x 500mmL and elbow pipes with bending radii of 450mm of the same diameter, both with flanges on both ends.

(b)Test condition

Each test condition is shown in Table 4. In this experiment, the flow speed is set slightly faster than the sedimentation limit flow speed as shown by the example of silica stone of 89mm dia. loop in Fig. 8. These curves are computed from Durand's law. Average wear loss and maximum wear loss were measured by measuring the weight loss and the thickness loss of specimen respectively. The water absorption of polyurethane elastomer was corrected by the use of a damy specimen which was immersed in water at the same temperature during the same period of loop wear test.

(3)Results of Loop test

Table 5 shows loop test wear results. Wear loss of straight pipe is shown as average wear loss. Wear loss of bend pipe is shown as maximum thickness loss.

The results can be summarized as follows.:

(a)Using coarse silica stone slurry (condition No. 1), relative wear resistance of R T V lining is 14 times greater in straight pipe and 9 times in bend pipe compared to that of mild steel. Holes are caused in the case of mild steel of bend pipe. In the case of polyurethane lining however, no undesirable blister or breakage is found, although the gloss is degraded and the surface is somewhat roughened.

(b)Using medium grain silica stone (condition No. 2), wear resistance of R T V lining is 11 times in straight pipe and 7 times in bend pipe compared to that of mild steel. In bend pipe, R T V lining especially shows better wear resistance than thermosetting polyurethane lining.

(c)Using coarse coal slurry (condition No. 4), wear resistance of R T V lining is 26 times greater in straight pipe and 14 times greater in bend pipe compared to that of mild steel. Coal is less abrasive but more corrosive than silica stone slurry pH range. Corrosion greatly affects weight loss of mild steel pipe, therefore R T V lining that resists corrosion shows greater benefit when compared with silica stone slurry.

(d)Using fine silica stone (condition No. 3), fine coal (condition No. 5) and iron ore (condition No. 6), wear by slurries was slight and there was not enough wear loss to estimate wear resistance of R T V lining relative to that of mild steel.

(e)Although no considerable difference in the degrees of abrasion was found between R T V and thermosetting polyurethane on the straight part, R T V showed better abrasion resistance with less abrasion on the bending part.

(f)Degrees of abrasion for high-carbon steel and 1%Cr steel were comparable to that of mild steel.

(4)Water-permeation Adhesion of R.T.V.Lined Pipe

The adhesion strength of R T V elastomer after loop tests were all more than 20 kgf/cm at 90° peel-off. Deterioration of adhesion was not detected.

2. Service test in field line

In order to verify the practicality of R.T.V. lined pipe, we investigated service tests in three slurry pipelines in Japan. Table 6 shows the results. In every environment, R T V lined pipe showed good wear resistance and adhesion strength. The installation of R T V lined pipe in the zinc ore pipeline and the blastfurnace dust concentrate pipeline continues to be successfull.

DEVELOPMENT OF JOINT

The welding joint is the most reliable joint method with its high work efficiency, low cost, strength of the joint part in pipeline. However, the welding joint is not usable in the case of internal polyurethane lined steel pipe, because field repairs using conventional thermosetting polyurethane lined pipe are not feasible. Therefore, flange joints or mechanical joints such as Victaulic Joint are commonly used for polyurethane lined pipelines. However, high cost in machining and installation of flanges and joints give rise to economical problems. The drawbacks of those joints are considered excessive abrasion due to local level difference and water permeation to the joint part. In order to solve these problems, we developed a new welding joint with R.T.V. method.

This repair method's main characteristic is feasibility of field lining.
We consider that it is the first practical method in the world. This method used a special inner core which is fitted to the inside of the pipe. Resin is injected into the repair gap between the steel pipe and the core, and is hardened to form lining.

1. Shape of repaired part

In order to examine the effect of the repaired shape on the abrasion resistance, various shapes were studied.

(1)Experiment method

Fig. 9 shows 7 kinds of samples (steel plates size 50 x 100 x 7t, lining thickness 3 to 6mm). The samples were prepared by the procedure shown in Fig. 10.

Abrasion test was conducted using a rotary abrasion machine with slag ballast under the following conditions.

* Slurry particle diameter	: 10 ~ 15mm
* Slurry concentration	: 60 wt% (tap water)
* Peripheral speed	: 3m/sec.
* Testing time	: 1200 hrs.

* Frequency of slurry replacement: every 24 hrs.

Degree of abrasion was measured at points 5mm away from the joint. The appearance was checked visually. Mild steel plate was used as the reference.

(2) Results of experiment

1. Mild steel plate was worn by 0.20mm in average and 0.5mm max. (center of the sample).

2. Urethane lining received only slight rubbing flaws and no defect such as peeling of repaired lining was found.

3. Edge of the stepped part was worn to become round, but no local scraping was caused.

From these results, it is considered that the standard shape of the joining part is vertical because a special shape will make the actual repair work difficult.

2. Actual repairing experiment

Loop test was conducted to verify the practical serviceability of the repaired part. The sample was mild steel pipe with welded joint at the center and repaired lining length was 300mm. The lining thickness was within the range from 5 (target thickness)- 1mm to 5 + 2mm. Both the initial adhesion strength between the repaired part and the steel surface and the initial adhesion strength with the existing lining were 10 kgf/cm or over. The samples were tested in silica stone slurry (average particle size 3mm, concentration 15wt%, flow rate 4.0m/sec.)

No peeling of the repaired lining and concentrated abrasion were found after a 3 - month test. According to the results of this test both wear resistance and adhesion strength of repaired area were sufficiently satisfactory with regard to practical durability.

The technique and equipment for practical pipeline construction are in the process of development.

CONCLUSION

The newly developed pipe demonstrates excellent performance with various types of slurries, and possesses the following advantages for practical use. :

(1) Application of the lining can be performed easily at room temperature, enhancing productivity, as shown in Table 1.

(2) Because joints can easily be lined in the field, welded joints can be applied with this lining.

(3) The service life of the pipe greatly exceeds that of conventional alternatives, as shown in Table 5.

ACKNOWLEDGEMENT

The authors wish to gratefully acknowledge Mr. Yousaku Nagai and Mr. Chisei Murayama (Daiki Engineering Co., Ltd.), Mr. Tatsuaki Nakarai and Tatsuhiko Kobayashi (Mitsubishi Chemical Ind. Ltd.) for their kind cooperation and advice on the preparation of this paper.

REFERENCES

1. Oba, S., Ota, K. and Kawashima, T. :

" Analysis on Slurry Pressure Drop by various Fluid Models "
Proc. 9th Hydrotransport, Paper D2. Organized by BHRA, Cranfield. (October 17th to 19th, 1984).

2. Pitman, J.S. :

" Polyurethane for the Corrosion Protection of Pipework "
3rd Internal Conference on the Internal and External Protection of Pipes, Paper D2. Organized by BHRA, Cranfield. (1979)

3. Pentz, W.J. and Krawiec, R.G. :

" Hydrolytic Stability of Polyurethane Elastomer "
Rubber Age, pp. 39 - 43. (December, 1975).

4. Meier, J.F. and Rosenblatt, G.B. :

" Hydrolytic Stability of Cost Polyurethane Missile Launch Seals "
Elastomerics, pp. 21 - 26. (November, 1979).

Table 1 Comparison of Manufacturing Conditions of RTV Lined Pipe and Thermosetting Polyurethane Lined Pipe

		RTV Lined Pipe	Conventional Thermosetting Polyurethane Lined Pipe
Pretreatment	Prepolymer melting	R. T. *	80℃
	Curing Agent melting	R. T.	100~120℃
	Mixing temperature	R. T.	100~120℃
	Mixing ratio * *	3 / 2	12 / 1
	Primer application	2 layer	1 layer
	Baking		100℃
	Drying	R. T.	
	Pre-heating of Pipe	R. T.	100~110℃
Casting		R. T.	100~110℃
		20~30 min	30~60 min
Curing		R. T.	100~110℃
		1~2 hours	3~4 hours
Post Cure (Aging)		R. T.	R. T.
		1~2 weeks	1~2 weeks

R. T. * : Room Temperature
Mixing ratio * * : Prepolymer / Curing agent

Table 2 Specimens and Materials Used in Closed Loop Test

Specimen				Material			
Straight Pipe :	diameter (O.D.) (mm)	89,114	RTV Lined Pipe :	lining thickness (mm)	4
	thickness (mm)	4		hardness (Shore A)	80
	length (mm)	500	Thermosetting :	lining thickness (mm)	4
90° Bend Pipe:	diameter (O.D.) (mm)	89,114	Polyurethane	hardness (Shore A)	80
	thickness (mm)	4	Lined Pipe		
	bending radius (mm)	450	Steel Pipe : Mild Steel		
				thickness (mm)	8

Table 3   Chemical composition and mechanical properties of sample steels

Kind of steel	Chemical composition (Wt %)										T. T. (kgf/mm²)	Y. S. (kgf/mm²)	E ℓ (%)	Hardness Hv
	C	SO	Mn	P	S	Cu	Na	Cr	Mo	Nb				
Mild	0.15	0.19	0.49	0.015	0.008						44	31	36	133
	≦0.25	≦0.35	0.30~0.90	≦0.040	≦0.040	—	—	—	—	—	≧38	≧22	≧30	
1% Cr	0.24	0.25	0.50	0.005	0.003	0.01	0.02	1.00	0.57	0.025	76.3	70.9	28.5	243
	0.15~0.35	≦0.35	≦1.00	≦0.300	≦0.015	≦0.30	≦0.10	0.80~1.60	0.15~1.10	≦0.050	≧70.3	63.3~73.8	≧18.5	
High carbon	0.48	0.30	0.93	0.020	0.008	—	—	—	—	—	69.2	44.2	41	187
	—	—	—	≦0.040	≦0.060	—	—	—	—	—	≧52.7	38.7~56.2	≧24	

Table 4   Testing Conditions of Loop Test

NO	Specimen Size (O.D.) (mm)	Slurry	Particle Size (mm)		Concentration Cw (wt %)	Velocity (m/sec)	Temperature (°C)	pH	Period (days)
1	89	silica stone	d_{50} $d_{max.}$	3 6	32	4	20~36	8.1~8.3	90
2	89	silica stone	d_{50} $d_{max.}$	1.5 2.5	45	4	5~19	8.1~8.2	90
3	114	silica stone	d_{50}	0.5	50	3	15~35	7.8~8.2	120
4	89	coal	d_{50} $d_{max.}$	4 10	35	4	6~34	7.7~8.1	90
5	114	coal	$d_{max.}$	2.4	45	3	25~50	8.3~8.5	90
6	114	iron ore concentrate	$d_{max.}$	0.15	50	3	25~45	7.3~7.6	90

Table 5   Result of wear loss

Testing Condition	Specimen	Wear Loss (mm/year)			Relative Wear Resistance		
		R T V	Thermo-Setting P.Urethane	Steel Pipe	R T V	Thermo-setting P.Urethane	Steel Pipe
No. 1	Straight	2.12	2.35	30.2	14	13	1
	Bend	23.6	30.7	201.4	9	7	1
No. 2	Straight	1.64	1.64	18.7	11	11	1
	Bend	19.8	37.4	132.5	7	4	1
No. 3	Straight	0	0	2.53	*	*	1
No. 4	Straight	0.29	0.31	6.6	26	25	1
	Bend	1.5	1.5	21.3	14	14	1
No. 5	Straight	0	0	1.46	*	*	1
No. 6	Straight	0	0	1.66	*	*	1

*: Impossible to estimate the relative wear resistance

Table 6　Results of Field Test

NO	Slurry Condition				Specimen (Pipe)	Wear Rate	Wear resistance
	Slurry	Particle Size (mm)	Cw (wt %)	Velocity (m/sec)			
1	Dredging Sand	d_{50} 0.15 $d_{max.}$ 20 1.9 million ton (1700 hours)	60	5.0	750mm (I.D.) 1000mm long steel thickness 12mm RTV lining 7mm	Steel : 4.7mm/year RTV : 0.7mm/year	1 7
2	Blastfurnace Slag	$d_{max.}$ 25	45	4.2	200mm (I.D.) steel 10mm Natural Rubber lining 20mm RTV lining 10mm	Steel : 〉30mm/year life-3 months Rubber : 10mm/year life-2 years RTV : 1.5mm/year predicted· life-5 years	1 3 20
3	Blastfurnace Dust ·Concentrate	d_{50} 0.2～0.3	70	5.0	100mm (I.D.) steel 8.6mm RTV lining 5mm	Steel : 2.30mm/year life-3 years RTV : No failure after 4 years	1 —

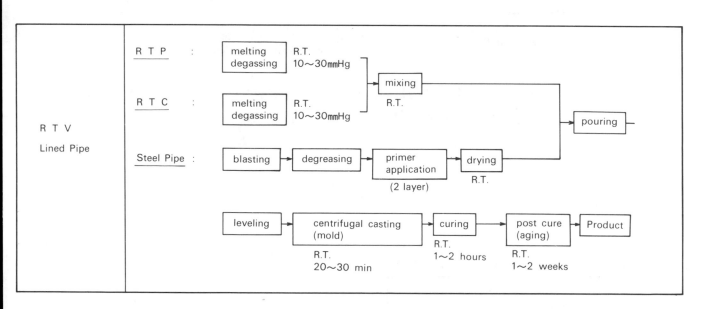

Fig. 1　Manufacturing Processes of RTV Lined Pipe

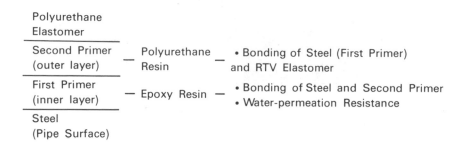

Fig. 2　Structure and Function of Double Layer Primer System

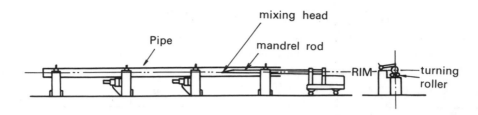

Fig. 3 Schematic Drawing of Pouring Method

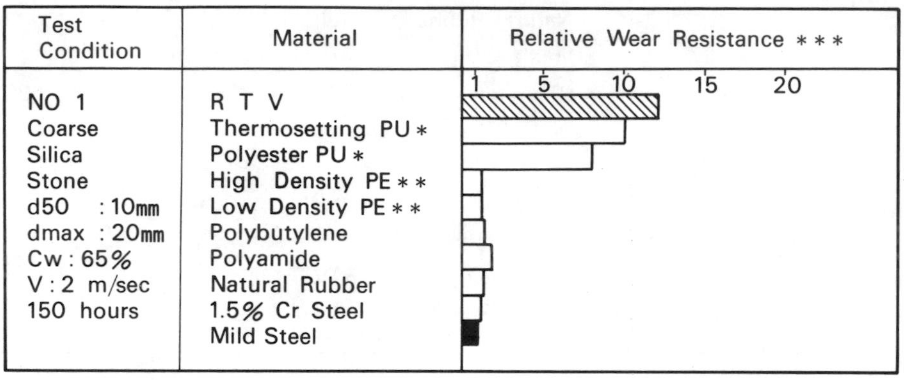

Test Condition	Material	Relative Wear Resistance ***
NO 1 Coarse Silica Stone d50 : 10mm dmax : 20mm Cw : 65% V : 2 m/sec 150 hours	R T V Thermosetting PU * Polyester PU * High Density PE ** Low Density PE ** Polybutylene Polyamide Natural Rubber 1.5% Cr Steel Mild Steel	1 5 10 15 20

PU *: Polyurethane , PE **: Polythylene

Relative Wear Resistance ***: $\dfrac{\text{average wear loss of wild steel}}{\text{average wear loss of other material}}$

Fig. 4 Chart of Relative Wear Resistance to Mild Steel by means of Rotary Wear Test

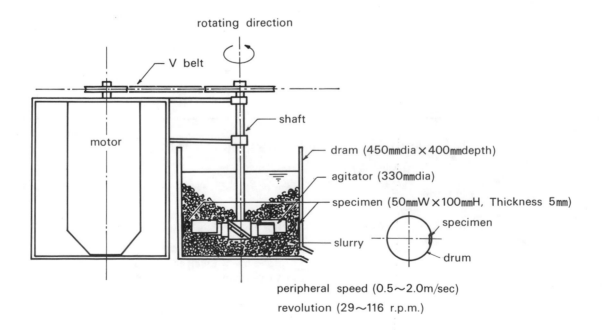

Fig. 5 Schematic Drawing of Rotary Wear Tester

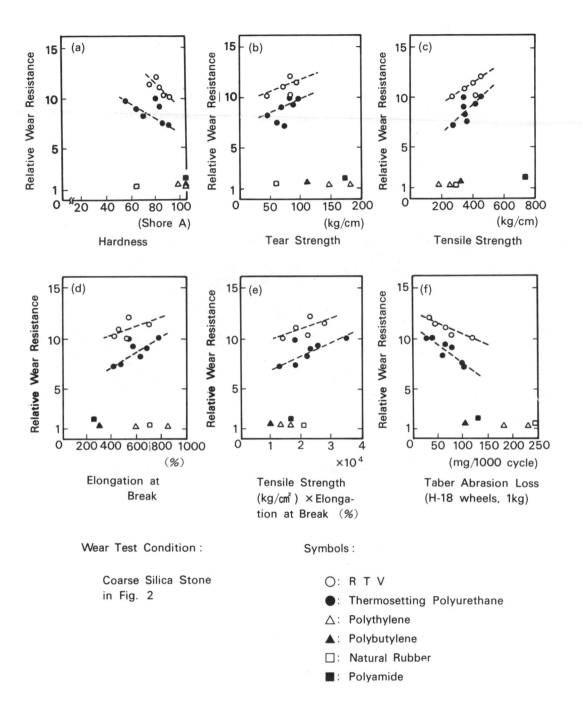

Fig. 6 Relation between Wear Resistance and Mechanical Property

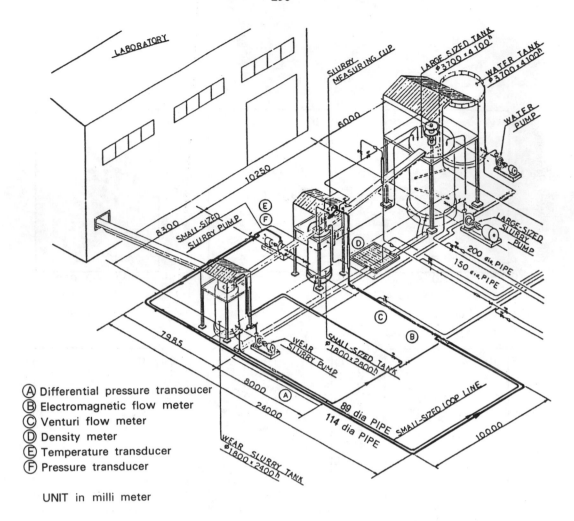

A Differential pressure transoucer
B Electromagnetic flow meter
C Venturi flow meter
D Density meter
E Temperature transducer
F Pressure transducer

UNIT in milli meter

Fig. 7   Schematic drawing of loop

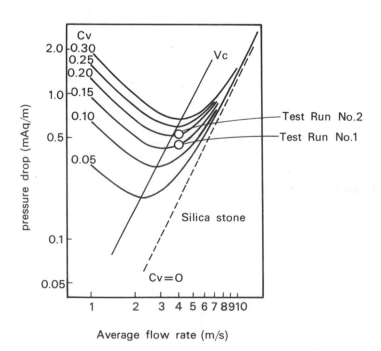

Fig. 8   Slurry Flow Condition
(Silica stone 89mm dia)

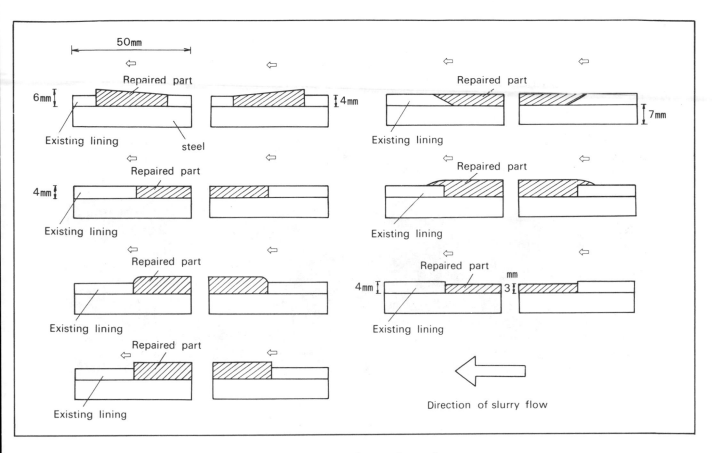

Fig. 9 Samples for abrasion test

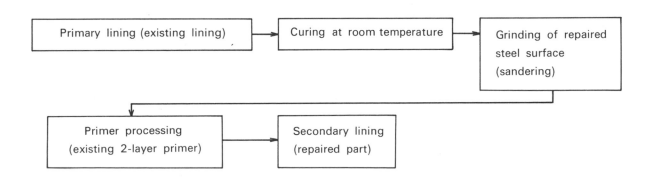

Fig. 10 The procedure of repairing

Photo. 1   View of pipeline

10th International Conference on the
Hydraulic Transport of Solids in Pipes

HYDROTRANSPORT 10

Innsbruck, Austria: 29-31 October, 1986

PAPER G5

HYDROTRANSPORT OF STOWING MATERIALS IN
POLISH COAL MINES

J. Palarski

Professor in the Department of Mining at the
Silesian Technical University, Poland.

SYNOPSIS

In Polish coal mines over 30 million m^3 of
materials are transported hydraulically to stow
headings. Problems connected with the preparation
of stowing mixture, its gravatational transport in
pipelines, and dewatering in the workings are
discussed in the paper. Technical parameters of
automated stowing installations are given and the
latest monitoring systems and measuring devices
are described. The paper also presents a method of
designing stowing installations for the transport
of different waste materials.

NOMENCLATURE

A - section, m^2

$$a = \frac{\gamma_s - \gamma_o}{\gamma_o} = \frac{\rho_s - \rho_o}{\rho_o} \quad , -$$

C_v - volume concentration, -

D - pipe diametr, m

d - grain diametr , m

g - acceleration of gravity, ms^{-2}

H - total height of installation, m

h - depth of a section of installation, m

L - total length of installation, m

l - length of a section of installation, m

p /P/ - pressure in point, Pa

ΔP_i - total losses of hydraulic flow
    of mixture from the beginning of
    the installation to point "i"
    on this section, Pa

Δp - unit hydraulic losses, Pa m^{-1}

t_p - time of rinsing installation, s

w - sedimentation velocity, $m.s^{-1}$

v - flow velocity, ms^{-1}

Q - efficiency, $m^3 s^{-1}$

α - angle of inclination of pipeline, ^{o}

γ - specific gravity, Nm^{-3}

ρ - specific density, kg m^{-3}

$f, f_1', k, r, \zeta, \zeta_1$ - computing
                  coefficients

λ - resistance coefficient

INDECES

b - barometric pressure

i, j - index of addition, section of
    installation

L - installation

m - mixture

s - solid particles

o - water

od - reference

p - particles in loose state,
    stowing material

max - maximum

min - minimum

kr - critical

INTRODUCTION

Mining of coal seams occurring under important
buildings and in difficult geological conditions
in thick seams requires the use of stowing to fill
old workings. Since the beginning of the 20th
century Polish mining has been using hydraulic
stowing which consists in hydraulic transport of
granular material to an empty space after mining
has finished; after water is separated from the
granular material the material becomes the support
for the roof. The development of coal output in
Poland from 1955 to 1985 according to the kind of
roof control is presented in Fig. 1. Stowing was
most extensively used from 1965 to 1970. In 1985
about 15% of coal was mined from backfill longwall
faces, which constitutes 30 mln t (the saleable
output being 192 mln t). To fill empty spaces
after mining 32 mln t of sand and 7 mln t of rock
were transported in 1985. Moreover, about 1 mln t
of other waste materials among them industrial
smoke-box dusts were used underground. To
transport this 39 mln m^3 of materials pipelines of
a total length of 1200 km are used. On account of
great abrasion about 300 km of pipes are changed
each year.

The total length of a network of stowing pipelines
in a coal mine can be from a few kilometers to 30
km. Water used in hydraulic transport as a liquid
carrier flows in a closed circuit, and its losses
are made up from mine water settling tanks.
Because of this, separate settling tanks are made
underground to clarify fill water which is then
pumped to the surface.

The widespread use of stowing is the result of its
technological simplicity and also of the fact that
it supports the roof very well, which is connected

with small deformations of the surface and rockmass and that the costs of stowing are low. Fig. 2 presents some examples of costs of hydraulic stowing in different coal mines (the costs being divided into labour, material and energy costs and amortization charges). It can be seen in Fig. 2 that the costs of hydraulic stowing are almost half the costs of pneumatic stowing. However, hydraulic stowing causes greater technological problems in a working face, especially in building a fill stopping. Although, the process of filling the working and draining off water from a longwall face is difficult to mechanize, as opposed to the almost fully mechanized process of getting coal in a face.

CHARACTERISTIC OF STOWING INSTALLATIONS

A stowing installation comprises (Fig. 3):
-installations on the surface producing stowing material and called stowing plant,
- series of pipelines, in which stowing materials with water are transported gravitationally,
- stoppings limiting the space which is to be filled, gutters to drain water off and settling tanks, i.e. installations located in the production section,
- pump chamber, water pipelines and water tanks. According to the method of unloading, the tanks are divided as follows (Fig. 4):
- slope tanks,
- shaft tanks,
- slit tanks,
- silo tanks.

Material is washed out from a tank with the help of monitors (Fig. 4a), nozzles (Fig. 4b) or is removed with the help of feeders (Fig. 4c) or falls under gravity on feeding devices. Feeding devices pass the material on to screens, next the material falls into channels where it is mixed in predetermined proportions and flows to a chamber which ends with a hopper. In order to obtain a mixture of required concentration or in cases of failure water can be added directly into the hopper.

Depending on the type of stowing material, each tank can store from 1500 to 5000 m^3 of stowing material. In a stowing plant there are also monitoring and measuring devices, pumps, crushers for breaking rocks of size larger than the clearance and a network of water pipelines. In modern stowing plants the process of producing stowing mixtures is fully automated.

HYDRAULIC TRANSPORT OF STOWING MIXTURES

The mixture at the correct concentration (the mixture being produced in a stowing plant) flows under gravity in pipelines to the space designed for stowing.

The efficiency of such transport depends on:
- such parameters of the installation as: diameter of pipes, the difference of height between the inlet and outlet of the installation, space arrangement of the pipeline,
- properties of the mixture, and in particular its concentration, grain-size distribution, and density of solid particles.

Moreover, there is a possibiltity of increasing the efficiency of an installation by adding

polymers to a mixture or by installing an additional pump on the horizontal part of the pipeline. Steel pipes, steel pipes with a basalt lining, or with a rubber lining, and rubber pipes of 150 mm or 185 mm outer diameter are used in Polish mining.

In order to determine the optimal parameters for hydraulic transport of stowing materials in pipelines measurements have been carried out both in test rigs and industrial installations. These measurments and theoretical considerations have enabled us to work out a method for designing stowing installations. The modified Bernoulli equation of the real liquid has been the initial equation for the theoretical consideration. As the measurements in the industrial installations have proved the mixture can be treated as a homogenous liquid if it moves with the velocity:

$$v \geqslant 1,3\ v_{kr}$$

Then the density of the mixture is almost constant in the whole installation irrespective of the inclination of a pipeline and solid particles fill the full section of the pipe. The results obtained confirm also the observations of abrasiveness of pipes which are worn evenly along the whole circumference and not only in its lower part (Fig.5). In Eq. 1 terms:

$p_b(A)$ and $\dfrac{v^2 \rho_m}{2g}$ can be omitted as small

in comparison with other terms (Fig. 6). Pressures determined by this equation have been compared with the results of pressure distribution measurements in the installation, and it has been stated that the flow of the mixture can be described by Eq. 2. In order to determine the efficiency of hydraulic transport one should know the equation of unit energy losses in inclined pipelines in which different types of stowing materials flow. As the result of measurements of energy losses in test stand for the grain-size distribution and different concentration in pipes of 120, 150, 185, 200, 225, and 250 mm diameter, Equations 3 & 4 have been determined. After substituting Eqs. 3 & 4 to the formula 2 the so-called equations of the flow mixture (Eqs. 5 & 6) and of the pressure distribution in the installation Eqs. 7 and 8 have been obtained. Calculating from Eqs. 5 and 6 the mean flow velocity we can easily determine the efficiency of the flow and pressure at every point of the installation.

In any stowing installation the mean flow velocity should be greater than the critical velocity but at the same time it should be smaller than a certain boundary value above which pipe wear rapidly increases. This velocity for pipes and stowing materials used in Polish mines should not be greater than about 10 m/s. In a stowing installation, where the flow velocity is high or the pressure is so high that it is close to the tear strength of the applied pipes, an additional hopper or lay-by can be built in to reduce the pressure or velocity (Fig. 7). In the case when the additional hopper is used the gravitational flow of the mixture begins from the surface of the mixture in this hopper to its outlet. The flow between the main hopper and the additional one has the character of free fall.

A lay-by is a series of pipes with bends in which the mixture loses energy of movement. The location of an additional hopper and a lay-by and its length can be calculated from the above mentioned equations.

Every break in the work of the installation requires flushing. Flushing of the pipeline is not a long-lasting process but it does cause problems at the longwall face due to excess water and significantly increases the cost of stowing. Hence the minimising of flushing time is one of the most important problems.

The minimum flushing time can be calculated from Eq. 9 and Fig. 8. In fully automated stowing plants there is a danger of pipe blockages in the event of the failure of a feeder. Then only water is fed into the installation and even at increased efficiency it is not always possible for water to transport solid particles from the pipeline. The critical time is when there is water in a vertical pipe and solids in a horizontal line. Then pressure of the column of water may be insufficient to overcome the resistance of the flow of the mixture in the horizontal pipe. This is why when designing a stowing installation, the highest concentration should be determined from the condition that there is water in a vertical pipe and mixture in a horizontal one. In such a situation pipe blockages are avoided.

A stowing installation in a coal mine usually connects a stowing plant with a number of longwall faces, which means that a stowing pipeline can branch in different directions, and constitute a network of pipes in a mine (Fig. 9). In a stowing shaft there are at least two pipelines and usually four, one of them being a reserve. In horizontal roadways a pipeline can fork and there are possibilities of connecting different pipelines. Therefore stowing pipelines in a mine constitute a network of the length from a few to 30 km. Most longwall faces with caving are connected to a stowing pipeline. Then the stowing pipeline can be used to convey filling material in case of fire. About 50% of fires are put out with the help of stowing systems in Polish coal mines.

Designing a stowing network in a changing spatial arrangement of workings is a very difficult process, especially because some parts of pipelines are common, some branches are in headings, some in inclines often along great dips or rises. In these districts there is a danger of breaking the continuity of flow, of cavitation and drawing in air on loose connections. This is why the process of designing - which comprises computing geometrical parameters of a network, properties of a mixture, pressure distribution, wear properties of pipes, number and order of the changes of pipes, parameters characterizing draining devices and the process of flushing the installation, and also economic analysis - is done with the help of a computer.

MANAGEMENT OF STOWING WATER

Stowing water flowing out of a fill stopping is directed to local settling tanks made in the stowing or directly in coal (Fig. 11). After cleaning water flows to settling tanks near the bottom of the shafts (Fig.12) from where it is pumped onto the surface to surface settling tanks and from where it returns to a stowing plant. Flocculants are used in settling tanks in order to speed up the process of settling impurities. So far the use of flocculants in stowed spaces has not given any positive results. There are technological problems connected with the localization of feeders and with the very way of adding flocculants. The activity of flocculants is also very limited because of significant turbulence and currents in the stowing area.

AUTOMATION OF THE PROCESS OF STOWING

Parameters of the work of a stowing pipeline in a network change almost all the time because of changes of the location of the stowed longwall face, and changes in the length of pipes, and because of the use of different stowing materials, etc. This is why in order to use automated control of hydraulic transport in the whole network, the network should be equipped with an appropriate measuring and controlling system, and data obtained from it should be processd, and controllable quantities should be corrected (controllable quantities being only the concentration of the mixture, the type of material, and the spatial arrangement of the pipeline). Within the framework of this paper it is impossible to discuss the whole system of automatic control together with safety devices, but we would like to characterize two of the most important stages of this process, namely the process of identification of the stowing installation and the system of safety devices. On account of changes in the location of the longwall face during mining and of the installation it is necessary to identify geometrical parameters on which the choice of input controllable parameters depends. Identification resolves itself into determining the ratio L to H on the basis of the measurements of water flow velocity during every flushing cycle. The ratio L/H can be determined on the basis of the Bernoulli equation and the equation of the resistance of turbulent water flow in the pipeline, and known diameter of pipes and flow velocity.

In order to eliminate very dangerous and expensive failures such as pipe blockages, different kinds of safety devices are used. Nevertheless, there are situations which cannot be foreseen, but which can lead to to pipe blockages and bursts of pipes. To limit the consequences of such failures special devices as shown in Fig. 13 are used in the installation. A rapid change of pressure accompanies every pipe blockage and burst. To limit the range of such a blockage which can lead to choking of the installation over a few to several hundred meters or to limit uncontrollable flow of the mixture to mining openings it is enough to place pressure sensers along the installation and short pipes with valves which direct the mixture into prepared tanks. The signal of a rapid change of pressure from a pressure senser causes valves to open in an established order and with the appropriate delay. The mixture flows into tanks. After repair the installation is ready to work once again. It is worth noting that the traditional process of cleaning the installation after a blockage can last from a few hours to up to a few weeks especially if a vertical pipeline is blocked.

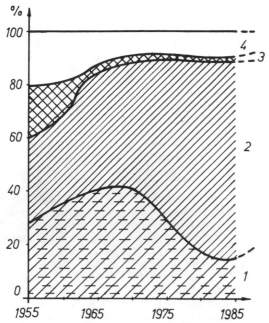

Fig. 1. Development of coal output in Polish coal mining according to the way of stowing: 1, hydraulic stowing; 2, pneumatic stowing; 3, caving; 4, other types of stowing and ways of mining.

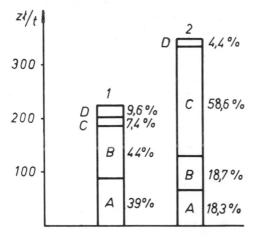

Fig. 2. Cost of hydraulic (1) and pneumatic (2) stowing: A, labour costs; B, material costs; C, energy costs; D, amortization charges.

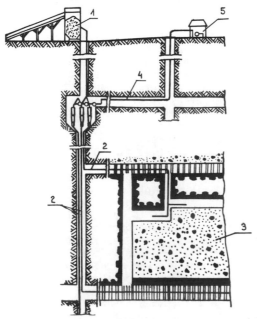

Fig. 3. Diagram of a hydraulic stowing installation: 1, material tank; 2, stowing pipeline; 3, stowing; 4, water pipeline; 5, pump chamber.

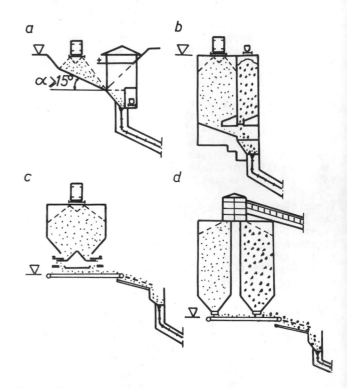

Fig. 4. Types of settling tanks: a. slope tank; b, shaft tank; c, slit tank; d, silo tank.

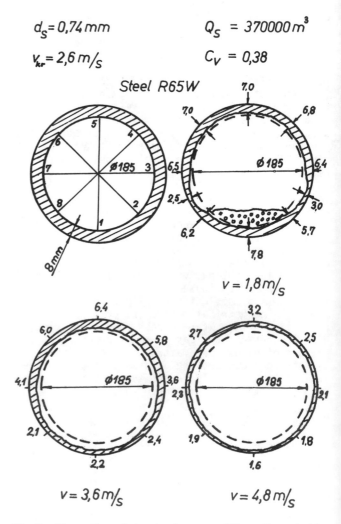

$d_S = 0,74\,mm$ $Q_S = 370000\,m^3$

$V_{kr} = 2,6\,m/s$ $C_V = 0,38$

Fig. 5. Observations of pipe abrasiveness at different flow velocities.

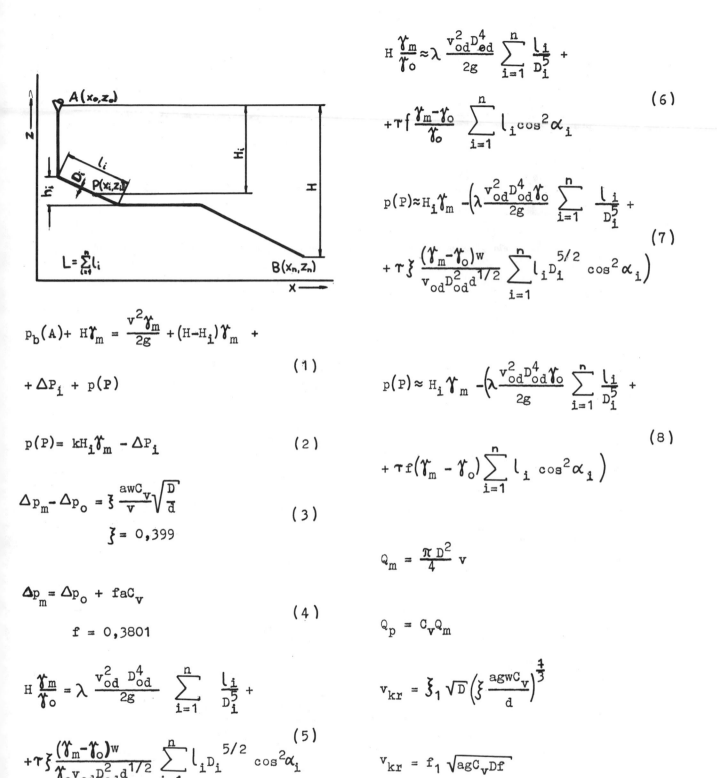

$$H \frac{\gamma_m}{\gamma_o} \approx \lambda \frac{v_{od}^2 D_{od}^4}{2g} \sum_{i=1}^{n} \frac{l_i}{D_i^5} +$$

$$+ \tau f \frac{\gamma_m - \gamma_o}{\gamma_o} \sum_{i=1}^{n} l_i \cos^2 \alpha_i \tag{6}$$

$$p(P) \approx H_i \gamma_m - \left(\lambda \frac{v_{od}^2 D_{od}^4 \gamma_o}{2g} \sum_{i=1}^{n} \frac{l_i}{D_i^5} + \right.$$

$$\left. + \tau \zeta \frac{(\gamma_m - \gamma_o)w}{v_{od} D_{od}^2 d^{1/2}} \sum_{i=1}^{n} l_i D_i^{5/2} \cos^2 \alpha_i \right) \tag{7}$$

$$p(P) \approx H_i \gamma_m - \left(\lambda \frac{v_{od}^2 D_{od}^4 \gamma_o}{2g} \sum_{i=1}^{n} \frac{l_i}{D_i^5} + \right.$$

$$\left. + \tau f (\gamma_m - \gamma_o) \sum_{i=1}^{n} l_i \cos^2 \alpha_i \right) \tag{8}$$

$$p_b(A) + H\gamma_m = \frac{v^2 \gamma_m}{2g} + (H - H_i)\gamma_m +$$

$$+ \Delta P_i + p(P) \tag{1}$$

$$p(P) = k H_i \gamma_m - \Delta P_i \tag{2}$$

$$\Delta p_m - \Delta p_o = \zeta \frac{awC_v}{v} \sqrt{\frac{D}{d}}$$

$$\zeta = 0,399 \tag{3}$$

$$\Delta p_m = \Delta p_o + faC_v$$

$$f = 0,3801 \tag{4}$$

$$H \frac{\gamma_m}{\gamma_o} = \lambda \frac{v_{od}^2 D_{od}^4}{2g} \sum_{i=1}^{n} \frac{l_i}{D_i^5} +$$

$$+ \tau \zeta \frac{(\gamma_m - \gamma_o)w}{\gamma_o v_{od} D_{od}^2 d^{1/2}} \sum_{i=1}^{n} l_i D_i^{5/2} \cos^2 \alpha_i \tag{5}$$

$$Q_m = \frac{\pi D^2}{4} v$$

$$Q_p = C_v Q_m$$

$$v_{kr} = \zeta_1 \sqrt{D} \left(\zeta \frac{agwC_v}{d} \right)^{\frac{4}{3}}$$

$$v_{kr} = f_1 \sqrt{agC_v Df}$$

Fig. 6. Way of computing parameters of hydraulic transport of stowing materials.

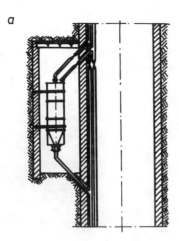

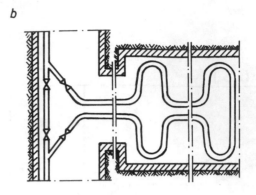

Fig. 7. Schematic diagram of an additional hopper (a) and a
lay-by (b).

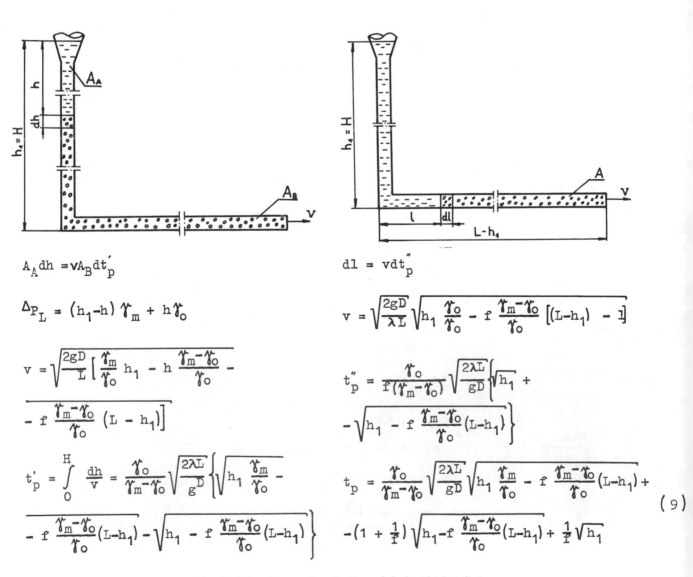

$$A_A dh = vA_B dt_p'$$

$$\Delta_{P_L} = (h_1 - h)\, \gamma_m + h\gamma_o$$

$$v = \sqrt{\frac{2gD}{L}\left[\frac{\gamma_m}{\gamma_o}\, h_1 - h\,\frac{\gamma_m - \gamma_o}{\gamma_o} - f\,\frac{\gamma_m - \gamma_o}{\gamma_o}\,(L - h_1)\right]}$$

$$t_p' = \int_0^H \frac{dh}{v} = \frac{\gamma_o}{\gamma_m - \gamma_o}\sqrt{\frac{2\lambda L}{gD}}\left\{\sqrt{h_1\,\frac{\gamma_m}{\gamma_o} - f\,\frac{\gamma_m - \gamma_o}{\gamma_o}(L - h_1)} - \sqrt{h_1 - f\,\frac{\gamma_m - \gamma_o}{\gamma_o}(L - h_1)}\right\}$$

$$dl = vdt_p''$$

$$v = \sqrt{\frac{2gD}{\lambda L}}\sqrt{h_1\,\frac{\gamma_o}{\gamma_o} - f\,\frac{\gamma_m - \gamma_o}{\gamma_o}\left[(L - h_1) - l\right]}$$

$$t_p'' = \frac{\gamma_o}{f(\gamma_m - \gamma_o)}\sqrt{\frac{2\lambda L}{gD}}\left\{\sqrt{h_1} + -\sqrt{h_1 - f\,\frac{\gamma_m - \gamma_o}{\gamma_o}(L - h_1)}\right\}$$

$$t_p = \frac{\gamma_o}{\gamma_m - \gamma_o}\sqrt{\frac{2\lambda L}{gD}}\sqrt{h_1\,\frac{\gamma_m}{\gamma_o} - f\,\frac{\gamma_m - \gamma_o}{\gamma_o}(L - h_1) +} \\ -(1 + \frac{1}{f})\sqrt{h_1 - f\,\frac{\gamma_m - \gamma_o}{\gamma_o}(L - h_1)} + \frac{1}{f}\sqrt{h_1}$$

(9)

Fig. 8. Way of computing the time of rinsing the installation.

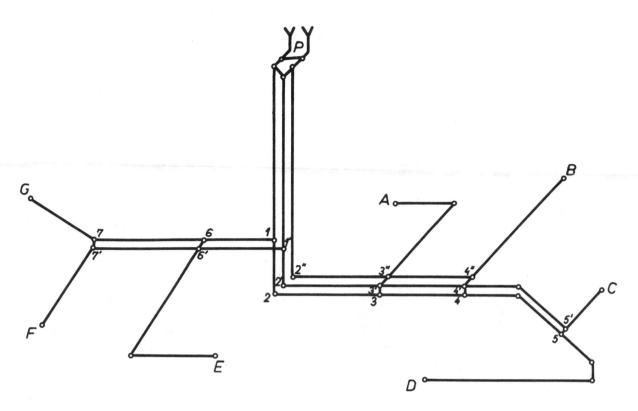

Fig. 9. Schematic diagram of a network stowing pipelines.

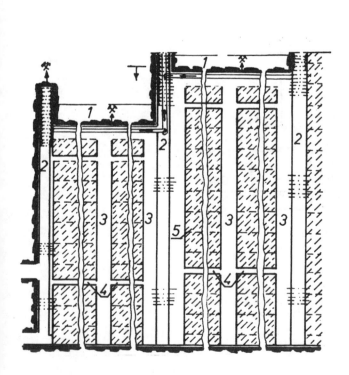

Fig. 10. Schematic diagram of settling tanks made in stowing: 1, longwall face; 2, roadway; 3, transverse settling tank; 4, longitudinal settling tank; 5, stowing.

Fig. 11. Schematic diagram of settling tanks in coal: 1, longwall face; 2, roadway; 3, stowing; 4, chamber settling tanks.

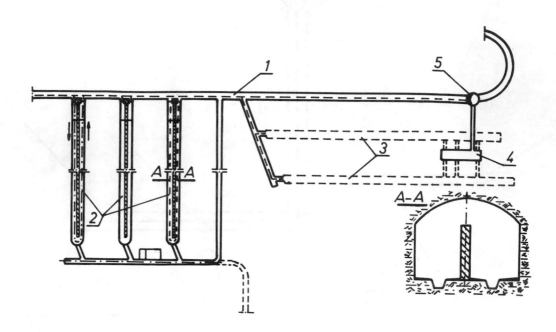

Fig. 12.   Schematic diagram of settling tanks near shafts: 1, roadway with a gauton; 2, settling tank; 3, water tank; 4, pump chamber.

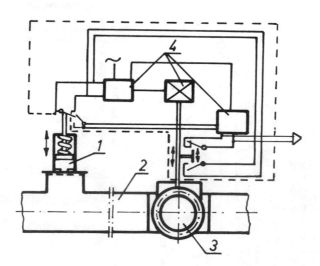

Fig. 13.   Safety devices against pipe blockages: 1, pressure pick-up;
2, pipeline; 3, valve; 4, system of controlling and propulsion.

10th International Conference on the
Hydraulic Transport of Solids in Pipes

HYDROTRANSPORT 10

Innsbruck, Austria: 29-31 October, 1986

PAPER H1

MEASUREMENT OF WALL SHEAR STRESSES FOR
HIGH CONCENTRATION SLURRIES

B.E.A. Jacobs and A. Tatsis
BHRA, The Fluid Engineering Centre.

SYNOPSIS

Considerable interest is being shown at present in the high concentration transport of coarse solids such as coal.

The prediction of the hydraulic gradients required to pump the materials is important both for technical and economic reasons. It is considered that the prediction techniques are not fully established. To rectify this a machine has been designed which will measure the shear forces between the slurry and pipe wall. The concepts lying behind the design are discussed and a description of the machine is given.

NOMENCLATURE

C	concentration by volume
f	coefficient of friction
g	acceleration due to gravity
H	depth beneath slurry surface
h	immersion of top of viscometer bob beneath surface
p	pressure
S	relative density
ℓ	viscometer bob length
ρ	density of water
τ	wall shear stress

1.   INTRODUCTION

One of the limiting factors of conventional hydraulic transport of solids in the turbulent flow regime is the upper particle size that can be carried without excessive power consumption or wear of the pipes. For many materials this problem is overcome by grinding it to a small particle size. Frequently, however, the end user may not require the material in finely divided form and for these situations stabilized or dense-phase pumping has been proposed. In stabilized flow the rheological properties of the carrier fluid are such that the coarse material is maintained in suspension. Dense phase transport is the limiting case of a sliding bed where the slurry fills the complete pipe cross section. For both these methods the potentially low operating velocity should reduce both the power consumption and wear rate.

Although attempts have been made to predict the flow of these slurries, by the assumption of Bingham plastic properties for concrete, for example, or the use of a friction factor for dense-phase slurries, it is considered that at present the flow mechanism at the pipe wall generating the shear stress is imperfectly understood. As part of a research project aimed at elucidating the flow behaviour of high concentration slurries an apparatus has been designed termed a shearometer, capable of simulating and measuring wall shear stresses in pipe flow. The device consists essentially of a disc which is rotated against the retarding force produced by a transversely loaded annular ring of slurry.

2.   REVIEW OF THE LITERATURE

It is the intention of this section of the paper to review briefly the literature describing the transport of coarse material, that is slurries containing particles up to approximately 50 mm nominal diameter. In particular, those papers dealing with slurries at a concentration of 35% by volume and above will be considered as this represents the upper limit at which settling slurries of a narrow particle size range would normally be transported. This value was achieved during the course of the well known work of Durand and Condolios (Ref. 1).

Much of the relatively recent work on coarse particle slurries has been concerned with the hydraulic transport of coal. For a coal having a relative density of 1.35, the above value of volumetric concentration is approximately equivalent to 42% by weight. Concentrations by weight are frequently quoted in the literature.

The first paper to create real interest in this field was that by Elliott and Gliddon (Ref. 2) who pumped coal up to 12 mm at concentrations up to 60% by weight. It was observed that the mixes were stable above a concentration of 55% by weight, that they could be pumped in laminar flow and behaved as Bingham fluids. A considerable amount of further work has been carried out since then with the practical aim of pumping coarse coal either for transport between mine and user or for short distance applications within the mine complex.

The Continental Oil Company (Ref. 3) pumped -63 mm material at concentrations up to 60% by volume. Below 35% by volume the hydraulic gradients increased linearly with concentration but between 35% and 42% the pressure loss decreased, to be followed by further increases as the concentration was raised still further. The highest concentrations were reached with a significant amount of fine material present in the mixtures. The Saskatchewan Research Council also pumped coal at up to 50% by volume and demonstrated that surface phenomena were important (Ref. 4).

Other workers who have experimented with coal slurries include Prettin and Gaessler (Ref. 5), Thomas (Ref. 6), Haas et al (Ref. 7), Sakamoto et al (Ref. 8) and many workers associated with the German coal mining industry (Refs. 9-11).

Some investigations have considered materials other than coal. For example, limestone aggregate and colliery spoil (Ref. 13), rock particles and PVC in Bentonite (Ref.14), granite chippings and mixtures with coal fines (Ref. 15) and concrete (Ref. 16). In all these investigations the authors recognised the need to describe the flow in terms different from those for heterogeneous settling slurries for the higher concentration mixes they investigated. Carleton et al (Ref. 13) found that a sliding bed model best described their experiments. Masuyama et al (Ref. 14) developed a correlation based on coarse particles suspended in a non-Newtonian carrier and the work of Boothroyde et al (Ref. 15), in which granite chippings were mixed with coal fines, also notes that the fines formed a non-Newtonian carrier fluid. Granite chippings without fines were best described by an equation which approximated the effect of fluid flow plus particle friction.

A most significant piece of work, which attempted to make use of the fine fraction to support coarser particles in a stabilized flow, is by Lawler et al (Ref. 17). They estimated that a 35% carrier with 65% coarse coal at an overall concentration of 70% by weight produced the optimum mixture. They were able to correlate the results by use of Metzner and Reed's generalised approach for laminar fluids.

Duckworth et al (Ref. 18) have carried out further work in which the properties of the carrier fluid have been studied together with the effects of addition of coarse material to the carrier on the rheological properties of the mixture. They found that they were able to correlate the pressure drop for a given flow rate by means of the Buckingham equation derived from the assumption that the mixture was a Bingham plastic. More advanced correlation techniques will be used in the future such as the Herschel-Bulkley flow model. It was demonstrated that coal mixtures up to 70% C_w could be pumped in this mode of transport.

Other work by BHRA on behalf of B.P. and briefly reported in Ref. 19 has shown that coarse slurries up to 83% by weight could be pumped.

In summary, it can be said that the above papers describe experiments in which higher concentrations were achieved than for normal single sized settling slurries by means of wider particle size distributions. In some cases, the concentrations were above the loose packing density of single sized spheres. The hydraulic gradients were seen to be affected by the fine material present which, when added to water, tended to form non-Newtonian mixtures.

3. THEORETICAL PREDICTIONS

As discussed above various investigations have shown that the flow of some high concentration slurries can be described on the assumption that they behave as Bingham plastics. At first sight it might appear that such an analysis is only applicable when the coarse particles are transported without significant mechanical interaction between them. When there is a high proportion of large particles they may interlock to form a plug. Bingham plastics, however, form plug flows by means of their yield stress. It may be fortuitous therefore that the plug formed by in-

terlocking particles can bury itself within the plug formed by a homogeneous Bingham fluid, thus allowing this type of analysis to provide a prediction technique (Ref. 20).

Fig. 1 shows a number of flow regimes that can exist where a large proportion of the pipe cross section is occupied by slurry. Fig. 1a shows the situation of a sliding bed flow where mechanical friction of the sliding bed/pipe wall interface provides the majority of the retarding force. Fig. 1b shows the case where the properties of the carrier fluid, in conjunction with the geometrical boundary conditions between the coarser fraction of the slurry and pipe wall, enable a hydrodynamic boundary layer to develop which lubricates the sliding layer. Fig. 1c shows the case where the boundary layer has thickened, with an associated change in the cross sectional shape above the bed. Finally Fig. 1d indicates the condition where the lubricating layer surrounds a fully formed plug (Ref. 16). Due to the influence of gravity the plug would be expected to be offset towards the bottom of the pipe.

The flows indicated in Figs. 1a and 1d can be evaluated by sliding bed and Bingham plastic analyses respectively. The importance of the flows shown in Figs. 1b and 1c is that they represent cases where a high concentration of solids, including coarse particles, is being carried at a lower hydraulic gradient than that required by a sliding bed but the carrier fluid lacks the fluid properties required by a fully stabilized slurry. It is also possible that a slurry stable under static conditions would become destabilized during flow and tend towards a sliding bed.

Because of the importance of the nature of the boundary flow subsidiary equipment is being designed by BHRA which will enable the extent of the boundary to be measured. This has importance both from the theoretical and practical aspects of high concentration flows.

Some of the earlier work on correlation of slurry flow properties centred on sliding bed flows (Ref. 21). Small diameter test loops were employed to determine the nature of correlating functions which could later be used for scale-up purposes. More recently, Wilson, in a series of papers (Refs. 22-24) has developed a mechanistic analysis of this type of flow in which the hydraulic gradient is determined by both the sliding bed friction at the pipe wall and the hydraulic shear forces due to the flow at the bed-liquid interface.

4. EXPERIMENTAL METHODS

Experimental techniques were described in Ref. 22 for determining the coefficient of friction between the sliding bed and the pipe wall. This involved measuring the angle at which a pipe containing a bed of the slurry under investigation had to be tilted for the solids to slide relative to the pipe walls. For the case of coarse solids in a relatively viscous carrying medium (compared with water) this technique presents some problems since it could be expected that a hydrodynamic lubricating layer will be formed between the bed and pipe wall. The formation of this bed will not be instantaneous with the start of movement but will develop with time as the forces cause acceleration of the bed and layer flows until equilibrium is reached. During this period the layer thickness may change.

In the case of mechanical friction the shear forces will be proportional to the normal force between the particles and the pipe wall. For a lubricated sliding action the relationship is more complicated as explained above and the forces will be dependent on both the normal force and velocity. If the tilting tube method is employed to measure the shear forces it must be sufficiently long for the layer to develop. As the shear forces are dependent on both normal forces and velocity, various tube diameters will be required and slurry velocity will need to be controlled. The complexity of the apparatus required is comparable with that of a closed loop slurry test rig.

An alternative method of measuring the shear is to use a rotating cylinder viscometer. For a conventionally sized machine, however, the ratio of particle size to cylinder diameter, for the slurries under consideration, is likely to be so large as to affect the interaction between the moving surface and solid-liquid mixture. The results produced, therefore, will not be fully applicable to pipe flow. For the case of a settling slurry there will be a gravitational effect which will change the shear forces dependent upon the depth of the rotating cylinder beneath the slurry surface, see Fig. 2. This is potentially a useful feature which might enable one to measure shear stresses for different amounts of submergence of the pipe wall below the slurry liquid interface. It is clear, however, that a large diameter cylinder is required to overcome the variations in force generated by the large particles and this in turn implies a gravitational effect from top to bottom of the cylinder. If the depth of the cylinder is reduced the end effects becomes more apparent.

In order to overcome these problems a special machine has been designed in which a horizontal disc is rotated in a bath of slurry, as shown in Fig. 3. The disc is driven at a controlled speed by means of a variable speed motor. An optical sensor enables rotational speed to be measured. As the surface velocity on the disc varies with radius, causing changes of shear stress, the slurry to be investigated is contained within an annular channel whose width is small compared with the disc radius. The coarse fraction of slurry is contained within the annulus and is prevented from escaping by setting it at a small clearance above the disc. This will not prevent the water and fine fraction from leaking out so the disc is immersed in water and fines (or other carrier fluid that is being used) until the height of this liquid is the same as that in the channel, thus setting up a balance of hydrostatic forces. Flat vertical plates within the channel hold the slurry in position and prevent it rotating as a solid mass with the disc. A flat ring is placed within the annular channel and rests on the coarse particles. By loading the ring it is possible to simulate the forces between the coarse particles and the disc that would be generated by gravity due to the depth of the pipe surface beneath the slurry. It is possible that part of this loading will be taken by the annular channel thus affecting the loading on the disc. To overcome this effect the channel support is strain gauged to determine the vertical force. The tangential force on the ring is measured by a spring balance which enables the shear stress to be calculated for a given disc velocity and depth of slurry. Thus it is possible, by carrying out a series of tests at different values of the submergence, h, to determine the shear stress around the pipe periphery, Fig. 4, for the sliding bed part of the flow. Various theoretical assumptions can be made, or separate tests carried out to determine the shear stresses between the top surface of the bed and the carrier fluid.

The above data will enable the hydraulic gradient to be established and repeat tests will determine the functional relationship with velocity.

5. EXPERIMENTAL PROCEDURE

At the time of writing the rotary shearometer is approximately 95% complete. Delays caused by unexpected technical problems in its construction have prevented us from reporting any experimental data obtained from it. It is anticipated, however, that data will be available by the time of the conference.

To illustrate the effect of a lubricant on the frictional forces at a moving surface some tests were carried out with a rotating cylinder viscometer immersed in a slurry composed of bentonite, 2 mm nominal diameter sand and water in the ratio of 1:31:12.7 by mass respectively. The ratios by volume were 1:27.4:28.6 respectively. The solids concentration by weight was 71.6% and the equivalent volumetric concentration was 49.8%. No large particles were used because of their excessive influence on the viscometer bob, as described earlier.

The slurry was contained in a cylindrical vessel. As shear only took place over a small region, the diameter of the vessel was unimportant from this aspect.

The experiment consisted of measuring the torque required to turn the viscometer bob at a constant speed at various depths beneath the surface. Values of shear stress on the bob were calculated from its dimensions.

The slurry was freshly mixed prior to the experiment.

6. RESULTS AND DISCUSSION

The results are shown in the form of a graph (Fig. 5) in which bob wall shear stress is plotted against immersion depth. It can be seen that shear stress is a non-linear function of immersion.

A sliding bed slurry can be regarded as a mixture of solids and fluid in which that part of the solids weight which is unsupported by buoyancy causes additional pressure forces which behave as if they were hydrostatic (Ref. 22). Thus, the pressure, p, due to the solids is given by

$$p = \rho (S_s - S_f) \, g \, HC$$

where

ρ is the density of water
S_s is the relative density of the grannular solids
S_f is the relative density of the fluid
g is the acceleration due to gravity

H is the height beneath the surface
C is the concentration by volume of the solids.

The shear stress on the moving surface is given by

$$\tau = pf$$

where f = friction factor between the surface and the settling slurry.

For a bob of length ℓ, immersed a distance h (see Fig. 2), an average depth can be taken of $h + \ell/$.

$$\therefore \quad \tau = \rho \ (S_s - S_f) \ g \ (h + \ell/2) \ Cf$$

(Ignoring the shear stresses from the carrier fluid) and

$$\frac{d\tau}{dh} = \rho \ (S_s - S_f) \ g \ Cf \quad :- \text{ a constant}$$

$$= \tau/(h + \ell/2)$$

Based on these assumptions the gradient $d\tau/dh$ has been plotted on Fig. 5 employing a friction factor of 0.29 which has been found to be applicable to a sand water mixture under sliding bed conditions. It is seen immediately that the shear stresses measured in this experiment are much lower.

Measurements of torque on the viscometer bob enable the friction factor to be determined experimentally for the slurry under test. For zero immersion f = 0.12. Using this value of friction factor a second value of gradient $d\tau/dh$ has been plotted.

The tangent cuts the wall shear axis at the same point as the experimental curve simply as a result of the method of calculation. It is seen that the measured stresses rise at a much lower rate than those predicted and that they are considerably lower than to be expected on the basis of friction of a settling slurry in water. The shear stresses are comparable with those quoted by Ede (Ref. 25) for concrete at low sliding velocities.

For a pure fluid the shear stress would not increase with depth, thus the indication is that the mixture has characteristics somewhat between those of a pure fluid and a settling slurry. A possible explanation is that the yield stress of the bentonite tends to hold the loose packed particles in position relative to each other. The shear stress measured would then be a combination of the yield stress of the bentonite plus a component of friction. On lowering the viscometer bob some compaction of the particles could take place which would tend to raise the frictional component and raise the shear stress. This is in agreement with experiment and also with the observation that on reducing the immersion the shear stresses were reduced below their previous values for a comparable depth. At higher rotational velocities a stream of bentonite was seen emerging around the bob spindle at the slurry surface suggesting that some mechanism exists for separating this from the mixture.

Clearly the above experiment corroborates the evidence that it is possible to reduce the wall shear stresses of a dense phase slurry below that of a settling slurry in water. The formation of a lubricating layer can only be inferred from these experiments and the accurate measurement of the shear forces must await the completion of the shearometer.

7. CONCLUDING COMMENTS

Some of the differences between sliding bed flows and certain types of non-Newtonian slurries have been discussed. In particular the border line region separating these two flow regimes is thought to offer the potential of coarse particle transport whilst not requiring the carrier fluid to be capable of supporting these particles under conditions of zero transport velocity. Hydraulic gradients are expected to be less than for sliding bed flows.

The technical requirements to enable the wall shear stresses to be measured have been discussed and the rotary shearometer, designed to measure these forces, has been described.

As the construction of the machine is incomplete it has not been possible to present experimental data other than those which illustrate some of the concepts behind its design. Some preliminary data will be available at the conference.

8. ACKNOWLEDGEMENTS

The work described in the paper forms part of a project aimed at improving the understanding and utilisation of slurry transport. The work is supported by Industry and the Department of Trade and Industry.

Thanks are due to the Council and Chief Executive of BHRA for permission to publish this paper.

9. REFERENCES

(1) Durand, R. and Condolios, G.
"The hydraulic transport of coal and solids in pipes". Colloqiuim on Hydraulic Transport, National Coal Board, London, (1952).

(2) Elliott, D.E. and Gliddon, B.J.
"Hydraulic transport of coal at high concentrations". Proc. 1st Int. Conf. on the Hydraulic Transport of Solids in Pipes". Paper G2, BHRA, The Fluid Engineering Centre, September, 1970.

(3) McCain, D.L. and Umphrey, R.W.
"Continuous underground coal haulage by hydraulic transport". Proc. 3rd Int. Conf. on the Hydraulic Transport of Solids in Pipes". Paper A6, BHRA, The Fluid Engineering Centre, May, 1974,.

(4) Gilles, R., Husband, W.H.W., Small, M. and Shook, C.A.
"Some experimental methods for coarse coal slurries". Proc. 9th Int. Conf. on the Hydraulic Transport of Solids in Pipes". Paper A2, BHRA, The Fluid Engineering Centre, October 1984.

(5) Prettin, W. and Gaessler, H.
"Bases of calculation and planning for the hydraulic transport of run of mine coal inpipelines according to the results of the

hydraulic plants of the Ruhrkohle A.G.".
paper E2, Hydrotransport 4, BHRA, The Fluid
Engineering Centre, May, 1976.

(6) Thomas, A.D.
"Coarse Particles in a heavy medium:-
turbulent pressure drop reduction and
deposition under laminar flow".
Hydrotransport 5, Paper D5, BHRA, The Fluid
Engineering Centre, May, 1978.

(7) Haas, D.B., Husband, W.H.W. and Shook, C.A.
"The development of hydraulic transport of
large sized coal in Canada". Hydrotransport
5, Paper H1, BHRA, The Fluid Engineering
Centre, May, 1978.

(8) Sakamoto, M., Uchida, K. and Kamino, Y.
"Transportation of coarse coal with fine
particle water slurry". Hydrotranpsort 8,
Paper J2, BHRA, The Fluid Engineering
Centre, August, 1982.

(9) Klose, R.B. and Mahler, H.W.
"Investigations into the hydraulic transportation behaviour of ore and coal suspensions
with coarse particles". Hydrotransport 8,
Paper D2, BHRA, The Fluid Engineering
centre, August, 1982.

(10) Gies, R. and Geller, F.J.
"Pressure gradients and degradation at the
hydrotransport of coarse washery shales".
Hydrotransport 8, Paper D4, BHRA, The Fluid
Engineering Centre, August, 1982.

(11) Maurer, H. and Mez, W.
"Hydraulic transport of coarse-grain
material in the hard coal mining industry
and experimental test results".
Hydrotransport 8, Paper J3, BHRA, The Fluid
Engineering Centre, August, 1982.

(12) Harzer, J. and Kuhn, M.
"Hydraulic transportation of coarse solids
as a continuous system from underground
production face to the end product in the
preparation plant". Hydrotransport 8, Paper
J4, August, 1982.

(13) Carlton, A.J., French, R.J., James, J.G. and
Broad, B.A.
"Hydraulic transport of large particles
using conventional and high concentration
conveying". Hydrotransport 5, Paper D2,
BHRA, The Fluid Engineering Centre, May,
1978.

(14) Masuyama, T., Kawashima, T. and Noda, K.
"Pressure loss of pseudo-plastic fluid flow
containing coarse particles in a pipe".
Hydrotransport 5, Paper D1, BHRA, The Fluid
Engineering Centre, May, 1978.

(15) Boothroyde, J., Jacobs, B.E.A. and
Jenkins, P.
"Coarse particle hydraulic transport".
Hydrotransport 6, Paper E1, BHRA, The Fluid
Engineering Centre, September, 1979.

(16) Loadwick, F.
"Some factors affecting the flow of concrete
through pipelines". Hydrotransport 1, Paper
D1, BHRA, The Fluid Engineering Centre,
September, 1970.

(17) Pertuit, P., Tennant, J.D., Lawler, H.L. and
Cowper, N.T.
"Applicationof stabilized slurry concepts of
pieline transportation of large particle
coal". 3rd Int. Conf. on Slurry Transportation, Slurry Transport Association, March,
1978.

(18) Duckworth, R.A., Pullum, L. and
Lockyear, C.F.
"The hydraulic transport of coarse coal at
high concentration". Journal of Pipelines,
3, 1983.

(19) Brookes, D.A. and Dodwell, C.H.
"The economic and technical evaluation of
slurry pipeline transport technqiues in the
international coal trade". Proc. 10th Int.
Conf. on Slurry Technology, Slurry
Technology Association, March, 1985.

(20) Tattersall, G.H. and Banfill, P.F.G.
"The rheology of fresh concrete". Pitman
Advanced Publishing Program, 1983.

(21) Newitt, D.M., Richardson, J.F., Abbot, M.
and Turtle, R.B.
"Hydraulic conveying of solids in horizontal
pipes". Trans. I. Chem. E., Vol. 33, 1955.

(22) Wilson, K.C., Streat, M. and Bantin, R.A.
"Slip model correlation of dense two-phase
flow". Hydrotransport 2, Paper B1, BHRA,
The Fluid Engineering Centre.

(23) Wilson, K.C.
"Co-ordinates for the limit of deposition in
pipeline flow". Hydrotransport 3, Paper E1,
BHRA, The Fluid Engineering Centre.

(24) Wilson, K.C.
"A unified physcially-based analysis of
solid-liquid pipeline flow". Hydrotransport
4, Paper A1, BHRA, The Fluid Engineering
Centre.

(25) Ede, A.N.
"The resistance of concrete pumped through
pipelines". Magazine of concrete research,
November, 1957.

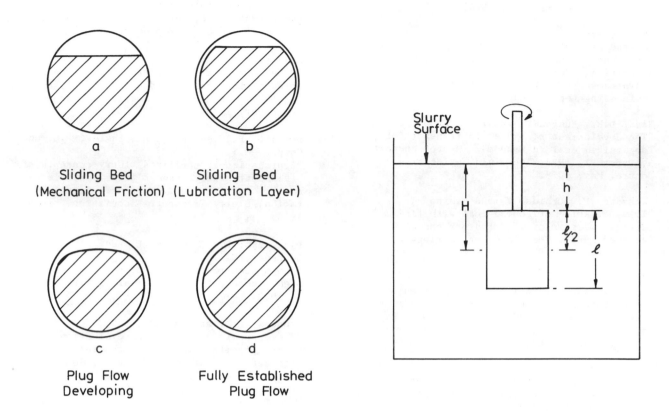

FIG.1 POSTULATED FLOW REGIMES FOR
CONCENTRATION SLURRIES.

FIG. 2 ROTATION VISCOMETER BOB IMMERSED
IN SLURRY BATH.

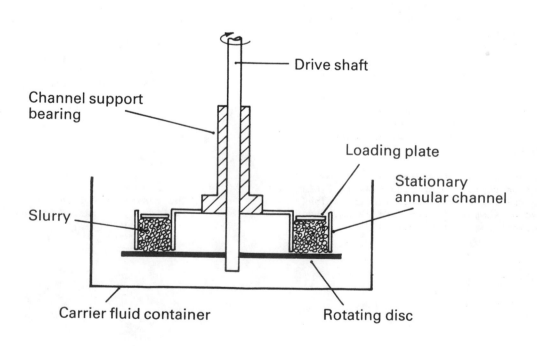

FIG. 3 ELEMENTS OF SHEAROMETER

273

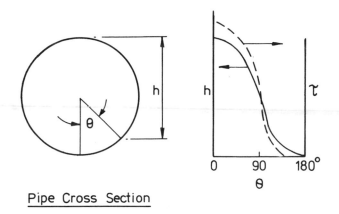

Pipe Cross Section

FIG.4 QUALITATIVE EXAMPLE OF WALL SHEAR
STRESS VARIATION.

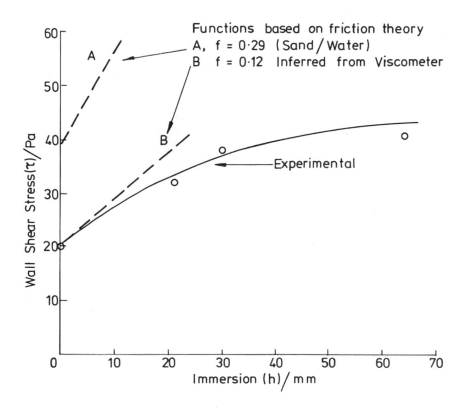

FIG. 5 EFFECT OF IMMERSION ON MEASURED SHEAR RATE.

10th International Conference on the
Hydraulic Transport of Solids in Pipes

HYDROTRANSPORT 10

Innsbruck, Austria: 29-31 October, 1986

PAPER H2

OPTICAL AND ULTRASONIC INSTRUMENTS FOR MIXTURE
FLOW INVESTIGATION

O. Scrivener, H. Reitzer, A. Hazzab, A. Idrissi

Institut de Mécanique des Fluides, UA CNRS,
Université Louis Pasteur de Strasbourg

SYNOPSIS

The purpose of this work was to investigate
solid-liquid flows in pipes on a fundamental
point of view so as to try to understand the
problems encountered in the design of industrial
size plants. Test experiments with model
suspensions have shown the existence of critical
regimes of concentration probably related to the
rheological behaviour of the mixture but also to
the local parameters of the flow: velocities,
distribution of concentration. Specific
instruments to investigate these parameters were
developed: a pulsed ultrasound Doppler
velocimeter to measure the velocities of both the
particles and the surrounding fluid, a system for
local concentration measurement based on light
absorption measurement. Results obtained for test
measurements with fine or coarse sized particles
are presented and have show the feasibility of
these techniques.

NOMENCLATURE

Y	distance from pipe wall
D	pipe diameter
I	transmitted light intensity
I_o	zero concentration intensity
D_p	particle diameter
U	particle velocity
$\bar{U}$	flow rate velocity
C_v	volumetric transport concentration
U_{max}	maximum particle velocity
R_o	radius of pipe
J_m	pressure loss for mixture flow
J_o	pressure loss for water flow

INTRODUCTION

There is nowadays no doubt the hydraulic
transport of solids being an effective and non
polluting methods. But, a look through the
specialized literature, shows to the reader that
the published results of experimental
investigations into the dependence of pressure
losses on the mixture velocity, solid
concentration distribution reveal great
quantitative discrepancies. It seems obvious that
the problems are related to a lack of information
on the experimental conditions and on the
fundamental results on the interaction of the
particles and the flow. The empirical or
semi-empirical models proposed are usually based
on specific experiments with specific solid
materials. This means that physical parameters,
like particle shapes or pipe wall material, are
not taken into account.

After several contacts with laboratories working
in the field of application of hydrotransport we
were convinced that more fundamental data are
needed to be able to understand better the flow
of the mixtures. Our first interest was to study
these flows in a laboratory size experimental rig
with well characterized solid materials and to be
able to obtain simultaneously the most important
parameters of the flow: friction losses, velocity
profiles and local concentration distribution.

One of the main problem we encountered is the
development of measuring techniques suitable in a
non-homogeneous flow. Classical methods, like use
of probes for velocity measurements, are not
suitable due the perturbation brought by the
probe and biaising of their response related to
the impact of the particles in the flow. Our
primary interest was to develop measuring
techniques for velocity and local concentration
measurements. For the first, we developed an
ultrasound velocimetry technique starting from a
system initially designed for medical use, this
means for small particles in small vessels and
weak difference of acoustic impedence. For the
second, we kept-in a method based on laser light
absorption.

EXPERIMENTAL FACILITIES

The test pipe used for the friction losses and
the velocity measurements is shown in figure 1.
The pipe loop consists of glass tube sections
20 mm in diameter. The mixture circulates
continuously round the circuit, moved by a
variable speed vortex pump. Special care was
taken for the design of the mixing tank so as to
control continuously the volume and the transport
concentrations. This tank is double and the
coarse particles, which sediment in the inner
tank, are reinjected in the flow by the mean of a
screw pump with variable speed. The pipe loop is
equipped with an horizontal and two vertical,
respectively upward and downward, test sections.

A suspension of fine particles was obtained using
particles either of starch or titanium oxide or
silt with a mean diameter of 10 m. Glass beads
were used by way of coarse particles (0.1 to 2mm

in diameter). The volumetric concentration of the particles was in the range of 1 to 10%. Higher concentrations were not obtained due to a blockage of the pump. Samples of narrow size distribution were used in order to avoid the problem of size effects, in particular on the scattering of the ultrasound waves.

HEAD LOSSES

The pressure losses are measured in the horizontal and vertical test sections at different flow rates and for different diameters and concentrations of the particles. The results are not far different from those found by most of the authors (figure 2) showing the tendancy to reach a minimum for the pressure gradient corresponding to the critical deposit velocity as obtained by use of the relation proposed by Zandi. This paper being mainly devoted to the experimental techniques, more details about the influence of the physical parameters of the flow and of the solid material can be found in other references (1).

Nevertheless, we noticed an interesting change in the behaviour of the suspension in horizontal or vertical flow when the volumetric concentration reaches a critical value close to 3% (figure 3). The same effect was observed by plotting the factor Fl of Durand and is probably related to the drastic increase of hydrodynamic interaction and shocks between the particles at higher concentrations. Similar observations were made by Parzonka (2) in plotting the results of several authors in different flow situations. This change of behaviour correlates probably with a change of rheological behaviour of the mixture and to the local characteristics of the flow: local velocities and distribution of the particles.

PULSED ULTRASOUND VELOCIMETRY

Theoretical predictions of pressure losses require better knowledge of effects of the particle on the velocity profiles of the liquid and the dispersed solid phase. Previous local velocity measurements relied primarily on classical methods involving e.g. pressure probes (Sillin et al.(3), Ayukawa et al (4)) or flow visualization. These methods are badly adapted to measurements in non-homogeneous flows due to the conditions of experiments. On the other hand ultrasound techniques were developed for blood flow measurements in medical diagnostics. Such a system was initiated by Peronneau (5) and allows the measurement of velocity profiles in blood vessels, using pulsed ultrasound waves. Kowalevski (6) also used a similar system to study the flow of concentrated liquid or solid suspensions in a pipe. In both cases the accoustic properties of the two phases are the same or roughly ajusted. Recently Garbini et al (7) published a theoretical analysis on the use of a pulsed ultrasound velocimeter for turbulence measurements.

In the present work we developed a pulsed ultrasound velocimeter adapted for measurements with coarse particles up to 2mm in diameter and of large acoustic impedance compared to that of the surrounding fluid.

Fundamentals of the system

Like for Laser Doppler Anemometry in ultrasound velocimetry a frequency shift is detected when a continuous ultrasound wave is reflected by a moving particle. This frequency is related to the velocity of the particle V, the incident ultrasonic frequency fe, the sound velocity c in the fluid and the angle between the incident beam and the velocity vector :

$$f = \frac{2 V f_e \cos\theta}{c}$$

This relationship shows that, just like for Laser Doppler Anemometry, ultrasound velocimetry is an absolute method of velocity measurements, which requires only the knowledge of the probe geometry and the frequency of wave emission in order to calculate the velocity. In the pulsed system, a burst of ultrasound of finite duration Te is launched in the fluid by the piezoelectric transducer and repeated at a given frequency Fd = 1/Td (figure 4). During the time interval between two pulses, a part of the transmitted energy is reflected by particles whose acoustic impedance is different from the surrounding fluid. The reflected signal is detected by the same probe.

The time T between pulse emission and reception of the scattered signal determines the location of the measuring point. The "window" or reception gate duration Tr determines the length of the measuring volume. Due to the flat shape of the disc of piezoelectric material used as transducer the ultrasound beam diverges slightly. This divergence can be reduced by focusing the beam.

In our system the probe was designed as a Perspex cylinder with a disc of piezoelectric material 2 mm in diameter glued at its flat edge and inserted in a block fixed at the pipe wall at an angle of 65°. The measuring volume is a cylinder of 2mm in diameter and 0.5 mm in length. The frequency of emission is 8 Mhz and the ultrasound bursts repeated with a frequency of 15 to 30 Khz. Owing to these characteristics the maximum depth is limited to 40 mm and the velocity to 2.9 m/s. These characteristics were choosen to give the best sensivity on the velocity measurements in our pipe but can be increased for other applications in changing the emmitted frequency, the repetition frequency or the duration of the window. Also, for measurements in large pipes, the transducer can be affixed on a probe, the measurement being far from the probe, the perturbation being considered as negligible.

The Doppler bursts received by the probe are processed by a data analyser. The signals are digitized at a variable sampling rate in the range of 20 to 50 Khz and stored on magnetic disks. The spectral density is then computed by Fast Fourier Transform and the successive individual spectra stored or accumulated and the noise removed by suitable filtering. The mean frequency and thus the mean velocity is computed by taking the first order moment of the probability density function or the root square of the second order moment. More than the velocity information, the Doppler burst contains an information about the dimension of the particle through the level of energy of the spectrum.

By setting a threshold on the energy of the individual spectra it was possible to measure simultaneously the velocity of the coarse particles and of the surrounding carrying fluid seeded with a few very small size tracers.

Velocimeter limitations

Beneath the limitations in velocity and depth inherent to the system some limitations are due to the scattering of the ultrasound due to the particles located on the beam path between the probe and the measuring volume. As a result from the random positions of the particles on the path of the beam, the phase of the diffused waves are random and the wave received by the probe is non-coherent (Frohly et al (8)) and an attenuation of a finite number of spectra can be observed depending from the measuring volume location and the size of the particles (figure 5). In order to determine the upper limit of the particle concentration we have simulated the path of a wave through a random distribution of spheres (9) and found that over 20% of the signal scattered by the particles located at 10 mm from the wall can reach the probe with a volume concentration of 5%.

The second limitation is due to a broadening of the spectrum very similar to that observed by i.e. George and Lumley (10) using Laser Doppler Anemometry. With the aim of evaluating the relevance of Ultrasound Velocimetry for turbulence measurements Garbini et al (7) have theoretically studied the Doppler Spectrum of a back scattered, pulsed, ultrasonic wave. They show that, in addition to the effect of turbulence velocity fluctuations, the broadening of the spectrum is related to the finite dimensions of the measuring volume (transit time error), to the velocity gradient across the measuring volume and to a "geometrical" effect. Contrarily to LDA, the ultrasonic system uses only one beam and the plane waves diffused by the particles when crossing the measuring volume are received by the transducer in a range of angles. The broadening of the Doppler spectrum causes a distortion of the turbulence spectrum, significative in the wall region, and, thus, makes difficult application of Ultrasound Velocimetry for turbulence measurements. For the present measurements we applied corrections mainly for the gradient and geometrical broadenings.

An other limitation of the system comes from the non negligible dimensions of the measuring volume which makes difficult precise location of the wall position and measurements very close to the wall.

VELOCITY MEASUREMENTS

Velocity profiles measurements were performed in the vertical upward and downward test sections with fine particle suspensions and with coarse sized glass beads. The results were recorded on half the diameter of the pipe at Reynolds number ranging between 14000 and 30000. The symmetry of the profiles was controled by a 180° probe rotation round the tube axis.

Velocity profiles with fine particles

Measurements were performed with starch, titane oxide and silt suspensions. For the three the particles diameter are ranging around 10 μm in diameter and the settling velocities are

negligible as compared to the mean flow velocity.

For starch suspensions, at a concentration of 0.1%, it can be seen (figure 6) that the experimental data at the lowest flow-rates are not significantly different from the profiles computed for water flow. At higher flow rates, the profiles diverge progressively from the Newtonian profiles: the velocities are higher in the turbulent core and lower toward the wall. This can be explained by the rheological behaviour of the suspension which was found to be slightly shear-thickening. The same effect was also observed with titane oxide suspensions at a concentration of 0.6% (figure 7). In the same figure we have plotted the results obtained with a suspension of silt at a concentration of 3.5%. A blunting of the profile is observed in the core of the flow. This blunting is related with the rheological behaviour of this suspension which was found to be plastic (constitutive law of Casson) (11). This result suggest that the assumption made by several authors saying that the yield shear stress can be neglected when computing the turbulent profile of a plastic fluid seems not to be valid.

Velocity profiles with coarse sized particles

Similar measurements, made with 1% in volume of suspended glass beads 1.4 and 2 mm are shown in figure 8. It is to notice that the measured velocities are the velocities of the solid particles and not of the mixture. Compared with the theoretical Newtonian profiles of fluids with same viscosity and density it can be observed that the velocity profiles of the coarse sized particles have a very similar shape but are shifted toward lower velocities.

The slip velocity, or difference between the velocity of the fluid and of the particle is not far from the settling velocity except for the lowest mean velocities. The slip velocity was also directly obtained by simultaneous measurements of fluid and particle velocities by setting an energy threshold on the individual Doppler spectra (figure 9).

For higher concentrations and 1 mm particles a blunting of the profile, as compared to the fluid profile, was put in evidence both for upward and downward vertical flow (figure 10). It is also observed that the diferences of the velocities in the two directions and same location in the pipe are lower than two times the settling velocity.

Settling velocities measurements

Settling velocities were measured in calm water in a tube of 20 mm in diameter filled up with water. The glass beads are launched continuously with a constant flow rate. The velocities of the particles were measured using the ultrasound probe located at several diameters from the entrance of the tube so as to measure the terminal settling velocities.

The mean values of the terminal settling velocity of 1 and 2 mm particles were measured by the choice of a reception gate duration corresponding to a measuring volume of half tube diameter length. The settling velocity versus the concentration is shown in figure 11.

It can be observed that, for the 2 mm spheres,

the settling velocities are higher than the theoretical values calculated by Valembois for single particles settling in an infinite medium. For the smaller particles the results corresponds to the theoretical values. In both cases the settling velocity decreases when the particle concentration increases. The results are in agreement with the model proposed by Maude and Whitemore (12) for the latter but diverge from about 16% for the larger particles.

To measure the local settling velocities the length of the measuring control volume was set to 0.7 mm. The results are shown in figure 12. It can be observed that for both diameters of particles the settling velocities are higher than the predicted values. An increase of the settling velocity is observed on the axis of the tube and in the wall region. For the latter this increase is higher than the expected values taking into account the wall corrections.

MEASUREMENT OF CONCENTRATION DISTRIBUTION

Different authors proposed to model the distribution of particle concentration in a flow using the mass diffusion equations for the fluid and for the particles. These models are generally based on axisymetric distribution of the velocities. However, Ayukawa (4) has shown that the profiles in a pipe can be strongly disymetric if the concentration is high. Moreover distorsion of the velocity profiles may be related to the spatial distribution of concentration.

Most of the existing instruments used for concentration measurements are instruments developed for single phase, mass or volume flow rate, calibrated for two phase flows. Only few instruments (like isokinetical sampling probes) were developed for local concentration measurements in the flow. These methods generally need the use of probes introduced in the flow and which perturbation cannot be neglected. An interesting technique was designed by Przewlocki, Michalik et al (13) for measuring the spatial concentration in a pipe by the way of -ray absorption. This method was not directly applicable to our measurements and we developed a similar system but based on laser light absorption.

Measuring technique

The method is based on the measurement of the light attenuation of a beam transmitted through the flow. The theory of geometrical optic (valid for particles larger then the wave length) shows that, when the particles are transparent with a refractive index close to that of the fluid, attenuation is related to the diffusion of the light by the particles. The transmitted light intensity is related to the concentration Cv of the particles, their diameter Dp and the length X of the beam the relationship:

$$I = I_0 \exp(-KC_V X)$$

$$K = f(D_p)$$

where Io represents the zero-concentration intensity. This equation can be approximated by a polynomial development:

$$C_V = \sum_0^n A_{jk} x^j y^{n-p}$$

In the system represented figure 13 a photodetector is set in regard to the beam of a 3 mw laser on the opposite side of the glass tube. The light is collected by an optical glass fiber in front of the photocell so as to reduce the angle of difused light collection. The system can be moved up and down by use of a step-motor so as to measure stepwisely the mean concentration for parallel beam path at different levels in the pipe. The different chords are regularly spaced from the top to the bottom of the pipe. To avoid the beam deviation when crossing the cylindrical dioptres of the pipe walls the test section was put in a parallel walls glass tank filled up with a liquid of refractive index matched to that of the glass tube. The same liquid, obtained by mixing two fluids of refractive index respectively lower and higher to that of the glass, was used as carrying fluid.

A first run is made with the carrying fluid alone to measure the zero-concentration intensity corresponding to each chord. After measuring the mean particle concentrations for the different chords the measuring head is turned to a new fix angular position. By measuring the mean concentrations for four angular position of the head it is possible to built-up a network of beam paths and to compute the corresponding coefficient of the polynomial equations. By solving this system of equations it is possible to obtain the local values of the concentration and the isoconcentration curves.

Experimental results

The system was calibrated with samples of glass beads with a narrow diameter distribution (figure 14). The results show the influence of the particle diameter on the calibration curves. It is to notice that the effects of the particle diameter or granulometry are often neglegted in the systems based on wave absorption proposed by several authors. Figure 15 shows the evolution of the mean concentration across the pipe in horizontal flow for the four angular positions of the measuring head. The corresponding results after computing the values to determine the local concentrations are presented in figure 16. It can be observed that the flow, in the present experimental conditions, is presenting a zone of high concentration near the lower part of the pipe.

The computing programm and the method were also tested by using data corresponding to an uniform concentration and the data published by Michalik (13).

CONCLUSION

An experimental investigation was carried out to study the relevance of ultrasound velocimetry to velocity measurements in coarse particle suspensions whose phases display high differences in acoustical impedance. Analysis of the performance of the system show its applicability, with a good accuracy, to fine and coarse particle measurements in a volume concentration range up to 6%. The velocimeter is limited in depth and concentration and gradient or geometrical broadening of the Doppler spectrum have to be taken into account for the measurements of mean velocities. In the present state of our investigation the system is less suited to measurements of turbulence fluctuations. In the same order of ideas we investigate the feasability of Laser Doppler Anemometry for coarse sized particles at high concentrations using methods like refractive index matching of the glass particles and the carrying fluid.

On the other hand we developed a method to measure the concentration distribution of the solids suspended in a liquid by a technique of light absorption measurements. The first results presented here are promising and show that this kind of measurements are suitable in laboratory experiments.

The purpose of this work was to study the feasability of the two experimental methods we developed for velocity and concentration measurements. The results obtained in both cases are promising but more systematic investigation, in vertical and horizontal flows, is needed before to try to interpret these first results, to try to understand the existence of critical values of the concentration, to contribute to the modelisation of the flow and to transpose the results to industrial size pipes.

ACKNOWLEDGMENT

This research programm was supported by a contract of the CNRS-PIRSEM. The authors thank MM. Rambeloniaina and Belibel for their contribution to this work.

REFERENCES

(1) RAMBELONIAINA H. V.; 1984: Etude de la perte de charge de mélanges liquides-solides en conduite horizontale et verticale
Thèse, Universite de Strasbourg

(2) PARZONKA W., KENCHINGTON J.M., CHARLES M.; 1981: Hydrotransport of solids in horizontal pipes: effects of solids concentration and particle size on the deposit velocity.
Can. J. of Chem. Eng., 59, p. 291

(3) SILIN I.A., PITTCHENKO V.F., OCHERETKO ; 1964: Etude expérimentale de la distribution des vitesses, des consistances et des grandeurs de grains dans la section de la conduite pendant l'écoulement des courants à deux phases (in russian)
Hydrotechnic and Hydromechanic, Ukr. Ac. Sci., p.63

(4) AYUKAWA E., KATAOKA H., HIRANO M. ; 1980: Concentration profile, velocity profile and pressure drop in upward solid-liquid flow through a vertical pipe.
Proc. Hydrotransport 7, SENDAI, p.E2, ed. BHRA

(5) PERONNEAU P., HINGLAIS J., PELLET M., LEGER F.; 1970: Vélocimètre sanguin par effet Doppler à émission ultrasonore pulsée.
L'Onde Electrique, 50, 5, p.3

(6) KOWALEWSKI T.A.; 1984: Concentration and velocity measurements in the flow of droplet suspension through a tube.
Experiments in Fluid, 2, p.213

(7) GARBINI J.L., FORSTER F.K., JORGENSEN J.E.; 1984: Measurement of fluid turbulence based on pulsed ultrasound technique, part 1 and 2.
J. of Fluid Mech., 118, p.445

(8) FROHLY J., PERDIGAO J.M., LEFEVRE J.E.; 1984: Propagation des ondes ultrasonores dans les milieux hétérogènes. Limitation à l'utilisation des techniques ultrasonores en controle non destructif.
Spectra 2000, 14, 13, p.31

(9) BELIBEL C.; 1985: Ecoulement en conduite de mélange solide-liquide. Champ des vitesses en écoulement vertical.
Thèse Université de Strasbourg

(10) GEORGE W.K., LUMLEY J.L.; 1973: Laser Doppler Velocimeter and its application to the measurement of turbulence.
J. of Fluid Mech., 60, p. 321

(11) FAM D., DODDS J., LECLERC D., SCRIVENER O.; 1986: Caractérisation rhéologique et écoulement en conduite de schlamms de phosphates.
Entropie, (to be published in 86)

(12) MAUDE A.D., WHITEMORE R.L.; 1978: A generalized theory of sedimentation.
The British J. of Appl. Phys., 9

(13) PRZEWLOCKI K., MICHALIK A. et al; 1979: A radiometric device for the determination of solids concentration distribution in a pipeline.
Hydrotransport 6, 83, p.105 and also private communication.

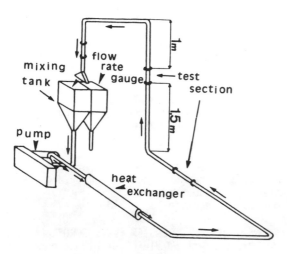

Fig. 1. Experimental set-up.

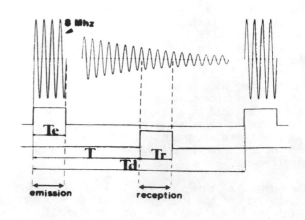

Fig. 4. Emission and reception of the ultrasound pulse.

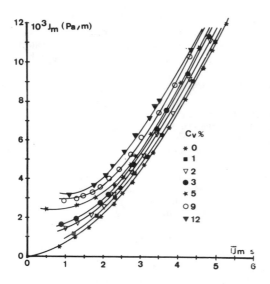

Fig. 2. Pressure gradient vs mixture velocity.

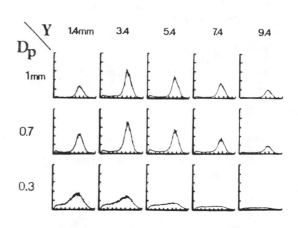

Fig. 5. Accumulated spectra for different depth and particle diameters; $C_v = 6\%$.

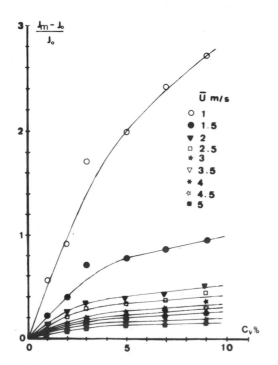

Fig. 3. Reduced pressure gradient vs concentration.

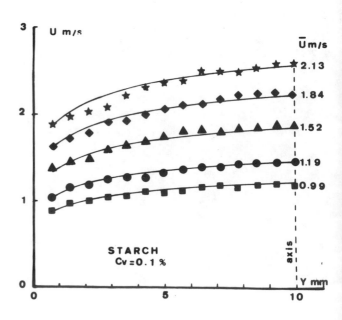

Fig. 6. Velocity profiles for fine particles in vertical flow; starch $C_v = 0.1\%$.

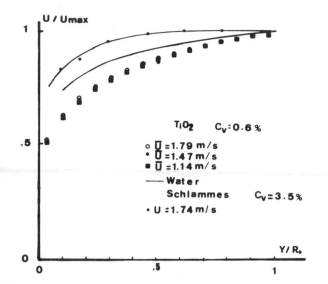

Fig. 7. Velocity profiles for fine particles in vertical flow, effect of the rheological model.

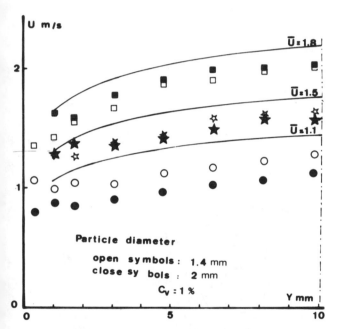

Fig. 8. Velocity profiles in vertical flow with coarse particles; glass beads $D_p = 1.4$ and 2 mm, $C_v = 1\%$.

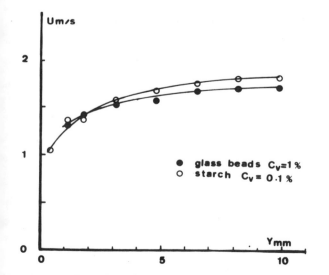

Fig. 9. Velocity profiles of the particles and surrounding fluid; glass beads $D_p = 1$ mm, $C_v = 1\%$.

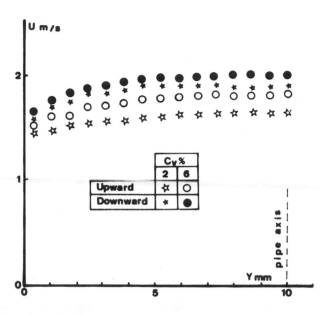

Fig. 10. Velocity profiles of the coarse sized particles; $D_p = 1$ mm, $C_v = 2$ and 6%.

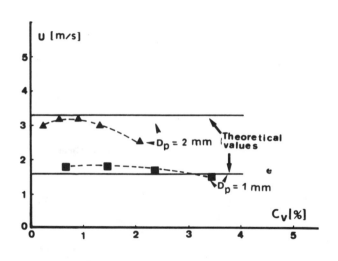

Fig. 11. Mean settling velocity.

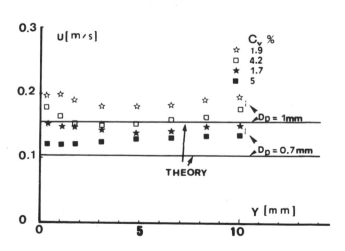

Fig. 12. Local settling velocity of coarse particles.

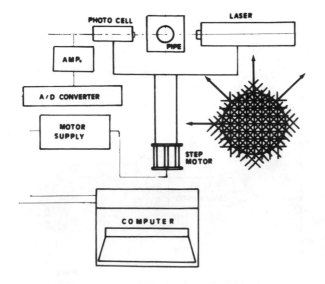

.Fig. 13. Experimental set-up for local concentration measurements.

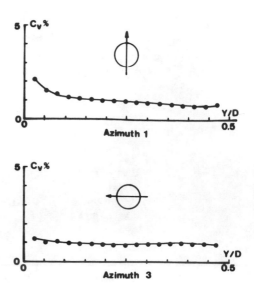

Fig. 15. Mean concentration across the pipe.

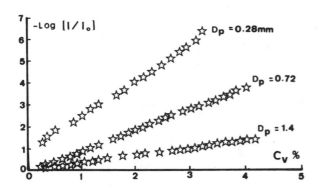

Fig. 14. Influence of the particle diameter on light absorption.

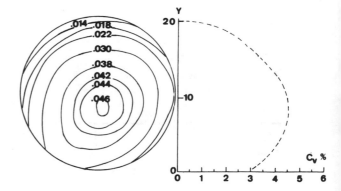

Fig. 16. Isoconcentration curves in a section of the pipe; $D_p = 0.2$ mm, $C_v = 4\%$.

10th International Conference on the
Hydraulic Transport of Solids in Pipes

HYDROTRANSPORT 10

Innsbruck, Austria: 29-31 October, 1986

PAPER J1

MODELLING THE EFFECTS OF NON-NEWTONIAN AND TIME-DEPENDENT SLURRY BEHAVIOUR

K.C. Wilson

Professor, Department of Civil Engineering,
Queen's University at Kingston, Ontario, Canada
K7L 3N6

SYNOPSIS

Fine-particle slurries tested in laminar flow often display non-Newtonian or time-dependent rheological properties, and in turbulent pipe flow such slurries can exhibit significant drag reduction. A recent analysis links this drag reduction to the scale of the dissipative micro-eddies which determine the thickness of the viscous sub-layer. This analysis is now extended from Bingham-plastic and power-law behaviour to a more general non-Newtonian model.

For time-dependent slurries the difficulty of predicting turbulent-flow friction from viscometric data indicates that it is best to scale up pilot-plant tests for turbulent flow. The proposed scaling relations, which maintain constant shear velocity, are generally applicable to non-settling slurries -- Newtonian, non-Newtonian, or time-dependent.

NOMENCLATURE

C	concentration of solids in slurry
D	internal diameter of pipe
d	particle diameter
f	friction factor
f_N	value of f for equivalent Newtonian flow
k_1, k_2	parameters in Eqs. 11,12
L	particle length
p	pressure
t	time
U	velocity (time-mean) in x direction
U_v	value of U at edge of viscous sub-layer
u_*	shear velocity [$u_* = \sqrt{(\tau_w/\rho)}$]
V	throughput velocity [discharge/pipe area]
V_N	value of V for equivalent Newtonian flow
y	distance from pipe wall
x	distance along pipe
α	ratio of integral under non-Newtonian and Newtonian rheograms
δ_v	thickness of viscous sub-layer
η	slope of high-stress rheogram asymptote
η_t	slope of the curve of τ **versus** dU/dy
λ	shape parameter in Eq. 8
μ	viscosity [$\mu = \tau/(dU/dy)$]
ξ	yield stress ratio [$\xi = \tau_y/\tau_w$]
ρ	density of pseudo-fluid
τ	shear stress
τ_i	intercept of high-stress rheogram asymptote
τ_w	shear stress at pipe wall
τ_y	yield value of shear stress
Ω	velocity correction term, see Eqs. 5,6

Subscripts

1,2	model and prototype conditions, used in scale-up
o	value at t = 0
∞	value approached at very large t

INTRODUCTION

Slurries with a significant fraction of fine particles often exhibit non-Newtonian or time-dependent rheological behaviour. Such behaviour alters frictional resistance, affecting the energy requirements and economics of a pipeline. For turbulent flow of pseudo-homogeneous mixtures both the velocity gradient in the viscous sublayer and the thickness of this sublayer are strongly influenced by the rheological properties of the pseudo-fluid. This influence often results in a decrease of fluid friction, so-called drag reduction.

A second, indirect, effect of rheological properties on energy consumption occurs when the slurry to be transported includes coarser solid particles in addition to the pseudo-fluid. Some of these particles can be supported by the pseudo-fluid, but the submerged weight of any remaining fraction of coarse particles must be transferred to the pipe wall through granular contact. As described in previous work (Refs. 6,15,16) transport of this contact load requires much more power than transport of suspended load; hence any reduction in the fraction of contact load will decrease energy consumption. This reduction can be accomplished by diminishing the fall velocity of the coarse particles, which in turn depends on the rheological properties of the pseudo-fluid composed of the carrier liquid and the fine particles. The question of fall velocity will be discussed later in this paper, but first it is necessary to consider slurries without coarse particles.

NON-NEWTONIAN TURBULENT FLOW

It is well known that certain chemical additives reduce the friction factor for turbulent flow to values well below those obtained for equivalent flows of water alone (Ref. 10), but it is not equally appreciated that the same effect can be achieved by inert rod-like particles with lengths in the order of hundreds of microns. A striking example is provided by the slurries of nylon fibres tested by Bobkowicz (Refs. 3,4). The largest effect was found for fibres with length near 1 mm and diameter approximately 20 microns. The results of Bobkowicz's tests with the largest concentrations of these fibres (2 percent and 4 percent by weight) are displayed on Fig. 1, which shows the ratio of the observed friction factor to that for Newtonian flow of water alone (for the same shear velocity) plotted against the shear velocity. It should be noted that the friction-factor ratio depends on fibre concentration as well as on shear velocity. As shown, the reduction of the friction factor can amount to more than 50 percent in favourable circumstances, implying equivalent reductions in the pressure drop and the power required for pumping.

The mechanism of drag reduction for non-Newtonian slurries has recently been the subject of an analysis by Wilson and Thomas (Ref. 17). Apart from velocity-profile changes accompanying density gradients (which are not applicable to the non-settling slurries of interest here), the principal mechanism of drag reduction in turbulent flow is associated with thickening of the viscous sub-layer. This phenomenon has been described by Lumley (Refs. 10,11), and is illustrated on Fig. 2, which graphs the distance from the wall, y, in the vertical direction and representative eddy size in the horizontal direction. The size of the largest eddies (the macro-scale of turbulence) is directly proportional to the distance from the wall, while the size of the smallest eddies (the dissipative, or Kolmogorov, micro-scale) has a known non-linear relation with distance from the wall, as indicated on the figure. At large values of y the inertial macro-eddies are much larger than the dissipative micro-eddies, and between them is a whole range of turbulent eddy sizes, shown shaded on the figure.

As the pipe wall is approached, the range of possible eddy sizes shrinks until the size of the largest and smallest eddies are equal. At this point y equals δ_v, the thickness (in a statistical sense) of the viscous sub-layer. The dashed line on Fig. 2 shows that increasing the size of the dissipative micro-eddies leads to an increase in the viscous sub-layer thickness. If other quantities are unaffected this, in turn, will produce a higher mean velocity for the same wall shear stress, giving a reduced friction factor.

Turbulent flow is inherently unsteady, and involves a series of rapid dissipative events, as eddies drop from the inertial to the dissipative size range. There is an analogy between this process and the dynamic loading of a structure, for which strain energy is determined from the integral of the stress-strain curve. As proposed by Wilson and Thomas (Ref. 17), the typical rate of energy dissipation of a turbulent eddy is determined by the integral of the curve of shear stress **versus** strain rate, i.e. the area beneath the rheogram.

Rheograms are based on viscometric measurements under laminar conditions, plotted as shear stress τ **versus** transverse strain rate dU/dy. (Of course it is equally valid to consider the shear stress as the independent variable, which determines the shear rate.) Figure 3 illustrates the rheogram shape displayed by many non-Newtonian materials, with no permanent deformation taking place until the applied shear stress exceeds a certain yield value, τ_y. At shear stresses greater than τ_y there is, in the general case, a non-linear relation between stress and strain rate, although this often approaches a

linear asymptote at high shear rates. As for a Newtonian fluid, the viscosity, μ, is defined as the ratio of τ to dU/dy. This represents the slope of a secant line joining the origin to any point of interest on the curve, as shown on Fig. 3. Another quantity associated with viscosity is the slope of a tangent to the curve of τ **versus** dU/dy, denoted by η_t and sometimes referred to as "tangent viscosity" or "incremental viscosity".

The ratio of the integrals under the non-Newtonian and Newtonian rheograms (for the same values of τ and dU/dy at the upper limit) is denoted by α. For the typical conditions illustrated on Fig. 3, the area under the non-Newtonian rheogram is significantly larger than that of the triangle which defines Newtonian behaviour, and it can be seen that for this type of behaviour α will lie between 1.0 and 2.0. The increased area under the non-Newtonian rheogram does not imply a greater rate of energy dissipation, since the energy cascade dictates the energy dissipation rate which will be imposed on the smallest eddies. Instead, the increased area indicates that, as far as the dissipative turbulent eddies are concerned, the non-Newtonian fluid acts as if it were a Newtonian fluid with viscosity $\alpha\mu$. Since the dissipation process imposes a constant-Reynolds-number relation on the smallest eddies, it follows that the product of eddy size and typical velocity for these eddies will be proportional to $\alpha\mu/\rho$.

At the interface between the inertial (logarithmic) layer and the viscous sub-layer the length scales of the largest and smallest eddies effectively coincide, as shown on Fig. 2. The same applies to the velocity scales of these eddies, which at this point are proportional to the shear velocity u_*, evaluated using the wall shear stress τ_w. As the product of velocity and eddy size varies with α, it follows that the minimum eddy size for a non-Newtonian fluid is larger than that for a Newtonian fluid (of the same μ) by the factor α. Based on the geometric relationship shown on Fig. 2, it can be seen directly that the same factor applies to the thickness of the viscous sub-layer, δ_v.

Within the viscous sub-layer conditions can be taken as essentially steady-state, so that the viscosity μ, (evaluated at the wall shear stress τ_w) is appropriate for a non-Newtonian fluid, and the velocity distribution, both Newtonian and non-Newtonian, is given by

$$U/u_* = \rho u_* y/\mu \qquad (1)$$

For the Newtonian case, the intercept with the logarithmic profile occurs at distance $\delta_v = 11.6\ \mu/\rho u_*$ and velocity $U_v = 11.6\ u_*$. For the non-Newtonian case, with its thickened viscous sub-layer, 11.6 must be replaced by 11.6 α. In the logarithmic zone the general form of the velocity profile is given by

$$U/u_* = 2.5\ \ln\ (y/\delta_v) + U_v/u_* \qquad (2)$$

where 2.5 is the inverse of the von Karman coefficient. Substitution of $U_v/u_* = 11.6$ and $\delta_v = 11.6\ \mu/\rho u$ yields the classic expression for the Newtonian case

$$U/u_* = 2.5\ \ln\ (\rho u_* y/\mu) + 5.5 \qquad (3)$$

The equivalent non-Newtonian expression (Ref. 17) becomes

$$\frac{U}{u_*} = 2.5 \ln\left(\frac{\rho u_* y}{\mu}\right) + 5.5 + 11.6(\alpha-1) - 2.5 \ln(\alpha) \qquad (4)$$

Since the area ratio α (like the viscosity μ) is evaluated for $\tau = \tau_w$, the last two terms of Eq. 4 will be constant throughout the pipe, and thus their contribution to the velocity ratio U/u_* will apply equally to the ratio of the throughput velocity V to u_*.

For convenience V_N will be used to represent the throughput velocity for equivalent Newtonian flow, i.e. flow with the same τ_w for a Newtonian fluid with μ corresponding to the non-Newtonian value at $\tau = \tau_w$. A final term, Ω, is included to represent any effect of a possible blunting of the velocity profile in the logarithmic or core regions of the flow. The expression is then written

$$V/u_* = V_N/u_* + 11.6(\alpha - 1) - 2.5 \ln \alpha - \Omega \qquad (5)$$

As the flow in the logarithmic and core regions is governed by inertia, not viscosity, non-Newtonian viscosity **per se** should have virtually no effect in these regions, implying a zero value for Ω. However it is known that non-Newtonians with a yield stress have a velocity profile with a flattened core, and in this case Ω should depend on τ_y/τ_w (for convenience this ratio will be denoted by ξ). One reasonable way of approximating this effect, described in Ref. 17, gives the equation

$$\Omega = -2.5 \ln(1 - \xi) - 2.5 \xi(1 + 0.5\xi) \qquad (6)$$

Substitution of Eq. 6 into Eq. 5 gives a relationship of a general nature, and this can readily be applied to those problems for which the wall shear stress τ_w is known in advance, permitting the area ratio α to be determined from the rheogram. Conversely, if the observed friction factor is available, together with the Newtonian one, the associated value of α can be calculated. These calculations were made for the nylon-fibre data of Bobkowicz, mentioned previously in connection with Fig. 1; and it was verified that the values of α for these data all fall in the range between 1.0 and 2.0, as expected.

Rheograms are often approximated by simple algebraic formulas. The most popular -- the Bingham and power-law models -- were discussed in Ref. 17. Unfortunately, these two-parameter models cannot give an accurate representation of behaviour such as that mentioned above in connection with Fig. 3. The Bingham model tends to be in good agreement with data at high shear stress, but does not reproduce the non-linear nature of the relation at lower stresses, although it does display a yield point. The power-law model (corresponding to a straight line when stress and strain-rate are plotted on logarithmic co-ordinates) tends to give a reasonable simulation in the mid-range of the variables, but shows weaknesses toward both ends of the range. At the low end it cannot reproduce the yield point, and it has a vertical tangent as dU/dy approaches zero (actual materials, with or without

a yield stress, typically have a non-vertical tangent at this location). At high shear stress the power-law formulation predicts a constantly-diminishing viscosity μ. As noted by Fredrickson (Ref. 8) and shown graphically by Thomas (Ref. 13), this progressive diminution of viscosity does not reflect reality. Moreover, the difficulties found for the Bingham and power-law formulations are not overcome by combining the two by addition of a yield shear stress to the power-law model -- the tangent is still vertical at dU/dy of zero, and the viscosity still decreases indefinitely with increasing shear stress.

PROPOSED NON-NEWTONIAN MODEL

Very simple functions, such as two-parameter models discussed above, were required before the computer came into common use. This simplicity is no longer necessary, and more complex modelling functions can now be considered. In selecting a function of this sort it is desirable to begin with the behaviour near the ends of the range. Thus, at sufficiently high shear stress the function should approach a straight line which can be defined by the relation

$$\tau = \tau_i + \eta \, dU/dy \qquad (7)$$

Here η is the limiting tangent viscosity, i.e. the slope of the high-shear asymptote on the rheogram, and τ_i is the intercept of this asymptote when projected back to the shear axis. The function itself will also intercept the shear axis; this occurs at the yield shear stress, τ_y, which in general will be different from τ_i.

To complete the mid-range of the rheogram, a function is required which passes through τ_y and approaches Eq. 7 at high shear stress. This requirement can be achieved by incorporating a term which has an exponential decrease in magnitude with dU/dy, i.e.

$$\tau = \tau_i + \eta \, dU/dy - (\tau_i - \tau_y)\exp[-\lambda \, dU/dy] \qquad (8)$$

As the Bingham model is a particular case of Eq. 8, it is clear that this equation has the potential to surpass the Bingham formulation in fitting rheologic data. In addition to the "Bingham" parameters τ_i and η, use of Eq. 8 requires evaluation of the true yield stress τ_y and the shape parameter λ.

In comparing the proposed equation with the other well-known model -- the power-law function, it should be recalled that the latter does not provide good agreement with data at the extremes of the range, and merely shows that much rheologic data approximate a straight line on logarithmic co-ordinates over a sizeable portion of the mid-range of the variables. The proposed model will be able to replace the power-law model provided it gives logarithmic plots which approximate straight lines for appropriately chosen values of the parameters (for substances without yield stress, τ_y can be set to zero). By testing various values of the parameters in Eq. 8, it was verified that most cases do give a good approximation to a straight line for the mid-range of logarithmic plots. Moreover, with other values of the parameters less usual behaviour can also be duplicated; for example the concave-upward rheogram for a Bentonite slurry obtained by Engelund and Wan (Ref. 7) can be represented by Eq. 8 with τ_y set greater than τ_i.

Up to this point, the proposed model for non-Newtonian flow has been presented only in the form suitable for dealing with flows having a constant stress field. In a pipe, however, the shear stress varies, and integration is required to obtain the velocity distribution and the discharge through the pipe (which determines the throughput velocity V). The results for laminar pipe flow are usually presented as a function (or graph) linking the wall shear stress τ_w with the flow number 8V/D. For the proposed model, the relation is not given in closed form but can readily be evaluated by numerical integration.

Figure 4 shows data obtained by the GIW Hydraulic Laboratory for phosphate slimes of a strongly non-Newtonian character. These data could not be represented in a satisfactory way by either Bingham or power-law models, but are matched closely by the new model, expressed by the equation

$$\tau = 55.0 + 0.0105 \frac{dU}{dy} - 33.0 \exp[-0.0135 \frac{dU}{dy}] \quad (9)$$

The evaluation of α was made previously for the Bingham and power-law models (Ref. 17). For the proposed new model α is given by

$$\alpha = \frac{1}{\tau}[2\tau_i + \eta \frac{dU}{dy} - \frac{2(\tau_i - \tau_y)}{(\lambda\, dU/dy)}[1 - \exp(-\lambda \frac{dU}{dy})]] \quad (10)$$

where τ is to be evaluated by Eq. 8.

Figure 5 is a plot of wall shear stress **versus** 8V/D, showing the fit line and the data points for laminar flow previously plotted on Fig. 4, together with associated data points obtained by the GIW Hydraulic Laboratory for turbulent flow of this slurry. Also shown on the figure are various calculated curves in the turbulent-flow range. The curve furthest to the left, marked "N", is based on the viscosity μ (i.e. $\tau/[dU/dy]$), evaluated using Eq. 9. It represents the first or "Newtonian" term of Eq.5, and predicts the velocity which would occur (for given τ_w) in the absence of drag reduction from sub-layer thickening. It can be seen that the observed velocities are about 20 percent in excess of this prediction, showing that significant drag reduction has occurred.

The curve furthest to the right, marked "WT", includes the terms in α and Ω of Eq.5, with α evaluated using Eqs.9 and 10, and Ω using Eq.6. It can be seen that, at least for conditions just above the laminar-turbulent transition, this curve predicts velocities significantly greater than those observed. The reason for this over-prediction may be found in the thickness of the portion of the flow dominated by viscosity, which is normally such a small function of the pipe radius that can be ignored in the calculation of throughput velocity (as was done in Eq.5). The case considered here is unusual in this respect. For wall shear stress less than 75 Pa, the viscous sub-layer thickness δ_v exceeds 5% of the pipe radius. In these circumstances both the viscous sub-layer and the buffer layer between it and the logarithmic portion of the profile should be taken

into account in predicting throughput velocity. Such calculations are not entirely straightforward, even for Newtonian flows. However, indicative computations have been carried out, based on the simple buffer-layer formulation of von Karman (Ref.12). The results are shown by the line marked "VB" on Fig.5. This line intersects the laminar line at the last experimental point for this type of flow, and is also in good accord with the two turbulent-flow observations at the largest values of τ_w. A more sophisticated modelling could well provide a closer fit to the remaining two experimental points, but the general agreement is clear.

It is widely held that the throughput velocity can be calculated by using the "tangent viscosity" η_t in place of μ in a Newtonian formulation (e.g. Eq.3). There is no physical rationale for this method, and thus it has not been plotted on Fig.5. If such a line were drawn it would pass close to the experimental point at $\tau_w \simeq 75$ Pa, and have a steeper slope than the curves shown, significantly under-predicting the throughput velocity of the highest two experimental points.

EFFECTS OF TIME-DEPENDENT RHEOLOGY

In addition to slurries which display non-Newtonian behaviour of the type described above, there are slurries which have rheograms that depend on the duration of testing. Such time-dependent behaviour can sometimes be expressed mathematically by a quasi-Newtonian relation in which the viscosity μ depends on the duration of the test, t , in accord with an exponential-decay law of the type

$$\mu = \mu_\infty + (\mu_o - \mu_\infty)e^{-k_1 t} \quad (11)$$

Here μ_o is the value of viscosity at t = 0 and μ_∞ is the value approached at extremely large test durations.

In other cases the behaviour of the time-dependent material resembles that of a Bingham plastic for which the yield stress varies between τ_{io} at t = 0 and $\tau_{i\infty}$ at very large test times, as given by the equation

$$\tau = \tau_{i\infty} + \eta\, dU/dy + (\tau_{io} - \tau_{i\infty})e^{-k_2 t} \quad (12)$$

It can be seen that the form of Eq. 12 is somewhat similar to that of Eq. 8. However, the evaluation of α is now made more complicated by the necessity of integrating over time as well as over strain rate. The appropriate time span for the eddies at the edge of the viscous sublayer is proportional to the time scale for micro-eddies, which is given by $\mu/(\rho u_*^2)$ for Newtonian flow and by $\alpha\mu/(\rho u_*^2)$ for the non-Newtonian case considered here.

There are great practical difficulties in the evaluation of time-dependent relationships such as Eqs. 11 and 12, not only because of the involvement of the unknown α in the time scale, but also because viscometers are generally not well suited for obtaining accurate measurements at the small durations that are of interest in these cases. Fortunately, the required time scale (and the associated value of α) depends only on the properties of the material and on the shear velocity u_*. This fact has led to the development of a scaling technique which will be described in the following section.

Time dependence has also been found in the recovery of rheologic parameters after the cause of deformation has ceased to act. This matter was investigated by Chu and Wang (Ref.5), who observed the fall of spheres through a non-Newtonian material which had previously been agitated. Particles with insufficient submerged weight to overcome the force produced by the yield stress of the material will not settle (Ref.1). Chu and Wang found that the maximum size of non-settling particles increased with the time elapsed since agitation, with an asymptotic approach to a constant value at large elapsed times. This behaviour clearly indicates a corresponding increase of yield stress with time. As with the other aspects of time-dependent flow mentioned above, the variation of yield stress with time will have an indirect effect on the quasi-steady-state turbulent flow in a pipe. Once again, the flow within the viscous sub-layer approximates the long-term or steady-state situation, while the adjacent turbulent eddies follow the dissipative, or Kolmogorov, scale.

Tests of the fall of particles in a material having a yield stress were also made by Highgate and Whorlow (Ref.9) and by Thomas (Ref.13). In their experiments the material was contained in the annulus between two coaxial cylinders with vertical axes, arranged so that one of the cylinders could be rotated to set up stress and deformation. It was observed that particles in the annulus fell more rapidly as the transverse shear between the cylinders was increased. When this finding is applied to pipeline flow of similar mixtures, it is seen that below a certain wall shear stress (with its associated shear velocity u_*) the coarse particles will be completely supported by the non-Newtonian carrier fluid; but at higher wall shear stress the particles will begin to settle. As mentioned previously, flows with settled or contact-load solids require high pumping power, and thus should be avoided.

As the boundary between settling and non-settling behaviour is determined by u_* (for specified coarse particles and carrier fluid), maintaining u_* constant will give similarity in this regard. The shear velocity also fixes the quantities μ and α for the non-Newtonian and time-dependent slurries discussed above, and thus it appears that scaling relations for turbulent flow should be based on constant u_*.

SCALING TURBULENT FLOW

For important pipelines involving turbulent flow of slurries, it is highly desirable to carry out pilot-plant tests under turbulent conditions. Usually the diameter of the test pipe is less than that of the prototype line, and scaling relations are employed to obtain prototype values. These relations are generally based on power-law approximations; an approach which follows the early work of Blasius (Ref.2).

For non-Newtonian flows, it is better to lay aside the power-law scaling formulation and reconsider the question of hydraulic similitude.

On the basis of the analysis put forward above it is found that, for all non-settling mixtures, similitude can be achieved by maintaining the shear velocity at a constant value. Constant wall shear stress, and hence constant u_*, implies a constant value of both the viscosity μ and the rheogram-area ratio α; leading to similitude in the behaviour of the viscous sub-layer. Likewise, for a material with a yield shear stress, the value of u_* fixes ξ (i.e. τ_y/τ_w) and hence produces a constant term Ω for the yield-effect correction in the core of the velocity profile. Thus, for given u_* all the terms on the right hand of Eq.(5) will be constant except the first, which can be written $2.5 \ln (\rho D u_*/\mu)$. This result is used in obtaining expressions for the throughput velocities V_1 and V_2 in pipes of diameters D_1 and D_2, respectively. On comparing the two expressions, based on the condition that u_* is the same in both pipes, it is found that

$$V_2 = V_1 + 2.5 \ u_* \ \ln(D_2/D_1) \tag{13}$$

The associated pressure gradients $(\Delta p/\Delta x)_1$ and $(\Delta p/\Delta x)_2$, must also be scaled, using a second relation based on the constant value of wall shear stress, i.e.

$$(\Delta p/\Delta x)_2 = (\Delta p/\Delta x)_1 \ (D_1/D_2) \tag{14}$$

The combined use of Eqs.(13) and (14) allow the scale-up of individual data points for turbulent non-Newtonian flow from one pipe size to another. Such separate scaling of velocity and pressure gradient on a point-by-point basis may at first appear unusual, but it is no different in practice from the scaling of, say, head-discharge curves for centrifugal water pumps.

The technique is illustrated on Fig. 6, using Thomas's data for a 7.5% kaolin slurry. This material has a non-Newtonian rheogram similar in form to Fig.3, and displays significant drag reduction (Refs.14,17). Figure 6 plots pressure gradient **versus** throughput velocity for two pipe sizes (D_1 = 18.9 mm, D_2 = 105 mm). The points from the smaller pipe have been scaled up individually using Eqs.13 and 14, and, as shown on the figure, these scaled points are in excellent accord with the observations from the larger pipe.

It should be noted that the experimental points do not plot as straight lines on the logarithmic co-ordination of Fig.6, and this is to be expected for non-Newtonian flow. The strength of the proposed scaling technique is shown by the fact that this non-linear behaviour is correctly scaled to the larger pipe size.

CONCLUSION

Fine-particle slurries often show non-Newtonian behaviour in tests under laminar conditions. The friction factor for turbulent flow of such slurries is also affected, primarily through an increase in the size of the small dissipative eddies which determine the thickness of the viscous sub-layer. In favourable circumstances this mechanism can reduce the friction factor to less than half the value for an equivalent Newtonian flow.

The analysis of this behaviour made previously for Bingham-plastic and power-law rheological models has now been extended to a more general model (Eq.8) which can fit a broader range of material characteristics.

Time-dependent rheological behaviour also occurs for some slurries, greatly increasing the difficulty of predicting the turbulent-flow friction factor from viscometric data. For these cases it is best to use pilot plant testing in the turbulent range, which can be scaled to prototype values by means of the constant-shear-velocity relations. These are given by Eqs. 13 and 14 and are generally applicable to non settling slurries -- Newtonian, non-Newtonian, and time-dependent.

ACKNOWLEDGEMENTS

The test data which form the basis for Figs.4 and 5 were provided by the GIW Hydraulic laboratory, and the writer wishes to express his thanks to Thomas W. Hagler Jr. Thanks also go to A.J. Bobkowicz and A.D. Thomas for making their data available.

REFERENCES

1. Ansley, R.W. and Smith, T.N.: "Motion of Spherical Particles in a Bingham Plastic", AIChE Jour., Vol. 13, No. 6, pp 1193-1196, Nov. 1967.

2. Blasius, P.R.H.: "Das Aehnlichkeitsgesetz bei Reibungsvorgaengen in Fluessigkeiten", V.D.I. Forschungsheft No. 131, 1913.

3. Bobkowicz, A.J.: The Effect of Turbulence on the Flow Characteristics of Model Fibre Suspensions, Ph.D. Thesis, McGill Univ., 1963.

4. Bobkowicz, A.J. and Gauvin, W.H.: "The Turbulent Flow Characteristics of Model Fibre Suspensions", Canad. Jour. Chem. Engng. Vol. 43, pp 87-91, Apr. 1965.

5. Chu, J. and Wang, Z.: "Settling Velocity of a Sphere in Homogeneous Bingham Fluid", Paper I-1, Proc. Internatl. Workshop on Flow at Hyperconcentrations of Sediment, IRTCES, Beijing, China, Sept. 1985.

6. Clift, R., Wilson, K.C., Addie, G.R. and Carstens, M.R.: "A Mechanistically-Based Method for Scaling Pipeline Tests for Settling Slurries", Proc. Hydrotransport 8, pp 91-101, BHRA Fluid Engineering, 1982.

7. Engelund, F. and Wan, Z.: "Instability of Hyperconcentrated Flow", Jour. of Hydr. Engng, ASCE, Vol. 110, No. 3, pp 219-233, March 1984.

8. Fredrickson, A.G.: Principles and Applications of Rheology, Prentice-Hall, Englewood Cliffs, N.J., 1964 (pp 22-23, p 87).

9. Highgate, D.J. and Whorlow, R.W.: "The Viscous Resistance to Motion of a Sphere Falling through a Sheared Non-Newtonian Liquid", Brit. Jour. Appl. Phys., Vol. 18, pp 1019-1022, 1967.

10. Lumley, J.L.: "Drag Reduction in Turbulent Flow by Polymer Additives", Jour. Poly. Sci. Macromol. Rev. 7, A. Peterlin (Ed.), Interscience, New York, pp 263-290, 1973.

11. Lumley, J.L.: "Two-Phase and Non-Newtonian Flow", Chap. 7 of Turbulence, P. Bradshaw (Ed.), Topics in Applied Physics, Vol. 12, Springer-Verlag, Berlin, 1978.

12. Reynolds, A.J.: Turbulent Flows in Engineering, Wiley, London, 1974, (pp 186-190).

13. Thomas, A.D.: "Settling of Particles in a Horizontally Sheared Bingham Plastic", Proc.1st Natl. Conf. on Rheology, Melbourne, Aust., 1979.

14. Thomas, A.D.: "Slurry Pipeline Rheology", Proc. 2nd Natl. Conf. on Rheology, Sydney, Aust., 1981.

15. Wilson, K.C.: "A Unified Physically-Based Analysis of Solid-Liquid Pipeline Flow", Proc. Hydrotransport 4, pp A1-1-16, BHRA Fluid Engineering, 1976.

16. Wilson, K.C.: "Analysis of Contact-Load Distribution and Application to Deposition Limit in Horizontal Pipes", Jour. of Pipelines, Vol.4, pp 171-176, 1984.

17. Wilson, K.C. and Thomas, A.D.: "A New Analysis of the Turbulent Flow of Non-Newtonian Fluids", Canad. Jour. Chem. Engng. Vol. 63, pp 539-546, Aug. 1985.

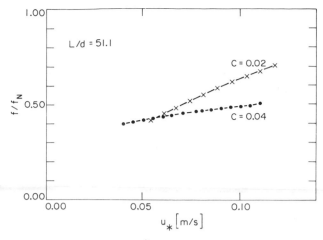

Fig. 1. *Friction-factor ratio for Bobkowicz's nylon-fibre slurries.*

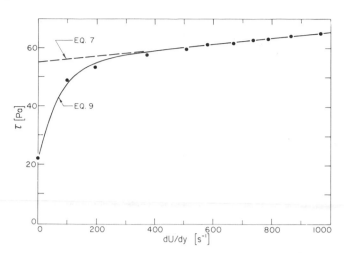

Fig. 4. *Rheogram for phosphate slimes tested at GIW.*

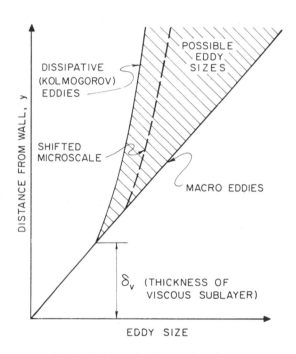

Fig. 2. *Eddy scales in turbulent flow.*

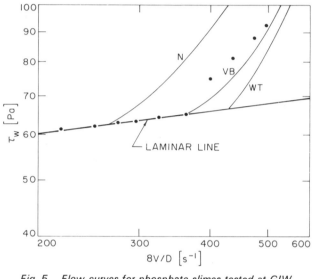

Fig. 5. *Flow curves for phosphate slimes tested at GIW.*

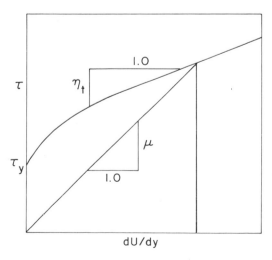

Fig. 3. *Representative non-Newtonian rheogram.*

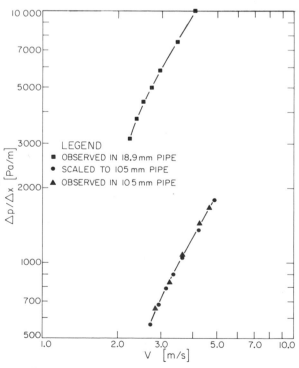

Fig. 6. *Scale-up for Thomas's 7.5% kaolin slurry.*

10th International Conference on the
Hydraulic Transport of Solids in Pipes

HYDROTRANSPORT 10

Innsbruck, Austria: 29-31 October, 1986

PAPER J2

COMPARATIVE RHEOLOGICAL CHARACTERISATION
USING A BALANCED BEAM TUBE VISCOMETER
AND ROTARY VISCOMETER

J H Lazarus and P T Slatter

Department of Civil Engineering,
University of Cape Town

SYNOPSIS

Kaolin Clay slurries and uranium mining tailing slurries of different concentrations were tested in the Balanced Beam Tube Viscometer (BBTV) and rheological characteristics were compared with rheologies determined in a rotary type viscometer. It was found that both the kaolin slurries and the uranium tailings slurries could be characterised by yield-pseudoplastic rheologies. The rheological parameters from both types of viscometers were used for laminar and turbulent flow predictions. In all cases the BBTV rheology proved to be a better indicator of friction head loss gradients in pipes than the rotary viscometer.

NOMENCLATURE

Symbol	Description	Unit
A	cross-sectional area	m^2
C	concentration	%
d	particle diameter	μm
D	tube diameter	m
f	Fanning friction factor	
i	hydraulic gradient	m/m
K	fluid consistency index	
K'	apparent fluid consistency index	
L	tube length	m
NR_e	Newtonian Reynolds Number	
n	flow behaviour index	
n'	apparent flow behaviour index	
p	pressure	Pa
R_e	non-Newtonian Reynolds Number	
r	radius at a point in the tube	m
S	relative density	-
S'	Rotary viscometer torque reading	
U	Rotary viscometer speed setting	-
u	point velocity	m/s
v	average slurry velocity	m/s

Δ	increment of	-
μ	dynamic viscosity	Pa s
ρ	density	kg/m^3
τ	shear stress	Pa
τ_y	yield stress	Pa

SUBSCRIPTS

o	at the pipe wall
m	mixture (slurry)
s	solids
v	volumetric

1. INTRODUCTION

Viscometer tests were carried out in the BBTV (Balanced Beam Tube Viscometer) and a Rotary Viscometer in order to determine the rheology of kaolin clay slurries and uranium mining tailings slimes. A range of tests were carried out using different concentrations and particle size ranges.

The yield-pseudoplastic Rheological Model was used to model laminar flow data. The results from the two viscometers did not agree (see Figs. 3 and 4) and are summarised in Table 1.

In order to determine which results were more meaningful, both laminar and turbulent flow predictions were made, using the above rheologies, and compared with tube flow data. The turbulent flow predictions were based on the theory of Torrance[3], Metzner and Reed[4] and Nikuradse[2].

An advantage of the BBTV is that it is, in fact, a miniature pipeline and valid turbulent flow data can be obtained using this instrument.

2. THEORETICAL PREDICTIONS

Test results where analysed using the yield-pseudoplastic rheological model characterised by

$$\tau = \tau_y + K\left[-\frac{du}{dr}\right]^n \qquad (1)$$

2.1 Laminar Flow

For a yield-psuedo-plastic fluid, the pressure drop across a length of pipe can be predicted from the equation

$$\frac{8v}{D} = \frac{4n}{K^{1/n}\tau_o^3}(\tau_o - \tau_y)^{\frac{1+n}{n}}\left[\frac{(\tau_o - \tau_y)^2}{1+3n}\right.$$

$$\left. + \frac{2\tau_y(\tau_o - \tau_y)}{1+2n} + \frac{\tau_y^2}{1+n}\right] \qquad (2)$$

For a given value of τ_o, v is calculated from equation (2) and i_m is calcualted from equation (3)

$$i_m = \frac{4 \tau_o}{D \rho_m g} \tag{3}$$

2.2 Turbulent Flow

2.2.1 Torrance

According to Torrance[3], for a yield-pseudoplastic fluid in turbulent flow, the friction factor (f) can be expressed in terms of the Reynolds Number:

$$(2/f)^{0,5} = \left(\frac{3,8}{n} - 4,17\right) + \frac{2,78}{n} \ell n \left[1 - \frac{\tau_y}{\tau_o}\right]$$
$$+ \frac{2,78}{n} \ell n \left[R_e \sqrt{f^{2-n}}\right] + 0,965 \left(5n - \frac{8}{n}\right) \tag{4}$$

where
$$R_e = \frac{8 \rho_m v^2}{K \left(\frac{8v}{D}\right)^n} \tag{5}$$

Equation(4) is solved for f using the Reynolds Number and i_m is calculated from

$$i_m = \frac{2 f v^2 S_m}{D g} \tag{6}$$

2.2.2 Metzner and Reed

Using the generalised approach of Metzner and Reed[4], K' and n' values are evaluated from the tangent to the curve of equ.(2) when plotted on logarithmic axes.

Dodge and Metzner[2] developed the following

$$\frac{1}{\sqrt{f}} = \frac{4,0}{n'^{0,75}} \log \left[R_e f^{(1-\frac{n'}{2})}\right] - \frac{0,40}{n'^{1,2}} \tag{7}$$

with R_e using K' and n' in equation (5) and

$$f = \frac{2 \tau_o}{\rho_m v^2} \tag{8}$$

For a given value of τ_o, K' and n' are evaluated. Equation (7) and (8) are solved for v and i_m is calculated as before from equation (6).

2.2.3 Nikuradse

For comparison, a simple approach was adopted by fitting a straight line to laminar flow data from the BBTV on the relevant pseudo-shear diagram. For the rotary viscometer, the line was fitted to the data points on the rheogram.

The slope of these lines were taken as the dynamic viscosity (μ) and turbulent predictions were made using Newtonian theory. Yield stress and rheogram curvature are therefore neglected.

According to Nikuradse[2]

$$\frac{1}{\sqrt{f}} = 4,0 \log \left(NR_e \sqrt{f}\right) - 0,40 \tag{9}$$

where $NR_e = \frac{\rho_m v D}{\mu} \tag{10}$

Equation (9) is solved and v and i_m calculated as before.

3. DESCRIPTION OF APPARATUS

3.1 Balanced Beam Tube Viscometer (BBTV)

Although the method of using this instrument has been considerably refined, the description is presented in ref. 1 and will not be repeated here.

The improved BBTV incorporates pressure tappings in the tubes, allowing direct measurement of wall friction loss. This avoids the inclusion of entrance losses, exit losses and hydraulic grade line curvature due to developing flow.

There are four pressure tappings in each tube, allowing the operator flexibility of pressure range. The pressure tappings are water filled and connected to the differential pressure transducer via solids collecting pods, and a flushing/bleeding facility.

The pressure tappings are located so that fully developed flow exists between them. The hydraulic effective length is thus equal to the physical distance between tappings (L).

Rheological parameters (yield stress (τ_y), fluid consistency index (K) and flow behaviour index (n)) are extracted from the BBTV data in the following manner.

A pseudo-shear diagram is plotted (Figs. 1 and 2) using the pseudo-shear rate (8v/D) as abscisa and wall shear stress ($D\Delta p/4L$) as ordinate. Data points in laminar flow only from both tubes are used and the best curve is fitted to the data by eye. At this stage, τ_y is read as the ordinate intercept.

Three co-ordinates are then read from the curve. These co-ordinates are used successively three times in pairs to solve equation (2) for K and n values. These values (τ_y, K, n) are then inserted in equation (2) to plot three curves through the data and the process is repeated until a satisfactory fit of equation (2) to the data points is obtained.

3.2 The Rotary Viscometer

A rotary viscometer is one in which the slurry to be tested is introduced into the gap between a rotating rotor and a fixed cup or beaker, known as the Searle system. The gap is annular, between two coaxial, concentric cylinders, one of which is stationary, while the other rotates. The rheology of the slurry can be determined from the resistance to rotation caused by the sample material. The quantities measured are torque and rotational speed. The rotor can be rotated at different fixed speeds. This instrument is an industry standard, and a detailed description is not included here.

The dual measuring head was used with a 50/500 scale. The rotor system used comprises a profiled rotor and a profiled cup representing the rotating and stationary measuring parts respectively. Rotational speed (shear rate) is varied by selecting a gear ratio, called a "U" value and torque measurements are measured on a graduated scale, called "S'" values. Rheological parameters are extracted from the rotary viscometer data as follows.

The speed and torque readings are converted to shear rate and shear stress respectively, and a rheogram is then plotted. The best fit of equation (1) is then used to obtain τ_y, K and n values.

Please note: The rotational speed and torque are converted to shear stress and shear rate respectively by the standard method given in the manual of the particular rotary viscometer which was used and which is an industry standard. The constants given in the manual were checked by an absolute method.

4. TEST PROCEDURE

The required slurries were prepared and tested in the BBTV as well as the rotary viscometer. Volumetric concentration (C_v) was controlled by adding or subtracting supernatant liquor as required. pH was controlled by adding sulphuric acid when necessary.

The particle size distributions were determined using a Malvern 2600/3600 Particle Sizer VF.6 .

5. DESCRIPTION OF MATERIAL TESTED

5.1 Kaolin

Kaolin from the Serina mine near Cape Town was obtained as a dry powder and mixed with water to the required concentration. The resulting pH was 6,8 and this was maintained throughout the tests. The particle size distribution is shown in Fig. 7.

5.2 Uranium mining tailings slurries

The slimes were received wet from the Rössing mine and the pH was very low (pH = 2,09) due to the acid leaching process used at the mine.

During the tests, the pH of the slurries rose from 2,09 to 3,43. In order to determine whether this change in pH could affect the rheology, tests were performed at pH = 1,94 and pH = 3,47. Sulphuric acid was used to reduce the pH. No change in rheology was detected.

In order to determine the effect of particle size on rheology, the slimes was separated into four particle size ranges using a SWECO Vibro-Separator.

The four particle size ranges were:-

$$d > 400\mu m$$
$$400\mu m > d > 250\mu m$$
$$250\mu m > d > 100\mu m$$
$$d > 100\mu m$$

The test procedure was to start with the d < 100μm fraction, and then reconstitute the total slimes in two further steps.

Slurry 1   d < 100μm C_v = 22,3%
The finest fraction was tested at C_v = 22,3%.

Slurry 2   d < 250μm C_v = 28,6%.
The 250μm > d > 100μm fraction was added so as to reconstitute the slimes up to the 250μm size. This resulted in a much higher volumetric concentration, since the fraction added had a very low moisture content. The mixed slurry was tested at this higher concentration.

The particle size distributions of these two slurries are shown in Figs. 8 and 9.

6. DISCUSSION OF TEST RESULTS.

The rheological results are summarised in Table 1. Experimental data is presented in figs. 1 and 2, the rheograms are shown in figs 3 and 4 (using equation (1)) and hydraulic gradients are shown as functions of velocity in Figs. 5 and 6.

The comparative rheograms show that there is a significant fundamental difference in the rheologies obtained from the two viscometers.

6.1 Observations during tests

Tests using the BBTV revealed that, in all cases, the onset of turbulence was characterised by a pulsating unstable flow, appearing to oscillate at a frequency of approximately 1 Hz. This phenomenon was apparent in the transducer readings, as well as by observation of the flow in the tubes.

The rotary viscometer was extremely difficult to read because the torque readings decay with time, in an exponential fashion. The "correct" reading is thus ambiguous. A possible explanation for this is that the action of the rotary viscometer leads to a centrifuge effect in a two phase mixture. The high shear rate readings are possibly erroneous due to centrifugal force also expelling overflow from the top of the rotor into the measuring gap, creating an end effect.

6.2 Observations from raw data

Figs. 1 and 2 show that all the data results in clearly defined laminar flow curves characterised by a yield-pseudoplastic rheology. The BBTV results indicate a laminar flow region, transition region and subsequent turbulent flow region.

6.3 Rheological comparisons

Compare the BBTV and rotary viscometers: Fig. 3

shows, for kaolin, that yield stresses (τ_y) are slightly higher for the BBTV whilst the fluid consistency index (K) is higher by a factor of about 20 for the rotary viscometer and the flow behaviour index (n) is about twice as high for the BBTV.

Fig. 4 shows a similar pattern for the uranium tailings. Yield stresses are similar for both viscometers, the K value is about five times as high for the rotary viscometer and the n value is about 40% higher for the BBTV.

6.4 A comparison of laminar flow predictions

Fig. 5 for kaolin shows that, for the BBTV, the experimental points are in excellent agreement with the theoretical predictions, validating the technique outlined in 3.1. The rotary viscometer rheologies are all higher than the pipe flow data points.

Fig. 6 shows better agreement between the two types of viscometers although Slurry 1, at $C_v = 22,47\%$, has data points higher than the rotary viscometer rheology.

6.5 Prediction of turbulent flow

Fig. 5, for the kaolin, shows that the predictions of Torrance and Metzner and Reed are in good agreement with experimental points for the BBTV while the rotary viscometer rheology is unable to predict turbulent flow.

The simple approach using Nikuradse was always too high, although the low concentration Kaolin and the Uranium tailings (Fig. 6) show good agreement for the BBTV rheology.

6.6 Laminar turbulent transition

The rotary viscometer gives no indication of the laminar turbulent transition. However, the BBTV indicates that the laminar–turbulent transition generally occurred at Reynolds numbers greater than 3000. It should be noted that for Newtonian flow, critical Reynolds numbers as high as 40 000 have been reported[5]. The figure of approximately 2000 represents the lower bound.

The BBTV has very little mechanical disturbance, and therefore high critical Reynolds numbers are to be expected. This would explain the unstable flow and the scatter shown by the smaller tube with uranium tailings (Fig. 6; BBTV; 22,47%). Examination of the larger tube results, on the same diagram, show that at higher Reynolds numbers (higher velocity), the flow "settles down" and is in good agreement with Torrance and Metzner and Reed.

7. CONCLUSIONS

1. The rotary type viscometer, in its standard form, is not suitable for the rheological characterisation of the slurries tested.

2. The BBTV is an instrument which is capable of correctly characterising the slurries tested.

3. The BBTV is capable of collecting turbulent flow data and indicating the laminar–turbulent transition.

4. The turbulent flow predictions of Torrance and Metzner–Reed show good agreement for the slurries tested in the BBTV.

5. The turbulent flow prediction of Nikuradse can be used as an approximation for fluids with a low yield stress and negligible rheogram curvature.

REFERENCES

1. LAZARUS, J.H. and SIVE, A.W.; "A Novel Balanced Beam Tube Viscometer and the Rheological Characterisation of High Concentration Fly Ash Slurries", 9th Int. Conf. on Hydraulic Transport of Solids in Pipes, Rome, 17–19 October, 1984, Paper E1.

2. GOVIER, G.W. and AZIZ, K.; "The Flow of Complex Mixtures in Pipes"; Van Nostrand Reinhold Company; 1972.

3. TORRANCE, B. McK., "Friction Factors for Turbulent non-Newtonian Fluid Flow in Circular Pipes"; The South African Mechanical Engineer; November 1963.

4. METZNER, A.B. and REED, J.C.; "Flow of non-Newtonian Fluids - Correlation of the Laminar, Transition, and Turbulent-flow Regions"; A.I.Ch.E. Journal; Vol.1, No. 4; December 1955.

5. SCHLICHTING, H; "Boundary layer theory"; McGraw Hill; Fourth edition; 1960; p. 376.

6. SLATTER, P.T. and LAZARUS, J.H.; "Research Report - Preliminary Rheological Investigation for Rössing Uranium"; Hydrotransport Research Unit, University of Cape Town; December 1985.

ACKNOWLEDGEMENTS

The authors wish to acknowledge the following:

1. Council for Scientific and Industrial Research.

2. Chamber of Mines

3. University of Cape Town

4. Rössing Uranium

5. Serina Kaolin.

BBTV Test Results

Rotary Viscometer Test Results

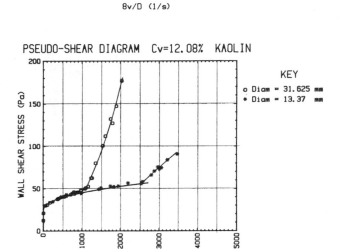

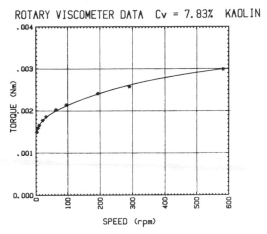

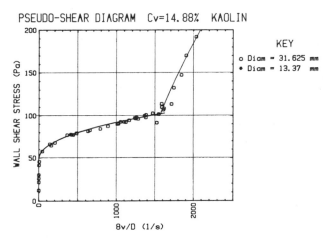

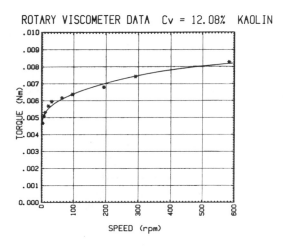

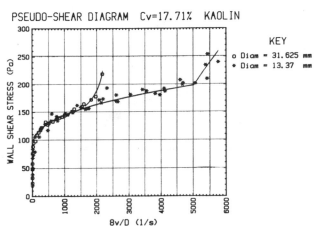

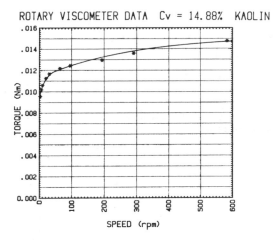

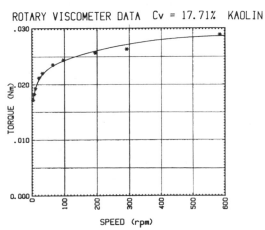

Fig. 1. *Raw data for kaolin using the Balanced Beam Tube Viscometer and the Rotary Viscometer.*

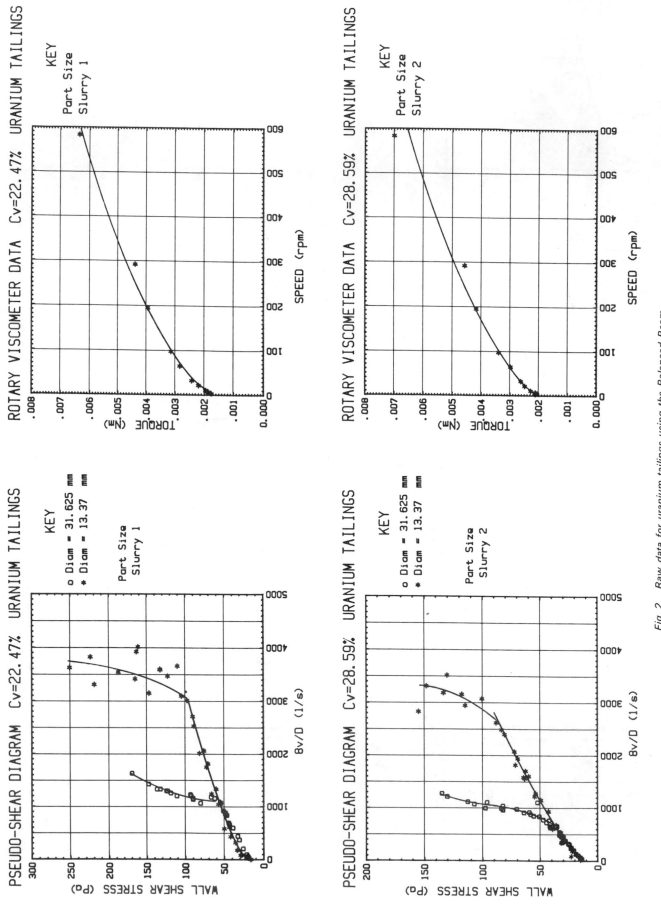

Fig. 2. Raw data for uranium tailings using the Balanced Beam Tube Viscometer and the Rotary Viscometer.

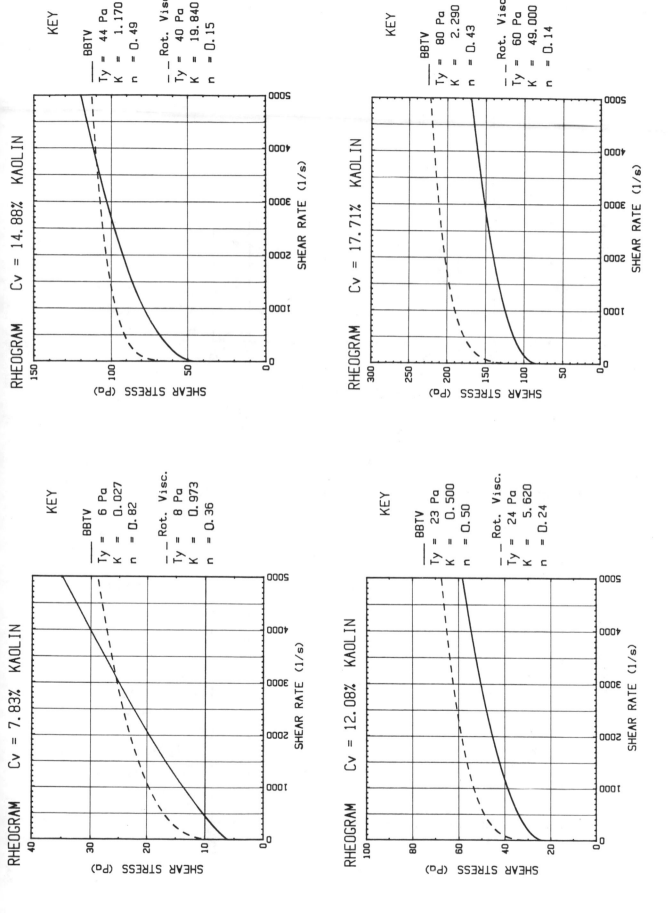

Fig. 3. Comparative rheograms for kaolin.

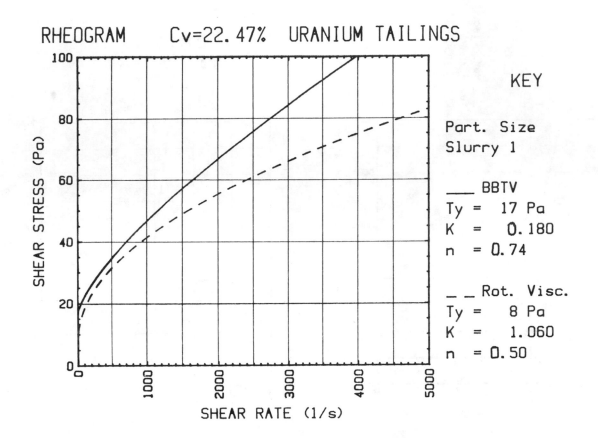

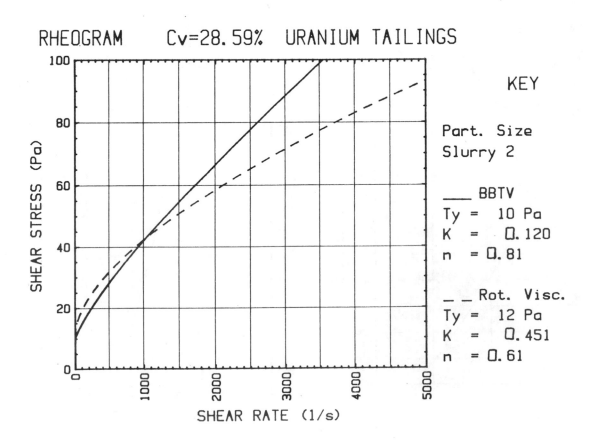

Fig. 4. Comparative rheograms for uranium tailings.

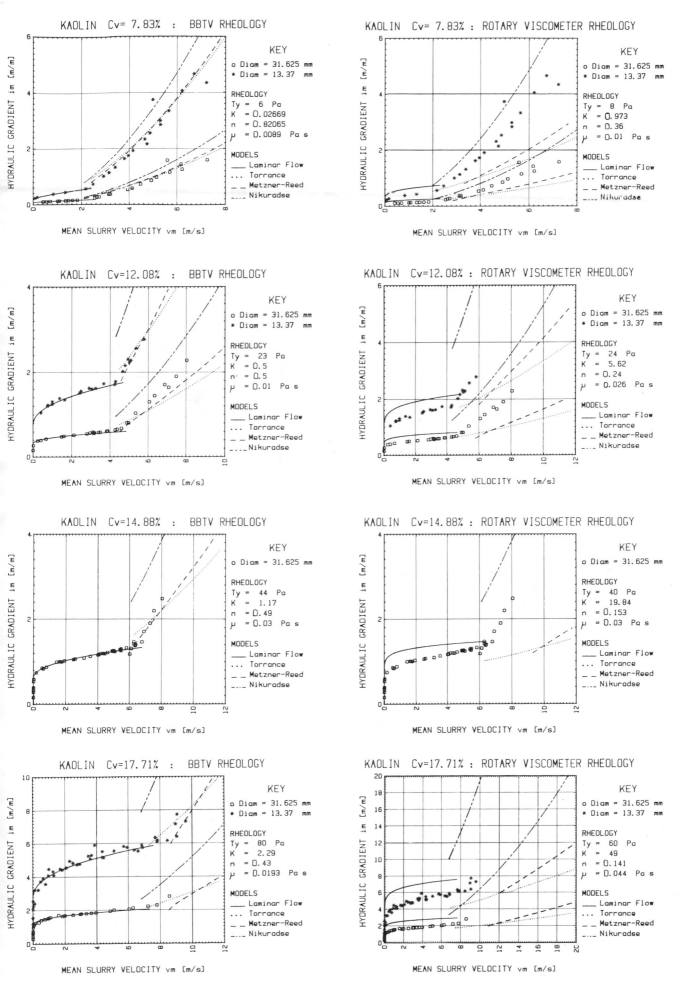

Fig. 5. Flow predictions: kaolin.

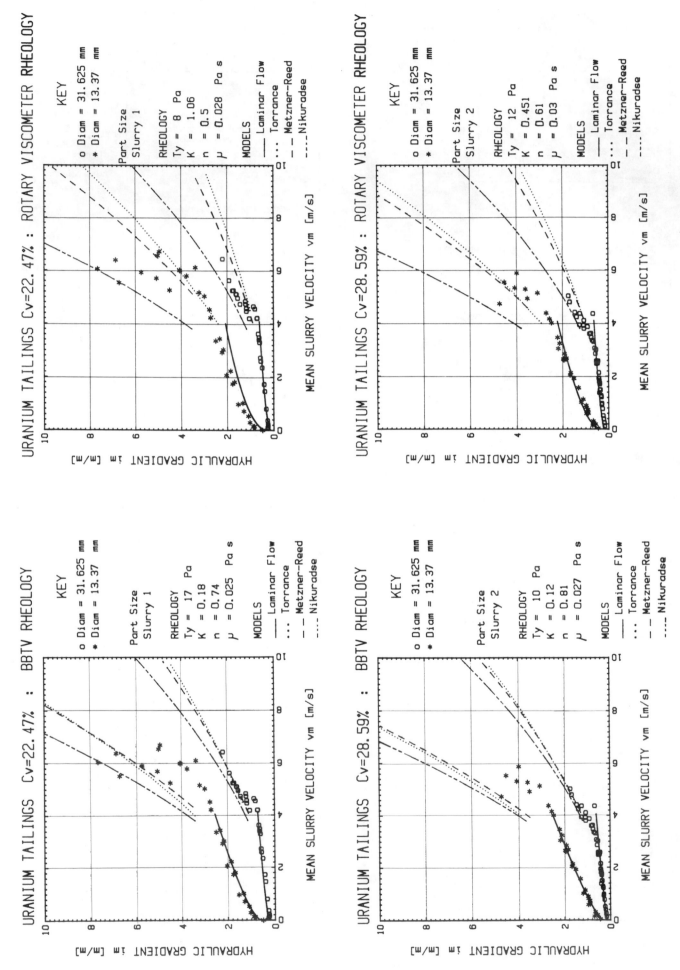

Fig. 6. Flow predictions: uranium tailings.

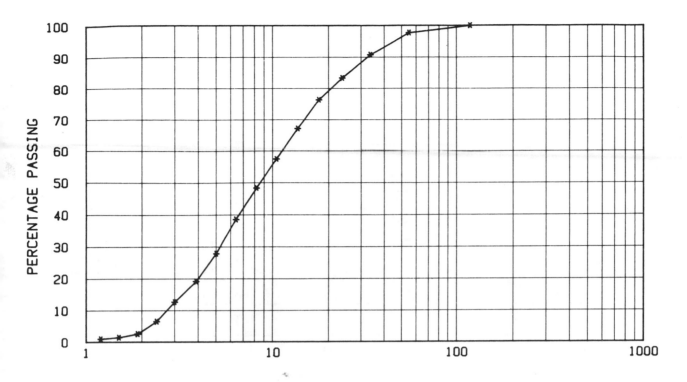

Fig. 7. Particle size distribution: kaolin.

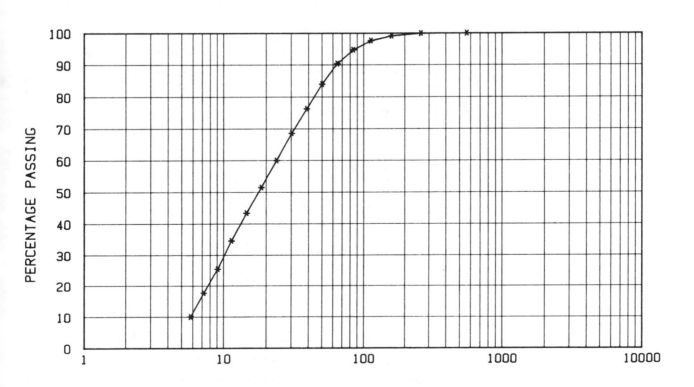

Fig. 8. Particle size distribution: uranium tailings, slurry 1.

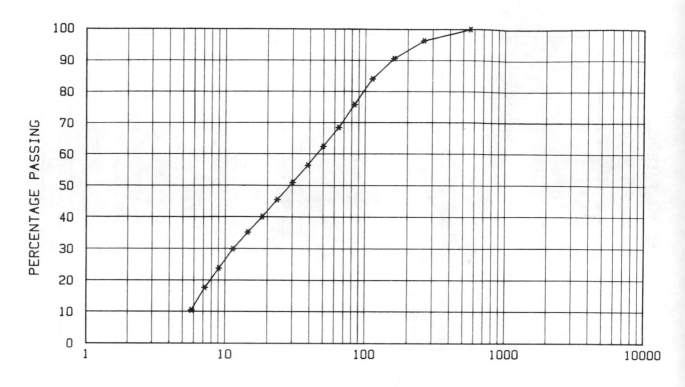

PARTICLE SIZE IN MICROMETERS

Fig. 9. Particle size distribution: uranium tailings, slurry 2.

SLURRY	VOLUMETRIC CONCENTRATION $C_v\%$	BBTV				ROTARY VISCOMETER			
		τ_y(Pa)	K	n	μ(Pa s)	τ_y(Pa)	K	n	μ(Pa s)
Kaolin	7,83	6	0,027	0,82	0,0089	8	0,97	0,36	0,010
"	12,08	23	0,50	0,50	0,010	24	5,62	0,24	0,026
"	14,88	44	1,17	0,49	0,030	40	19,8	0,15	0,030
"	17,71	80	2,29	0,43	0,019	60	49	0,14	0,044
Uranium Tailings Slurry 1	22,47	17	0,18	0,74	0,025	8	1,06	0,50	0,028
Uranium Tailings Slurry 2	28,59	10	0,12	0,81	0,027	12	0,45	0,61	0,03
Solids Relative Density : Kaolin 2,44 Uranium Tailings 2,75									

Table 1

10th International Conference on the
Hydraulic Transport of Solids in Pipes

HYDROTRANSPORT 10

Innsbruck, Austria: 29-31 October, 1986

PAPER K1

HYDRAULIC TRANSPORT OF COARSE PARTICLES WITH GAS INJECTION

D. Barnea, D. Granica, P. Doron and Y. Taitel

Faculty of Engineering,
Department of Fluid Mechanics and Heat Transfer,
Tel-Aviv University,
Ramat-Aviv 69978, Israel

ABSTRACT

The effect of gas injection on the flow of liquid-solid mixtures of coarse particles was examined. It is shown that the main effect of the gas addition is owing to the formation of gas pockets which reduce the pressure drop as well as the slurry superficial velocity. Based on this result the combined use of larger diameter pipes and gas injection is shown to be a practical procedure to reduce the pressure drop of a slurry flowing at a given rate.

NOMENCLATURE

A - pipe cross sectional area
C_S - slurry volumetric concentration
d - particle diameter
D - pipe diameter
N - gas/slurry flow rate ratio
P - pressure
Q_S - slurry volumetric flow rate
U_{SS} - slurry superficial velocity ($=Q_S/A$)
x - coordinate in the downstream direction
ρ_S - solid density
ρ_L - liquid density

INTRODUCTION

Hydraulic transport of solid particles is a well known method in chemical and mining industries. In the case of fine particles where pseudohomogeneous slurries can be assumed the pressure gradient is often predicted from the mixture properties using single phase correlations. However for coarse particle mixtures settling effects become important, resulting in significant frictional losses and high pressure gradients.

Settling slurries exhibit a variety of flow patterns in horizontal pipeline flow. The terminology varies in the literature. Parzonka et al. (1981) presented the four following flow patterns: Stationary bed, Sliding (or moving) bed, Heterogeneous suspension and Pseudohomogeneous suspension.

Experimental data on the typical pressure gradient behavior of fluid solid systems in horizontal pipes have been obtained by many investigators (Durand, 1953; Zandi & Govatos, 1967; Babcock, 1971; Carleton et al., 1978; Chhabra & Richardson, 1983; Noda et al., 1984 and Oba et al., 1984). The prediction of pressure drop is a complex problem and it is treated mostly via correlations of experimental data. Some of the empirical correlations claim to apply to all flow patterns for liquid solid systems (Newitt et al., 1955; Hayden & Stelson, 1971 and Turian & Yuan, 1977). Others are restricted to one or two flow patterns only (Durand, 1953; Zandi & Govatos, 1967; Babcock, 1971; Toda et al., 1979; Wani et al., 1983 and many others).

The theoretical prediction of pressure drop in slurry flow is quite complex. Some simplified physical models for the prediction of the flow patterns and the pressure drop were presented. Theoretical analyses were performed by Carstens (1969), Wilson (1976), Televantos et al. (1979), Shook et al. (1982), Roco & Shook (1984) and Noda et al., (1984). Recently an improved model which can be used to predict the pressure drop for any flow pattern was proposed by Doron et al. (1986) (Appendix A).

Very little work has been performed on three phase flow. Most of the works on three phase flow deal with gas injection to provide either pumping action in vertical slurry flow or to provide drag reduction in non-Newtonean suspensions of very fine particles. A comprehensive review on these two aspects of gas injection to slurry flow is included in Hetsroni (1982).

In the present work the effect of gas injection into slurries of coarse particles is studied and the possible benefit of such injection is examined.

EXPERIMENTAL SYSTEM

A schematic layout of the experimental system is shown in figure 1. The system consists of a 11 m long 50x50 cm cross section square steel frame supporting a transparent plexiglass pipe with an internal diameter of 51 mm (2 inches) which is leveled to an accuracy of ±0.1°.

The slurry feed pipe is of plexiglass with an internal diameter of 40 mm (1.5 inches) that slopes downwards at an angle of 2° to the inlet section (the feed pipe is always higher than the highest point of the test pipe).

The column is capable of being rotated around its axis from the horizontal position to 12 degrees in both directions, by means of a lead screw.

The system can be used in 3 modes of operation.
(1) As a slurry loop. In this case the slurry pump circulates only the slurry through the test section. The input concentration in this case is equal to the concentration in the slurry container.
(2) As a slurry loop with air injection. In this case air is injected to the test section and flows concurrently with the slurry. The air is vented to the atmosphere.

(3) An additional water loop can be added to the slurry loop. This water loop can be used to conveniently control the input concentration of the slurry.

The slurry is supplied from a 500 liter container, mixed by a 420 RPM mixer (with a 1.5 HP motor) and circulates in a closed loop through the system by two slurry pumps with rubber coated open impellers having a capacity of 15 m³/hr at 8 m head (using a 1.5 HP motor) and 30 m³/hr at 30 m head (using a 10 HP motor).

The slurry flow rate can be controlled by a butterfly control valve and by bypass lines, and is measured by a slurry magnetic flow meter. Concentration samplings are taken from the slurry supply line near the entrance to the test section.

Air is supplied from a central high pressure line passing through an oil separator and a pressure regulator, maintaining constant pressure at 2 bar. The air metering system consists of three gas rotameters with capacities of 2166, 182 and 18.4 liter/min. A mercury manometer is incorporated to ensure constant inlet pressure. Inlet temperature is measured by a thermometer located at the entrance to the test section.

The additional water is supplied from a 500 liter container that circulates in a closed loop through the system by a centrifugal pump having a capacity of 10 m³/hr at 15 m (using a 3 HP motor). To cover a wider range of flow rates there is an option to connect a larger pump with capacity of 40 m³/hr at 60 m (with a 20 HP motor). The liquid flow rate is controlled by five rotameters with capacities of 2x359.6, 75.7, 6.28 and 1.01 liter/min

The solid particles used are General Electric "Black Acetal" with a density of 1.24 gr/cm³ and diameter of 3 mm.

The pressure drop is measured using two Validyne differential pressure transducers (DP15 and DP7) with direct connection to a digital computer for data acquisition and reduction. The flow patterns were determined by visual observation.

RESULTS AND DISCUSSION

Figure 2 shows the pressure drop results as a function of the solid-fluid mixture velocity at various input solid concentrations. The results are typical to the pressure gradient behavior for slurry flow of coarse particles in horizontal pipes. At high superficial velocities, where the mixture is suspended, the slurry pressure drop is usually somewhat higher than that of the carrier liquid. Reduction of the mean velocity leads to bed formation and to pressure drops which are much higher than those of the pure liquid.

The solid lines present the results of a theoretical model that was recently developed by Doron et al. (1986). The theory compares favorably with the experimental data and predicts well the effect of mixture velocity and input concentration on the pressure drop.

Associated with each slurry concentration is a minimum pressure drop below which slurry flow can not exist. An attempt to overcome this situation is considered by injecting gas into the slurry line. The effect of gas addition on the pressure drop behavior is examined.

Figures 3 to 5 show the pressure drop vs. slurry superficial velocity for various ratios of gas-slurry flow rates. The pressure drop curves obtained for the case of air injection are shifted downwards and to the left relative to the case without gas. The physical meaning of this is obvious. Addition of gas results in intermittent flow of liquid and gas plugs flowing approximately at the same velocity. Consider for example a point on the curve of N=0. Adding gas at a ratio of 1:1, while keeping the total flow rate (slurry + gas) constant, does not change the velocity and the pressure gradient within the slurry slugs. However, since the slurry occupies only half the length of the pipe we expect the pressure drop as well as the slurry input superficial velocity to decrease approximately by a factor of two. Thus each point on the curve N=0 "moves" to an equivalent point on N=1 for which the pressure drop and the superficial velocity are lower by a factor of 2.

At relatively high slurry input rates addition of gas (for a given slurry flow rate) increases the pressure drop while at low flow rates to the left of the minimum lower pressure drops may be obtained. Thus addition of gas can be used to decrease the pressure drop when low slurry flow rates are needed.

The first order effect of the addition of gas is owing to the formation of gas voids that reduce the pressure drop as well as as the slurry flow rate (for constant total flow rate). In order to examine any additional effects of gas injection it is convenient to plot the data of figures 3 to 5 on normalized coordinates, namely $\frac{dP}{dx}(1+N)$ vs. the total mixture velocity $U_{SS}(1+N)$ where N is the gas slurry flow rate ratio (gas flow rate/slurry flow rate). The results are shown on figures 6-8. For the case of the low input concentration the normalized pressure drop curves are almost identical, with a slight tendency to higher pressure drops in the case of higher gas-slurry flow ratios. Gas-liquid slug flow is usually associated with higher pressure drop compared to pure frictional pressure drop at the liquid slugs. This is owing to the turbulent mixing in the front of the liquid slug that causes acceleration losses (Dukler and Hubbard, 1975). As a result, one would expect higher pressure losses than those observed in figure 6. However in the case of gas-slurry slug flow an additional effect takes place. The turbulent mixing behind the elongated bubbles enhances the radial transport of the particles in the bed. This effect may cause a relative reduction of the pressure drop and a shift of the minimum pressure drop towards lower values of the normalized slurry superficial velocity. Indeed, as seen in figure 6 the pressure drop curves with gas injection do not display a minimum in the range of the checked flow rates and it is obvious that the minimum is shifted considerably to the left.

A similar behavior is observed also for higher input concentrations (figures 7, 8). In these cases, however, the addition of gas results in higher normalized pressure drops indicating that the effect of radial floatation in slug flow is not as effective as in the case of low concentration.

The pressure drop behavior of gas-liquid-solid flow (figures 3-8) points out the potential benefit of gas injection to slurry flow.

For a single phase flow a common procedure to reduce the pressure drop is to increase the pipe diameter. This simple procedure is not applicable in solid-liquid mixtures. The solid lines on figure 9 show the effect of pipe diameter on the slurry pressure drop. The results presented in this figure are theoretical results obtained by using the predictive model of Doron et al. (1986). As seen, a wide range of operational conditions (below the doted line) are not accessible even with the use of large diameter pipes (contrary to the case of single phase flow). In other words, in slurry transportation, when operation is around the minimum pressure drop, it is not possible to reduce the pressure drop (for a given slurry rate) by using pipes of larger diameter. It can be shown that a combined use of a larger pipe diameter and gas injection may be a practical solution for reducing the pressure drop.

The first order effect of gas injection was shown to be due to the formation of the gas voids as demonstrated by the use of the normalized coordinates (figures 6-8). Using these results it is now possible to estimate the effect of gas injection on the pressure drop. The dashed lines in figure 9 represent the theoretical pressure drop for 25 cm diam. pipe with various gas-slurry ratios (each point on the solid curve "moves" by a factor of (N+1) towards lower values of pressure drop and slurry flow rate (where N is the gas/slurry ratio).

Consider for example the pressure drop of a slurry flowing in a 15 cm diam. pipe (point A on figure 9). It is possible to transport the same slurry flow rate at lower pressure drops by increasing the pipe diameter, for example to 25 cm, and adding gas at an appropriate ratio of, say 3:1 (point B). Thus the addition of gas enables, in fact, to cover the region below the dotted line that is not operable without gas injection.

SUMMARY AND CONCLUSIONS

Gas injection is shown to have first and second order effects.

The first order effect is owing to the formation of gas voids which reduce the pressure drop (as well as the slurry flow rate) for the same total (slurry + gas) input rate.

The second order effects are twofold. Pressure drop increases owing to acceleration losses associated with slug flow (intermittent flow of liquid and gas). On the other hand the turbulent mixing in the slug front enhances the particles floatation. This effect tend to reduce the pressure drop and causes the minimum in the normalized pressure drop to be shifted towards lower values of the total superficial velocity.

Finally it is shown that addition of gas combined with an increase in pipe diameter may be of practical use for reducing the pressure drop (for a given slurry flow rate).

REFERENCES

Babcock, H.A., "Heterogeneous flow of heterogeneous solids", in Advances in Solid Liquid Flow in Pipes and its Application, Zandi, I. Ed., Pergamon Press, 125-148 (1971).

Carleton, A.J., French, R.J., James, J.G., Broad, B.A. & Streat, M., "Hydraulic transport of large particles using conventional and high concentration conveying", Proc. of the 5th Int. Conference on Hydraulic Transport of Solids in Pipes, Hanover, West Germany, Paper D2, 15-28 (1978).

Carstens, M.R., "A theory for heterogeneous flow of solids in pipes", J. Hyd. Div., Proc. ASCE, 95, HY1, 275-286 (1969).

Chhabra, R.P. & Richardson J.F., "Hydraulic transport of coarse gravel particles in a smooth horizontal pipe", Chem. Eng. Res. Des., 61, 313-317 (1983).

Doron, P., Granica, D. & Barnea, D., "Hydraulic transport of coarse particles in horizontal pipe - Experimental and modelling", Int. J. Multiphase Flow, Submitted for publication (1986).

Dukler, A.E. & Hubbard, M.G., "A model for gas-liquid slug flow in horizontal and near horizontal tubes", I&EC Fundamentals, 14, 337-345 (1975).

Durand, R., "Basic relationships of the transportation of solids in pipes -experimental research", Proc. 5th Minneapolis Int. Hyd. Convention, 89-103 (1953).

Hayden, J.W. & Stelson, T.E., "Hydraulic conveyance of solids in pipes" in Advances in Solid liquid flow in pipes and its application, Zandi, I. Ed., Pergamon Press, 149-163 (1971).

Hetsroni, G., Ed., "Handbook of Multiphase Systems", Hemisphere publishing Co., McGraw-Hill Book Co (1982).

Newitt, D. M., Richardson, J.F., Abbott, M. & Turtle, R.B., "Hydraulic conveying of solids in horizontal pipes", Trans. Instn. Chem. Engrs., 33, 93-113 (1955).

Noda, K., Takahashi, H. & Kawashima, T., "Relation between behavior of particles and pressure loss in horizontal pipes", Proc. of the 9th Int. Conference on Hydraulic Transport of Solids in Pipes, Rome, Italy, Paper D4, 191-205 (1984).

Oba, S., Ota, K. & Kawashima, T., "Analysis on slurry pressure drop by various fluid models", Proc. of the 9th Int. Conference on Hydraulic Transport of Solids in Pipes, Rome, Italy, Paper D2, 163-190 (1984).

Parzonka, W., Kenchington, J.M. & Charles M.E., "Hydrotransport of solids in horizontal pipes: effects of solids concentration and particle size on the deposit velocity", Can. J. Chem. Eng., 59, 291-296 (1981).

Roco, M.C. & Shook, C.A., "A model for turbulent slurry flow", J. Pipelines, 4, 3-13 (1984).

Shook, C.A., Gillies, R., Haas, D.B., Husband, W.H.W. & Small, M., "Flow of coarse and fine sand slurries in pipelines", J. Pipelines, 3, 13-21 (1982).

Televantos, Y., Shook, C., Carleton, A. & Streat M., "Flow of slurries of coarse particles at high solids concentrations", Can. J. Chem. Eng., 57, 255-262 (1979).

Toda, M., Yonehara, J., Kimura, T. & Maeda, S., "Transition velocities in horizontal solid liquid two phase flow", Int. Chem. Eng.. 19, 145-152 (1979).

Turian, R.M. & Yuan, T-F., "Flow of slurries in pipelines", Amer. Inst. Chem. Eng. J., 23, 232-243 (1977).

Wani, G.A., Mani, B.P., Suba Rao, D. & Sarkar, M.K., "Studies on the holdup and pressure gradient in hydraulic conveying of settling slurries through horizontal pipes", J. Pipelines, 3, 215-222 (1983).

Wilson, K.C., "A unified physically based analysis of solid liquid pipeline flow", Proc. of the 4th Int. Conference on Hydraulic Transport of Solids in Pipes, Banff, Alberta, Canada, Paper A1, 1-16 (1976).

Zandi, I. & Govatos, G., "Heterogeneous flow of solids in pipelines", J. Hyd. Div., Proc. ASCE, 93, HY3, 145-159 (1967).

APPENDIX A:

A physical mechanistic model for the prediction of pressure drop and flow patterns in horizontal slurry pipe flow has been presented by Doron et al. (1986). The model is a two layer model: A stationary or a moving bed at the bottom of the pipe and a heterogeneous mixture of solid particles and carrier liquid at the upper part (see figure 10). The mathematical formulation consists of continuity equations for each phase, momentum equations for each layer and a diffusion equation which governs the dispersion of the solid particles in the upper layer:

$$U_h C_h A_h + U_b C_b A_b = U_{ss} C_s A \qquad (1)$$

$$U_h (1-C_h) A_h + U_b (1-C_b) A_b = U_{ss} (1-C_s) A \qquad (2)$$

$$A_h \frac{dP}{dx} = - \tau_h S_h - \tau_i S_i \qquad (3)$$

$$A_b \frac{dP}{dx} = - F_b + \tau_i S_i \qquad (4)$$

$$\epsilon \frac{d^2c}{dy^2} + w \frac{dc}{dy} = 0 \qquad (5)$$

where U denotes the mean velocity, C the mean concentration, c the local concentration, A the cross sectional area, S the perimeter, τ the shear stress, F_b the friction resistance force, ϵ the diffusion coefficient and w the terminal settling velocity of the solid particles. The subscript h denotes the upper dispersed layer, b the bed layer, i the interface and s the slurry.

For a given set of operational conditions, the simultaneous solution of the five equations yields the pressure drop, the mean velocity in each layer, the bed height and the solids concentration profile in the upper layer. The solution also yields the flow pattern of the slurry (stationary bed, moving bed, heterogeneous or homogeneous suspension).

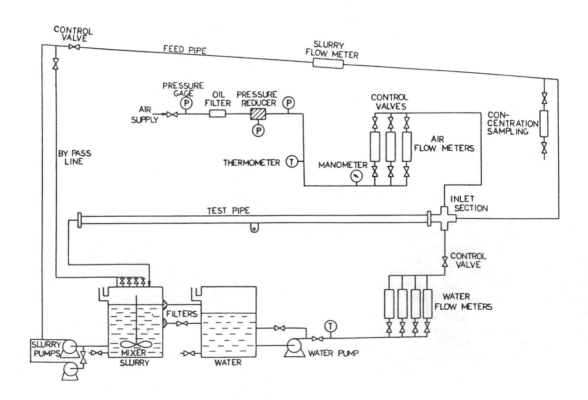

FIGURE 1: EXPERIMENTAL SYSTEM

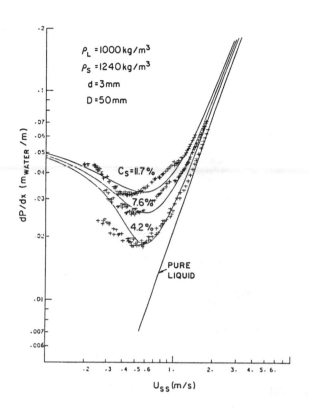

FIGURE 2: PRESSURE DROP IN SLURRY FLOW
+   Experimental data
—— Theory (Doron et al., 1986)

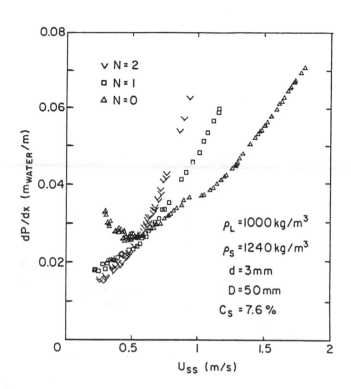

FIGURE 4: PRESSURE DROP IN SLURRY FLOW
WITH GAS INJECTION, C_S = 7.6 %

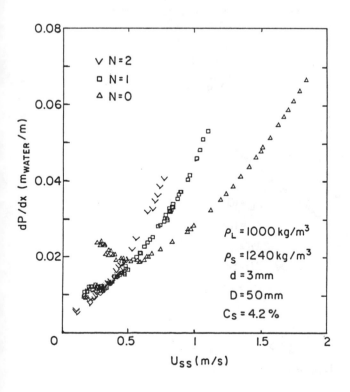

FIGURE 3: PRESSURE DROP IN SLURRY FLOW
WITH GAS INJECTION, C_S = 4.2 %

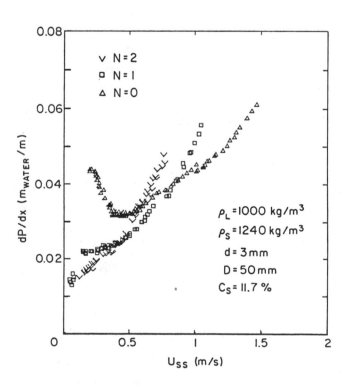

FIGURE 5: PRESSURE DROP IN SLURRY FLOW
WITH GAS INJECTION, C_S = 11.7 %

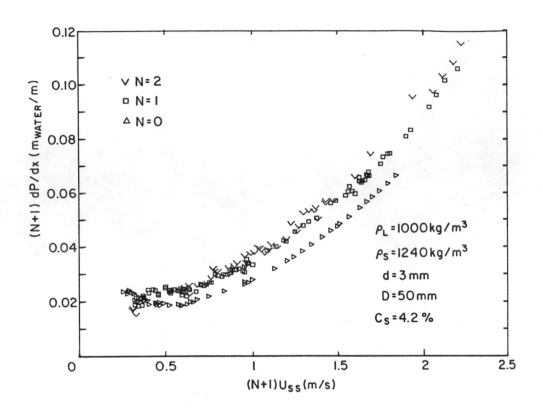

FIGURE 6: NORMALIZED PRESSURE DROP WITH
GAS INJECTION, C_S = 4.2%

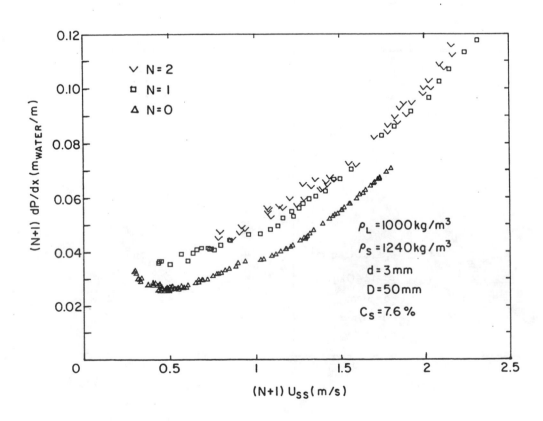

FIGURE 7: NORMALIZED PRESSURE DROP WITH
GAS INJECTION, C_S = 7.6%

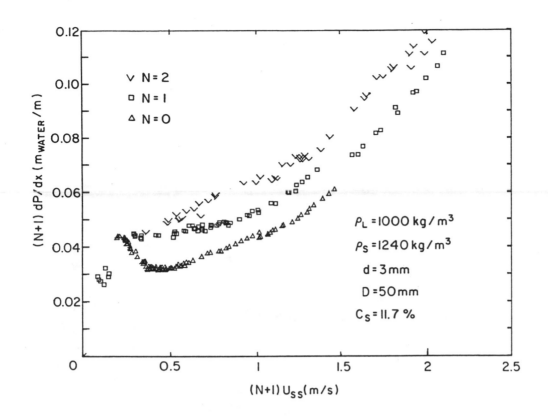

FIGURE 8: NORMALIZED PRESSURE DROP WITH
GAS INJECTION, $C_S = 11.7\%$

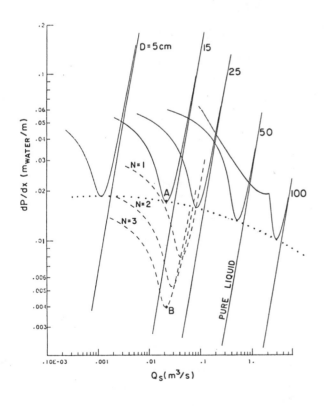

FIGURE 9: EFFECT OF PIPE DIAMETER AND GAS
INJECTION ON SLURRY PRESSURE DROP
$\rho_L = 1000 \text{ Kg/m}^3$, $\rho_S = 1240 \text{ Kg/m}^3$,
$d = 3 \text{ mm}$, $C_S = 4.2\%$

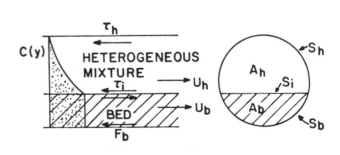

FIGURE 10: THE TWO LAYER MODEL

10th International Conference on the
Hydraulic Transport of Solids in Pipes

HYDROTRANSPORT 10

Innsbruck, Austria: 29-31 October, 1986

PAPER K2

TWO-PHASE SLURRIES FOR THE TRANSPORT OF
SOLID
PARTICULATE MATTER

N. Kundu and G. P. Peterson

Mr. N. Kundu, an Assistant Professor of
Engineering Technology, and Dr. G. P.
Peterson, an Assistant Professor of
Mechanical Engineering , are both
employed at Texas A&M University in
College Station, Texas.

SYNOPSIS

Slurry pipelines are used to carry solid
particles over distances of hundreds of
miles. A slurry pipeline requires vast
amounts of water, which in some areas may
not be readily available. Using foam as
a transport medium would require, at
most, half the amount of water. Due to
its superior particle suspension ability,
stability and viscosity, foam has a much
higher solid carrying capacity than
water. Solid particles transported by
foam can reduce or eliminate some of the
processes required for the slurry
preparation and drying of transported
materials. Presented here are the
experimental results supporting the
technical advantages of foam as transport
medium. Foam transport of solid
particles is technically sound,
economically attractive and
environmentally acceptable.

INTRODUCTION

Water is a necessary component in
countless agricultural and industrial
applications, as well as a vital factor
for human existence. It is also the
primary ingredient required in the
hydraulic transportation of solid
materials, where its availability along
with it's ability to hold solids in
suspension, together with its fluidity,
make it an efficient media for use in the
hydraulic transportation of solid
particulate materials. Pipelines are
currently used to carry solid particles
over hundreds of miles, with the vast
amounts of water consumed each day for

this purpose. Coal, gilsomite, copper
concentrates, solid wastes, and other
materials are just some of the materials
currently being transported by slurry
pipelines.

In areas where water is already scarce,
increasing emphasis is being placed on
water conservation. In these areas, care
must be exercised in the distribution and
use of water in order to prevent a
reduction in the local water table.
Desert areas in states such as Nevada and
New Mexico have already taken steps to
restrict the quantity of water
transported across state boundaries.
This presents a major problem since, in
traditional slurry systems, a reduction
in quantity of water results in a
proportionally smaller quantity of
materials that can be suspended and
transported. It is clear, that if the
hydraulic transport of solid materials is
to continue as a viable system, an
alternative medium for slurry
transportation is necessary, and a method
must be developed capable of reducing the
quantity of water required, and
increasing the overall efficiency. One
alternative is to use foam as the carrier
medium.

Foam is a two phase fluid, and is
created by introducing air into water
containing a foaming agent. When used as
the carrier fluid it will result in a
reduction in the quantity of liquid
required per unit quantity of solids
transported. This coupled with its
superior solid suspension capability make
it a desirable transport fluid[1].

Presented here are the results of a
preliminary investigation into the
feasibility of using foam as the carrier
medium in the transportation of solid
materials via pipelines.

FOAM AND ITS PROPERTIES

As the quantity of air introduced into
the water/surfactant mixture is
increased, the foam quality, which is
defined as the ratio of the gaseous
volume to the total volume,

$$\Gamma \, (\%) = \frac{\text{GASEOUS VOLUME}}{\text{TOTAL FOAM VOLUME}}$$

increases. This foam quality is the
principal factor that governs the
suspensionability and hence the transport
capabilities of foam.

Minssieux[1] reported that foam acts as a
Newtonian fluid up to a quality of 52%.
Up to this point, the bubbles are
dispersed uniformly in the liquid, and do
not come in contact with each other.
Blauer et. al.[2] reported that between
qualities of 52% and 74%, for laminar

flow, foam behaves like a Bingham plastic. In most materials handling applications however, the flow would be turbulent due to the high velocities resulting from the introduction of air. Initially, it appears that the introduction of air into the pipeline would result in a series of difficulties since, once the air is introduced, the fluid becomes compressible. However, the compressible nature of the foam actually assists in the overall performance. As the flow progresses, pressure losses in the pipeline increase the total volume, which in turn accelerates the two-phase fluid reducing the tendency for particles to settle out. Though this self-compensating increase in velocity is a very attractive feature, it must also be pointed out that the suspension characteristics of foam are viscosity dependent, and as the pressure decreases, it is accompanied by a corresponding decrease in the viscosity. Thondavadi and Lemlich[3] have reported that in addition to changes in velocity, the shape of the bubbles changes continuously from round to polyhedral as the quality increases.

Additional advantages resulting from the use of foam, are that the erosion of the pipeline would be reduced due to the cushioning effect of the foam, and that this type of process would be less susceptible to plugging than conventional slurry transport systems. In the event of failure or plugging, air introduced into the system could be used to restart the flow.

Pumping of mixtures containing solid particles present a problem however, triplex plunger type pumps have been used in the past to pump particles up to a maximum diameter of 3 mm and have been able to achieve satisfactory long term operation. In some cases, centrifugal pumps have been used to transport larger particles in a heterogeneous slurry, but these applications have been limited to fairly short distances[4].

Viscosity and Pressure Drop

The frictional pressure losses can be found by using effective viscosity of foam and the friction factor, as determined from the Moody Diagram and frictional pressure loss equation[8].

Using these methods the pressure loss in 440 km Black Mesa Pipeline would be 21014 KPa for water while that for 75% quality foam would be only 8337 KPa, and reduction of almost 60%.

Previous work by Blauer and Kohlhaas[4] as reproduced in figure 2 demonstrates pressure loss reduction for foam against water.

As is the case for traditional slurry pipelines, the pressure losses are quite large for long pipelines, and intermediate pressure boosts are required. This could be accomplished in foam pipelines through the use of intermediate compressor stations, where additional air is injected into the system to boost the pressure.

Settling Velocity

The settling velocity of particulate in foam is a function of the foam viscosity, interfacial surface tension between the liquid and gas phases, particle sizes and bubble size distribution, with smaller bubbles resulting in higher viscosities and a correspondingly higher solid transport capacity. At foam qualities higher than 74%, solid particles with diameters larger than the mean bubble diameter are suspended by a number of different bubbles. Particles with diameters smaller than the mean bubble diameter are retained within the bubbles and have a settling velocity of approximately zero.

EXPERIMENTAL STUDIES

In order to test the foam transport concept, a series of preliminary tests were conducted using the apparatus shown in figure 1. The test section consisted of a 13.72 meter section of 1.27 cm PVC pipe. Three sections of clear tubing were located at points within the flow loop, one at the pump suction, one at the pump discharge, and one at the midpoint of the loop. The purpose of these clear sections, was to visually observe the flow and detect the presence of stratified flow, and any settling or flow disruptions occurring within the loop. The system included an agitator, a .25 horsepower positive displacement pump, a foam generator, a .25 horsepower compressor, and the associated pressure gages.

All of the experimental tests were performed using a constant flow rate of 10.41 liter/min. with varying amounts of air introduced to obtain the different foam qualities required. Two different size ranges of zirconium oxide beads, 0.8 - 1.0 mm and 1.0 - 1.25 mm diameter, were used in the tests. Slurries with a predetermined solids concentration, were pumped through the foam generator, where air was introduced, creating the foamed mixture. Once the flow had been allowed to stabilize, the pressures at the beginning and end of the flow loop, and at a number of intermediate points along the flow loop, were measured and recorded.

Samples of the foam were taken throughout the tests. From these samples, the quantity of solids and water, along with the foam quality were measured. Based upon these measured values, the transport capacity was computed. (Tables 1,2 & 4)

Throughout the experiment the foam quality was maintained between 30% and 65%. Foam acts as a Newtonian fluid up to a quality of 52%[1], and between qualities of 52% and 74% foam behaves like a Bingham plastic for laminar flow[2]. However, high velocities due to the introduction of air makes the flow turbulent in materials handling applications.

In addition to the flow loop tests, a second series of tests were conducted to determine the settling velocity of particles in static foam. Foam of a fixed quality was prepared in a Waring blender and transferred to a graduated cylinder. Glass and zirconium oxide beads of different sizes and densities were placed in the foam and their settling velocities determined. (Table 3)

EXPERIMENTAL RESULTS

The results of the experimental test program are presented and separated into three different categories, pressure drop measurements, settling velocities, and transport capacity.

Pressure Drop Measurements

As mentioned previously a 13.72 m section of 1.27 cm PVC pipe was used for the experiment. A slurry flowrate of 10.41 liter/min was maintained throughout the experiment.

The relationship between the pressure drop and the solids concentration, as a function of the foam quality, is shown in figure 3. As illustrated, increase in the solids concentration result in a small increase in the pressure drop, while increase in the foam quality, results in significant increases in the pressure drop.

This increase in the pressure drop with increased foam quality is due to the increase in the flow velocity. Since the frictional pressure drop is a function of velocity squared, small increases in the flow velocity can result in substantial increases in the pressure losses. As additional air is added to the slurry to increase the quality, the total volume increases, thereby increasing the velocity.

A second phenomenon apparent from figure 3, is that as expected, the measured pressure drops are greater for particles of the same density but larger diameters.

Settling Velocity

Tests on a number of different sizes and densities of particles were conducted to determine the settling velocities as a function of particle size and density (Table 3). Figure 4 illustrates the effects of these two parameters on the settling velocity. Average particle sizes have been used for these graphs. For Example 0.9 mm for 0.8 to 1.0 mm, 1.25 mm for 1.0 to 1.5 mm and 2.25 mm for 2.0 to 2.5 mm range of particle diameter.

As expected, increases in the particle density result in an increase in the settling velocity for particles of the same diameter. However, as the foam quality increases, the settling velocity decreases, supporting the earlier statement that the particle suspension capability of foam is a function of the number of individual bubbles present and the mean bubble diameter.

Transport Capacity Tests

In an effort to compare the transport capacity of a foam system with a water slurry, a series of tests were conducted in which the foam quality was varied from 25% to 60%. Figure 5 illustrates the results of these tests. As shown, the transport capacity increases with respect to foam quality. Foam with a quality in the range of 55-60% was capable of transporting up to eight times solid material as pure water. Even at foam qualities 25-30%, the transport capacity of foam was up to four times the transport capacity of pure water, In addition to increases in the transport capacity with respect to foam quality, the transport capacity also increases with respect to particle size. As shown, at tank concentrations of 1.20 (%) and foam qualities in the range of 45-50%, particles whose diameters were between 1.00-1.25 mm were transported at a rate of 51 g/sec., while particles with diameters between 0.80-1.00 mm were transported at a rate of 41 g/sec.

CONCLUSIONS

The above experimental tests, although not conclusive, indicate that foam transport of solid particulate material is technically sound and may have several distinct advantages. Pressure losses in the pipeline could be compensated for by expansion of the gas bubbles, slurry holding facilities and mixing equipment may not be necessary at intermediate compressor stations, the cost of dewatering the solid material could be reduced, reductions in the required mass of liquid needed to transport a fixed amount of materials would result in an over all energy savings and finally, an up to fifty percent reduction in the water requirement could be achieved.

REFERENCES

1. Minssieux, L., "Oil Displacements by Foams in Relation to Their Physical Properties in Porous Media", Journal of Petroleum Technology, January,1974.

2. Blauer, R. E., Mitchell, B. J., Kohlhaas, C. A. , "Determination of Laminar, Turbulent and Transitional Foam Flow Friction Losses in Pipes", SPE Meeting, San Francisco, 1974.

3. Thondavadi, N. N., Lemlich, R., "Flow Properties of Foam With and Without Solid Particles", Ind. Eng., chem, Process Des, Dev., Volume 24, 1985.

4. Blauer, R. E. and Kohlhaas, C. A., "Formation Fracturing with Foam", SPE 5003, 194, or American Institute of Mining, Metallurgical, and Petroleum Engineers, 1974.

5. Wasp, E. J., Kenny, J. P., Gandhi, R. L., "Solid-Liquid Flow, Slurry Pipeline Transportation", Gulf Publishing Co., 1979

6. Ross, S., "Bubbles and Foam", Industrial Engineering Chemistry, October, 1969.

7. Kundu, N. K. and Peterson, G. P., "A Novel Approach for the Transmission of Solid Particulate Using Foam", ASME/ETCE Proceedings, February, 1986.

8. Mitchell, B. J., "Viscosity of Foam", Ph.D., Thesis University of Oklahoma, 1969.

Table 1

Test Results
Matl: Zirconium Oxide Beads 0.8-1.0 mm.
Density 3.8 g/ml

Tank Conc.* (%)	Pressure Drop (KPa/m)	Foam Quality (%)	Transport Capacity (g/s)
.03	1.9	45-50	12.2
.06	1.8	45-50	12.8
.09	1.6	25-30	19.7
	1.8	45-50	32.5
	1.3	60-65	35.4
1.20	1.5	24-30	30.0
	2.4	45-50	42.2
	1.3	60-65	45.8
1.50	1.4	25.30	29.3
	1.4	60-65	46.5
1.80	1.5	25-30	25.5
	1.9	60-65	58.9

Table 2

Test Results
Matl: Zirconium Oxide Beads 1.0-1.25 mm
Density 3.8 g/ml

Tank Conc.* (%)	Pressure Drop (KPa/m)	Foam Quality (%)	Transport Capacity (g/s)
.03	1.6	25-30	20.6
	2.1	45-50	25.6
.06	1.8	25-30	37.6
	2.1	45-50	31.4
.09	1.8	25-30	20.5
	2.1	45-50	40.2
1.20	2.0	25-30	45.5
	2.0	45-50	51.3

* Tank Conc. = Weight of solid / Weight of water

Table 3

Test Results
Settling Velocity

FOAM QUALITY 90%

Matl.	Density (g/ml)	Size (mm)	Settling Velocity (X 10⁻³cm/s)
Glass	2.5	1.0-1.5	3.53
Glass	2.5	2.0-2.5	4.13
Zirconium Oxide	3.8	1.0-1.25	5.41
Zirconium Oxide	3.8	0.8-1.0	4.66

FOAM QUALITY 70%

Matl.	Density (g/ml)	Size (mm)	Settling Velocity
Glass	2.5	1.0-1.5	5.43
Glass	2.5	2.0-2.5	6.21
Zirconium Oxide	3.8	1.0-1.25	6.94
Zirconium Oxide	3.8	2.0-2.5	5.49

FOAM QUALITY 33%

Matl.	Density (g/ml)	Size (mm)	Settling Velocity
Glass	2.5	2.0-2.5	13.6
Glass	2.5	1.0-1.5	10.2
Zirconium Oxide	3.8	0.8-1.0	28.9
Zirconium Oxide	3.8	1.0-1.25	32.9

Table 3 (cont.)

PURE WATER (FOAM QUALITY 0%)

Matl.	Density (g/ml)	Size (mm)	Settling Velocity (cm/s)
Glass	2.5	2.0-2.5	25.7
Glass	2.5	1.0-1.5	20.0
Zirconium Oxide	2.8	0.8-1.0	22.5
Zirconium Oxide	3.8	1.0-1.25	25.7

Table 4

Transport Capacity of Water Only

Material: Zirconium Oxide
Density: 3.8 (g/ml)
Size: 1.0-1.25(mm)

Tank Conc. (%)	Transport Capacity (g/s)
.03	5.5
.06	10.2
.09	14.2
1.20	20.0
1.80	30.5

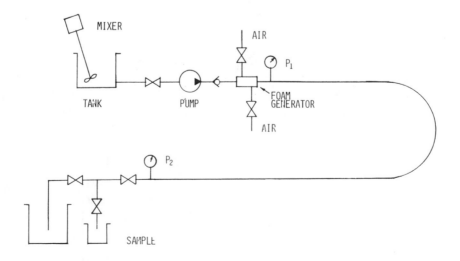

Fig. 1. Foam transport test system.

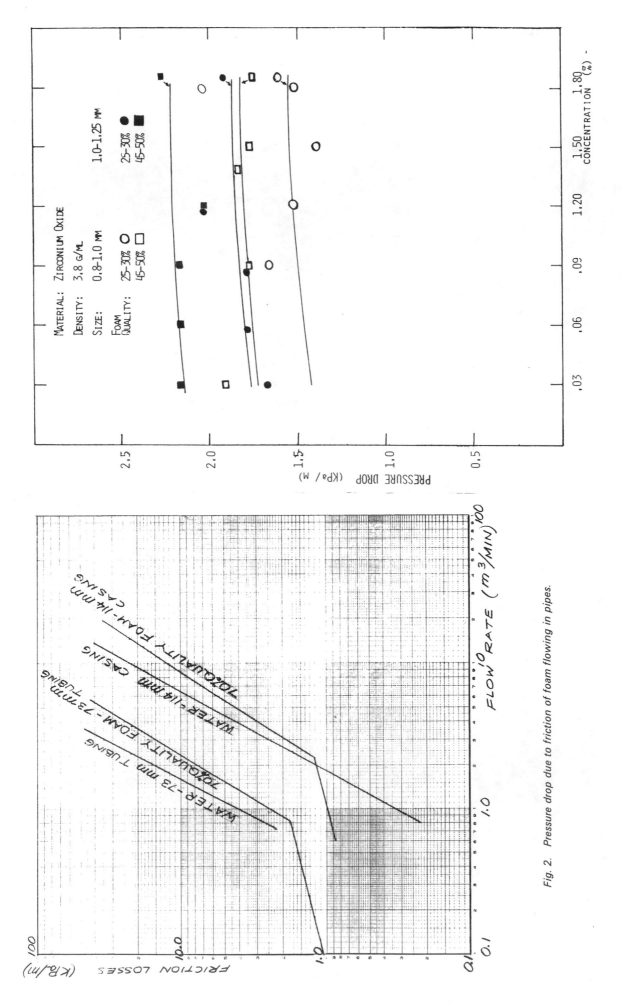

Fig. 3. Pressure drop versus tank concentration.

Fig. 2. Pressure drop due to friction of foam flowing in pipes.

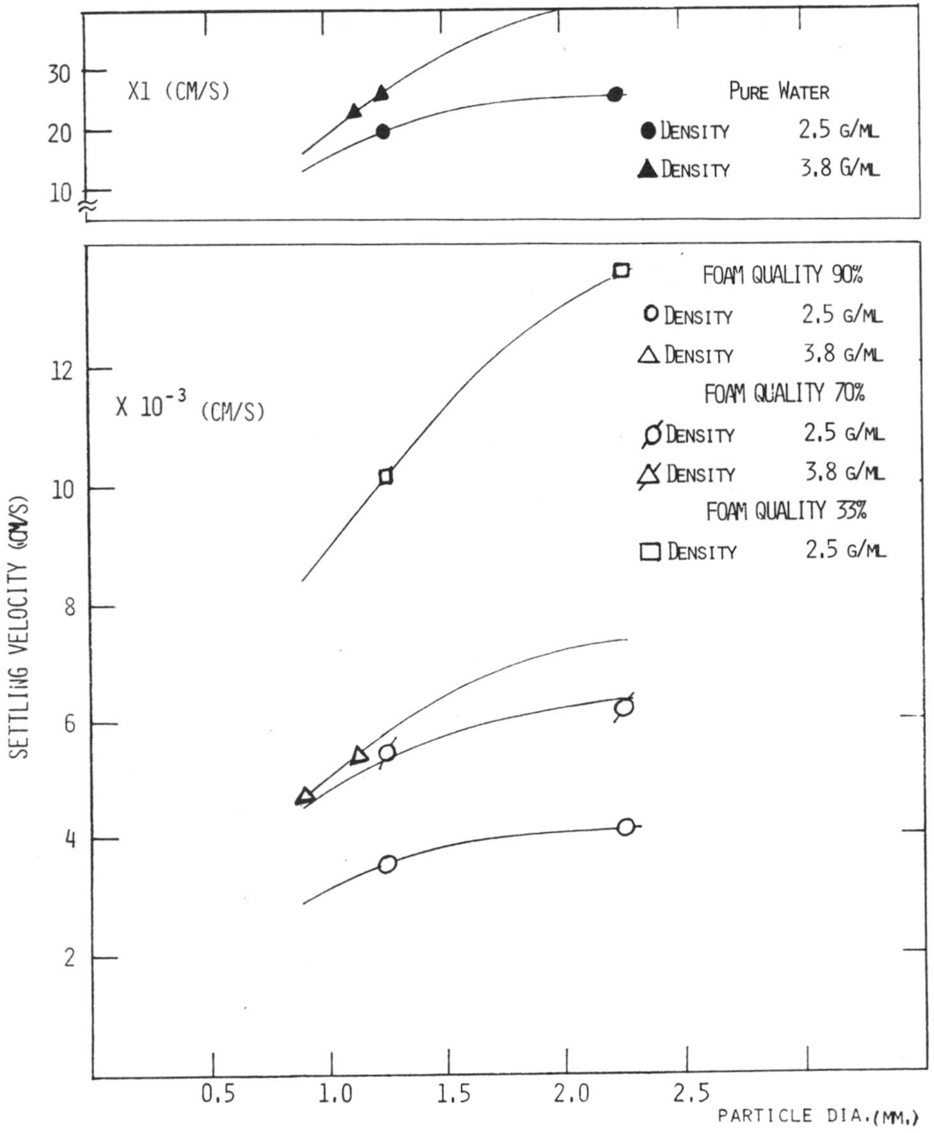

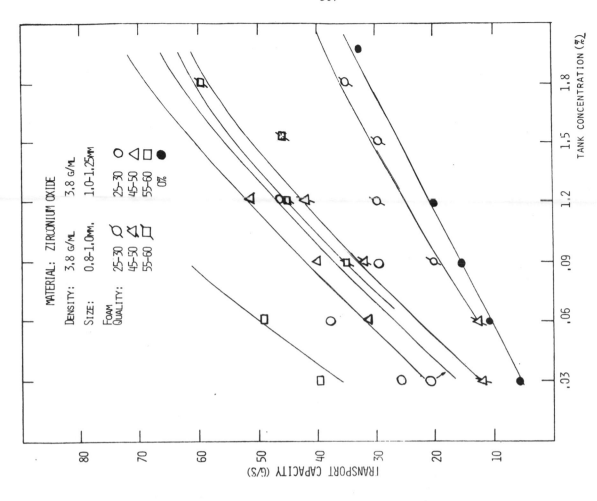

Fig. 5. Transport capacity versus tank concentration.

Fig. 4. Settling velocity versus particle diameter.

10th International Conference on the
Hydraulic Transport of Solids in Pipes

HYDROTRANSPORT 10

Innsbruck, Austria: 29-31 October, 1986

PAPER K3

A REVIEW OF TECHNIQUES FOR REDUCING ENERGY
CONSUMPTION IN SLURRY PIPELINING

N.I. Heywood

The author is in the Materials Handling Division
at Warren Spring Laboratory Stevenage, UK

SYNOPSIS

This paper reviews the methods available to reduce
head loss, and hence energy consumption, incurred
when slurries are pumped through horizontal pipes.
The methods are grouped according to whether they
are applicable to essentially "non-settling" or
settling slurries, whether they are effective in
the laminar or turbulent flow regimes and the
requirements, if any, for the laminar flow
behaviour of the slurry. The essential underlying
mechanisms for the techniques, together with their
advantages and disadvantages, are discussed. The
techniques covered include the addition of soluble
ionic compounds, polymers and soaps, the oscillation of either the pipe or the slurry flowrate
symmetrically or asymmetrically, the injection of
air into the pipeline and the use of helical ribs
or segmented pipe. In addition, attention is
given to three important particle property
effects: size, size distribution and shape.

All these techniques have been demonstrated
technically in the laboratory, often at pilot-
scale, but few have been taken up in industrial
installations. This is probably due in part to
the lack of awareness of these techniques by
design engineers.

INTRODUCTION

Over the last twenty years or so a number of
techniques for reducing head loss in slurry pipe
flow have been demonstrated, in many cases at both
pilot-plant and full-scales. These techniques are
capable of reducing frictional pressure drop
either by altering the slurry rheological
properties directly or by reducing the impact of
adverse flow properties indirectly.

Some of the more obvious ways of reducing slurry
viscosity levels directly include:

(a) reducing slurry concentration;
(b) increasing particle size, if Brownian motion
    and particle surface effects control
    viscosity levels;
(c) broadening the particle size distribution at
    constant total solids concentration;

(d) reducing the angularity of particle shape
    while maintaining the particle size distribu-
    tion and solids concentration essentially
    constant. Also, adding high aspect ratio
    fibres;
(e) adding deflocculants (soluble ionic
    compounds) to disperse flocculated slurries
    (important when a significant percentage, say
    5-10%, of solids are below approximately 5μm
    in size);
(f) addition of soaps;
(g) addition of high molecular weight polymer;
(h) injecting of water (or other suspending
    medium) into the flowing slurry periodically
    along the pipeline length.

Most of the above methods can alter a suspension
or slurry formulation irreversibly in the sense
that a significant cost arising from extra capital
and operating requirements may be incurred to
return the slurry to its original state (if it is
possible at all) once the slurry has been
transported through the pipeline.

A further set of techniques tends not to have this
disadvantage:

(a) oscillation of the slurry flowrate or
    pressure gradient;
(b) vibration or oscillation of the pipeline
    about the pipe axis while maintaining a
    constant slurry flowrate;
(c) injecting air (or other gas, if appropriate)
    into the pipeline to create a three-phase
    mixture which would generally be readily
    separated on discharge from the pipeline;
(d) use of spiral ribs on the inner pipe wall to
    reduce the deposit velocity of settling
    slurries, and hence power requirements;
(e) use of segmented pipe (a horizontal flat
    plate welded to the pipe bottom) to reduce
    deposit velocity and hence power
    requirements.

The methods listed above are effective under
certain conditions only. In determining whether a
method may be appropriate for a particular
application it is important to decide whether

(1) a settling or essentially "non-settling"
    slurry is to be pumped;

(2) turbulent or laminar flow conditions will
    prevail during pipe flow;

(3) Newtonian or non-Newtonian flow property is
    important, if the flow is in the laminar
    regime.

The methods can be classified essentially into
those applicable to the laminar pipe flow of "non-
settling" slurries (see Table 1) and the turbulent
pipe flow of settling slurries (see Table 2).

In the following sections, each of the important
influences on determining head loss and hence
methods of reducing power requirements will be
discussed. The review begins by considering how
modification of three important particle
properties can reduce slurry viscosity: particle
size, size distribution and shape.

EFFECT OF MODIFICATION TO PARTICLE PROPERTIES

Particle Size

Both from experimental observations and from theoretical considerations, the absolute size of particles has no effect on suspension rheology provided that the particles are well-dispersed in suspension and are influenced by hydrodynamic forces only. In practice, however, the effect of variation in particle size can be seen in rheological data for a variety of reasons.

Firstly, when particles are in aqueous suspension, an adsorbed layer of molecules on the particle surfaces causes the suspension to behave rheologically as though the volume of solids in suspension is greater than a value calculated from a knowledge of the weight and density of the particles. If the thickness of this adsorbed layer remains constant, and is independent of particle size, suspension viscosities will tend to increase with decreasing particle size, at constant particle volume concentration and shear rate. An adsorbed layer generally occurs in aqueous suspensions in which stabilisers are frequently used, but can also occur in non-aqueous suspensions.

Secondly, as particle size is reduced to micron sizes or below, Brownian movement forces play a significant part in determining suspension flow properties. Owing to particle collisions arising from Brownian motion, non-Newtonian behaviour is common and can consist of both shear-thinning and shear-thickening characteristics, as Wagstaff and Chaffey (1,2) have demonstrated. The effect of Brownian motion forces is similar to the effect of the presence of an adsorbed layer, in that suspension viscosity increases with decreasing particle size (3) and it is not clear at present how the two effects may be quantified separately, although the treatment by Woods and Krieger (4) shows promise. Figure 1 indicates how the suspension relative viscosity rises as particle size is reduced at constant total solids concentration.

Thirdly, as particle size is reduced to the order of a μm or below, Coulombic forces will become increasingly important, no matter how successful one is at suppression of the double layer of charge on the particle surfaces using appropriate types and concentration levels of electrolyte. The relative importance of these additional processes, besides hydrodynamic effects, can be assessed as a function of particle size by the method of Chaffey (5), who has defined and listed the relevant dimensionless groups and extended the pioneering analysis of Krieger (6).

It can be concluded therefore that the proportion of solids having a particle size below a few μm should be reduced if suspension viscosity is to be reduced. However, there are instances when a high viscosity, shear-thinning suspension is desirable to act as a suspending medium for the transport of coarse particles. The optimum solution is to reduce the proportion of fines while maintaining an essentially "non-settling" slurry.

Particle Size Distribution

The importance of particle size distribution in determining viscosity levels of essentially "non-settling" suspensions has generally been over-looked in the past. Yet there now exists ample theoretical evidence and a limited amount of experimental data which show that an economic advantage can be gained by adjusting particle size distribution. This usually means that, where particle surface effects are insignificant, a wide particle size distribution should be strived for. Then for a given solids concentration, viscosities will be reduced and hence head losses for laminar pipe flow will be minimised. The effect will be smaller for turbulent pipe flow and will depend upon the friction factor-Reynolds number relationship for the "non-settling" suspension. On the other hand, the effect of particle size distribution on head losses for settling slurries is far from clear.

Theoretical predictions of the viscosity reduction effect by Farris (7) suggest that its significance emerges only for slurries with total solids concentrations exceeding some 30% to 40% by volume solids. Figure 2 shows predicted viscosities for a Newtonian slurry consisting of a variable mixture of fine and coarse particles. If a slurry consists only of either fine or coarse particles, slurry viscosities are relatively high, but an appropriate mixture of the two particle fractions will lead to a minimum in viscosity. This minimum is likely to become more pronounced and its position dependent on the proportions of fine and coarse material, as the overall solids concentration is raised.

Such concepts and the Farris analysis can be extended to a slurry having many size fractions and also to continuous particle size distribution. The importance of theoretical predictions is that they are useful to determine approximately an optimal particle size distribution which can then be defined more precisely through experimentation.

Another way of looking at the particle size distribution effect is as follows. When the main concern is to maximise the dry solids rate through the pipe, this can be achieved by maximising solids concentration while keeping viscosity levels constant by adjusting particle size distribution. This conclusion obviously has a number of implications, including not only a reduction in cost per tonne km of dry material moved but also a reduction in water usage, important when water supplies are limited or when restrictions on water transport across state or national boundaries are present, such as those which occur in the USA and parts of Africa.

The basic concepts of the theory were first discussed by Eveson et al (8) and Mooney (9) in 1951 and Brinkman (10) and Roscoe (11) in 1952. More recently a different theoretical approach has been taken by Lee and Lee (12). Experimental evidence for the effect has been provided by Chong et al (13), Skvara and Vancourova (14), Parkinson et al (15) and Goto and Kuno (16). Investigations relevant to pipeline flows have been carried out by Ferrini et al (17) and Klose (18). However, despite these studies more experimental effort needs to be aimed at higher solids concentrations than those used up to now so that the limits of the Newtonian viscosity and even granular property can be studied. The importance of the granular property is particularly in evidence for compressible pastes and cakes which can be pumped over relatively short distances using high pressure positive displacement pumps originally developed for pipelining concrete.

This area is currently under investigation at Warren Spring Laboratory and a co-operative project with companies is planned to be launched shortly.

Particle Shape

It is well-established that reduction in particle angularity or anisometry generally leads to lower viscosities at a given overall solids concentration and while maintaining a similar particle size distribution, provided that only physical effects are occurring. Lower viscosities lead to lower head losses in laminar pipe flow. Clark (19) has provided ample evidence to support this conclusion. In addition, particles with high aspect ratios, typically polymeric or wood pulp fibres can exhibit two separate effects:

(a) a reduction in head losses for turbulent pipe flow when fibres are present in typically low concentrations, i.e. hundreds of ppm (see references 20-22);

(b) a reduction in head losses for both laminar and turbulent pipe flow when they are present in sufficiently high concentrations (but often no more than 1 to 5% by volume) to give an unsheared plug of fibrous suspension flowing in the central pipe core (see references 23, 24).

Drag reductions effected by fibre systems are usually rather lower than those obtained by polymer solutions. The advantage of a fibre-laden system is that the fibres do not generally degrade, unless they are particularly friable or subjected to very high shear stresses in tortuous flow fields, and hence the drag reduction effect does not diminish with time on progressive cycling within a pipeline loop or for flow in a long pipeline. A further advantage is that, because effects within the turbulent core of the pipe dominate the drag reduction process, in case (a) above the effect is not diminished on scaling up from small to large pipe diameters. With polymer solutions, the effect is diminished on scale-up. However, in case (b) it still remains to be established whether a pipe diameter effect exists or not.

The extent of the drag reduction increases with either an increase in fibre aspect ratio while holding solids concentration constant, or an increase in solids concentration at constant fibre aspect ratio; in the latter case, up to a maximum effect.

Both the laminar and turbulent flow behaviour of more concentrated fibre suspensions have some interesting features. Much effort has been devoted in the paper industry to the examination of the flow of paper pulp suspensions at weight concentrations typically from 0.5 to 3%. Although these concentrations may appear low, the highly fibrous nature of the pulp causes entanglement and the formation of large flocs and makes the material flow essentially as a plug, with a lubricating water annulus adjacent to the pipe wall. With these materials, unlike ordinary single-phase liquids, it is possible for the pressure drop-flow velocity plot to pass through both a maximum and a minimum in the laminar region. This behaviour may be attributable in part to changes in the width of the lubricating water annulus as the flow velocity is varied; this is governed by the degree to which water is

expressed from the matted plug of paper pulp as the hydrodynamic conditions vary. At higher flow velocities still, the pressure drop can fall below that for water flow at the same velocity, indicative of a drag reduction phenomenon, whose underlying cause is probably different from the similar effect produced by very low concentration fibres.

The consequence of such behaviour in the turbulent flow regime is that for many fibrous suspensions, a knowledge of their laminar flow properties either in a tube, pipeline or rotational viscometer will not aid in the estimation of turbulent head losses, nor in determining whether drag reduction will occur or not. In fact, there currently exists no method which allows this prediction. Thus two slurry systems may possess near identical laminar flow behaviour, but the fibrous slurry may exhibit drag reduction in the turbulent flow regime whereas the slurry containing non-fibrous particles normally would not show a drag reduction effect.

Work has been carried out on turbulent velocity profiles and turbulent radial and longitudinal intensities but is rather inconclusive and more effort is required here to obtain a proper understanding of the mechanism of drag reduction in fibre suspensions. Despite numerous experimental studies, conclusions drawn by various workers remain muddled and confusing. However, this of course should not prevent the exploitation of the drag reduction phenomenon in industrial applications. A precise understanding of the drag reduction mechanism is naturally not a prerequisite for the exploitation of the effect although predictions of the extent of the effect are almost impossible.

EFFECT OF ADDITIVES

Use of Soluble Ionic Compounds as Deflocculants

It has been known for some time that the addition of certain types of soluble, ionic compound to flocculated suspensions can result in substantial pressure drop reductions in pipe flow in the laminar flow regime. Usually, however, little effect is noticeable in the turbulent regime. The ionic compounds disperse the particles, thereby breaking up the flocculated particle network which had previously given rise to higher shear stresses in pipe flow. In order to disperse particles the charge on their surfaces needs to be of the same sign, and the higher the charge the greater the repulsive forces between particles.

It is only relatively recently that the reduction in pressure drop has been related to the degree of deflocculation in quantitative terms. Horsley et al (25-27) have carried out a number of studies in which the zeta potential on the particle surfaces, defined as the potential in the electrochemical double layer at the interface between a particle moving in an electric field and the surrounding liquid, has been measured and correlated with pressure drop reduction. The zeta potential is a measure of the repulsive forces acting on the particles and thus is useful in measuring the dispersing effects of chemical additives.

Figure 3 shows some typical head loss - slurry velocity data obtained by Horsley and Reizes (25) for the flow of a 43% by volume sand slurry (37% by weight of particles less than 10 μm) at different pH levels in the range 6.2 to 10.8; the pH was controlled by the addition of nitric acid and sodium hydroxide. The changes in pressure drop in the laminar regime were observed to be greater as the sand concentration was raised. Figure 4 shows a plot of head loss against zeta potential. This illustrates that the higher the zeta potential (i.e., the larger the negative value) the lower the pressure drop.

Sikorski et al (28) have studied the effects of chemical additives on a number of mineral slurries including drilling muds (thinners such as sodium acid pyrophosphate and sodium hexametaphosphate), phosphate rock slurries (caustic soda), limestone cement feed (combination of sodium tripoly-phosphate and sodium carbonate) and coal slurry (sodium tripolyphosphate, sodium dioctyl sulpho-succinate and sodium carbonate), while Shook et al (29,30) have studied the effects of various alkaline additives on the flow properties of coal-water slurries.

Addition of Soaps

Drag reduction using soap solutions which do not undergo irreversible mechanical degradation has been obtained using liquid and suspension systems.

Zakin et al (31) used various aqueous mixtures of l-naphthol and cetyl trimethylammonium bromide (CTAB) with sand slurries (particle size 0.9 mm and 1 mm). A 0.07% naphthol and 0.15% CTAB combination was found to give the lowest pressure drop reductions of up to 80% for flow through a 20 mm tube. Temperature and aging effects were also studied in the absence of sand particles. For quartz sand, 0.09% soap caused a 60% drag reduction of critical velocities, while doubling the soap concentration to 0.18% gave an additional improvement of less than 10%. It was also found that the critical velocities for the complex soap solution suspensions were about half those for the water-sand suspension and this was examined through settling rate data in graduated cylinders where settling rates were found to be much lower than for water-sand suspensions.

Bakelite sand suspensions showed a less marked drag reduction at 0.09% soap concentration while at 0.18% 40% drag reduction was obtained and at 0.225% soap concentration, 70%.

The significance of the Zakin et al study is that the soap does not appear to have been degraded mechanically. A critical shear stress was found with sand suspensions above which no drag reduction occurred, but on reducing the wall shear stress the same levels of drag reduction were returned to as previously. Thus the potential of complex soaps for drag reduction in suspension flow would appear to be greater than high molecular weight polymer and further work is required to assess whether the addition of this, or related, complex soap would give an economic as well as technical advantage.

Use of Soluble High Molecular Weight Polymer

Ever since the famous experiments of B.A. Toms in the late 1940's, it has been known that the addition of small amounts (10's to 100's ppm) of soluble, high molecular weight polymer to a liquid flowing in the turbulent regime will cause a reduction in the head loss. Typical polymers used to demonstrate this phenomenon have been polyacrylamides, polyacrylic acid and polyethylene oxide having molecular weights up to six million. One of the main drawbacks to the successful application of this phenomenon to numerous situations is that the polymer can degrade, often quite rapidly, under mechanical shear with a consequent reduction in average molecular weight of the polymer and hence reduction in effectiveness in reducing head losses. This is particularly acute in recirculating test flow loops which often incorporate many different types of fittings as well as repeated passage of the polymer solution through the pump. Degradation may prove to be much less of a problem in once-through applications, but can also occur through attack from micro-organisms present in water.

Despite the lack of industrial applications so far, the amount of research effort into this drag reduction effect for single-phase liquids has been immense during the last twenty years or so. More recently attention has been turned to the potential of polymers to reduce head loss in solid-liquid flows. Radin et al (32) have reviewed some of the work done in slurry flows up to 1975. Zakin et al (31) used polyethylene oxide and polyacrylamide with sand suspensions flowing in a 20 mm diameter pipe. These polymers gave very high drag reduction but excessive mechanical degradation occurred and the effect was lost in two minutes. However, they also used guar gum, a high molecular weight natural gum, which is much more stabled to mechanical degradation but which requires up to 100 times greater concentrations than polyethylene oxide for appreciable drag reduction.

Golda (33) recently undertook a systematic study of the effectiveness of six different polyacrylamides (all sold under the Dow Chemical tradename 'Separan') in reducing the head loss of a 2.91% coal slurry (mean particle size of 2.8 mm flowing in a 40 mm diameter pipe). A fixed polymer concentration of 80 ppm was used and mechanical stability tests were carried out by monitoring drag reduction initially and after 60 minutes. Separan AP45 performed best and consequently was further used in a series of tests to explore the effects of both polymer and coal concentration on the level of drag reduction. The degree of drag reduction was less for coal slurry than for water flow alone, but that in both cases, drag reduction passed through a maximum as polymer concentration was progressively increased; the polymer concentration giving the maximum drag reduction was higher for coal slurry than for water flow alone. The maximum drag reduction level also decreased progressively with increasing coal concentration.

Other studies include those by Pollert (34) who obtained a maximum drag reduction of 32% using polyacrylamide (Separan AP273) dissolved in flyash slurry, Kolar (35) who used mica and kaolin in polyacrylamide solution, and Sifferman and Greenkorn (36) who also used guar gum with sand suspensions flowing in three differently-sized pipes.

From the limited studies so far, it would appear that, despite the problem of mechanical degradation, the use of polymers may well facilitate substantial reductions in power consumptions.

AIR INJECTION INTO FLOWING "NON-SETTLING" SLURRY

For all Newtonian slurries in either laminar or turbulent flow and for all non-Newtonian slurries in turbulent flow, introduction of gas into a horizontal pipeline carrying slurry at a constant rate will invariably increase the pressure gradient at any point along the pipeline. This is because the average velocity is raised in the pipe, owing to the reduced mean cross-sectional area for liquid flow brought about by the gas presence, and this in turn creates higher shear stresses acting on the pipe wall which are responsible for the pressure drop.

In contrast, pseudoplastic non-Newtonian fluids, where the viscosity decreases with increasing flow (i.e. shear-thinning behaviour), can be in laminar flow concurrently with gas in a horizontal pipe and the presence of gas results in a decrease in average pressure gradient. This was first reported by Ward and Dallavalle (37). Some typical experimental results for shear-thinning flocculated kaolin suspensions, are shown in Fig. 5. Here the parameter ϕ_s^2 (termed the drag ratio in the figure) is plotted against the superficial air velocity, with the superficial slurry velocity in the pipe as a parameter. ϕ_s^2 is defined as the ratio of the pressure gradient along the pipe with gas injection to the pressure gradient for flow of suspension alone at the same suspension flowrate. It can be seen that provided the slurry flowrate is sufficiently low, i.e. the Reynolds number is low and in the laminar region, the presence of gas causes a reduction in the ϕ_s^2 parameter below unity, indicating a reduced pressure gradient over that for slurry flow alone at the same slurry rate. At any sufficiently low fixed slurry rate this reduction progressively increases with increasing gas rate until a maximum effect is reached. Thereafter the pressure gradient rises with further gas rate increases. The highly non-Newtonian, shear-thinning character of these kaolin suspensions is indicated by the flow curves in Fig. 6 taken from reference 40.

This phenomenon has important implications for pipelines carrying shear- thinning slurries. Three cases arise. In the design of a new pipe-line installation, the injection of air downstream from the slurry pump may result either in an overall operating cost saving or facilitate the transport of high viscosity paste-like material which a practical slurry pump could not otherwise achieve alone. The second situation arises when the capacity of an existing pipeline needs to be raised while retaining the same pump system. Air injection will reduce the frictional head against which the pump acts at the old slurry rate and the result will be an increase in slurry throughput for the same pump speed. Thirdly, an existing pipeline can be lengthened to transport the same slurry using the same pipe diameter.

In taking advantage of these benefits it is obviously important to be able to predict the maximum extent of drag reduction achievable for a given slurry and the air flow required to obtain maximum reduction in head loss.

It has been shown (40,43) that the maximum reductions could be collapsed onto a single curve either by plotting minimum values of ϕ_s^2 against the slurry Reynolds number with flow behaviour index n in the non-Newtonian power law mode as a parameter or by plotting ϕ_s^2 against a parameter

λ_c defined by

$$\lambda_c = \left[\frac{V_{SL}}{(V_{SL})_c}\right]^{1-n} \qquad (1)$$

where V_{SL} is the superficial slurry velocity and $(V_{SL})_c$ is the critical superficial slurry velocity at the breakdown of laminar flow for slurry flow alone.

Further work (42,46) has involved investigation of these effects in vertical pipe flow, where introduction of air always introduces a reduction in head loss because of the decrease in the static head component to the total head loss which arises due to a decrease in the density of the mixture.

Although gas injection into a pipeline has yet to be exploited widely and its benefits under some conditions yet to be appreciated, some examples of practical applications already exist. Three are cited here:

(a) a transfer system for asbestos slurry was designed at Warren Spring Laboratory (38,39) for a brake manufacturer. The system involving the injection of air into a 90 m long pipe carrying the slurry has been in operation in the client's factory for over ten years;

(b) the distance over which waste residue is transferred at a sugar-producing factory has been increased by some 50% by injecting air into the pipeline immediately downstream from the pump, while retaining the same pump;

(c) Putzmeister Ltd markets a 'MIXOKRET' pumping system (48) for transfer of concrete on building sites. The system consists of a single-shaft positive-action mixer which also acts as a conveyor. The pressure vessel lid to the mixer which is hinged on one side is closed with a quick-action clamp and air above the concrete is compressed, pushing the mix out of the vessel. Compressed air is added from a second line directly to the delivery line connection on the vessel outlet, giving "air cushions" which transport the mix in an intermittent flow through the hose line. The compressed air to the vessel and to the delivery line connection can be controlled separately. A discharge stand with pot or curved discharge device lets the conveying air escape at the end of the line allowing uniform discharge of the concrete.

Recently an attempt has been made (47) to define more precisely the economically beneficial operating conditions when using air injection. A simple method was presented for evaluating the power saving achieved and the volume of air required to obtain maximum power saving. There is no doubt that this is an important technique for head loss reduction when there exists no or little scope for reducing suspension viscosities by particle deflocculation, and hence turning a non-Newtonian, shear-thinning suspension into an often much lower viscosity Newtonian suspension. One of the main advantages of air injection is that the separation of the phases on discharge from the pipeline is relatively straightforward since the air normally flows in the form of discrete slugs separately from the suspension. Hence no alteration of suspension properties using this power saving technique occurs during pipeline transport.

DISTURBANCES TO SLURRY FLOW OR PIPE

Oscillation of Slurry Flowrate or Pressure Gradient

Oscillation of either the flow of slurry or the applied pressure gradient have been shown to result in a decrease in head loss under certain circumstances. When the fluid flow is oscillated then typically a sine wave pulsation is superimposed on a steady flow with the frequency lying in the range 0-10 Hz. Edwards and Wilkinson (49) have observed both experimentally and from theoretical considerations that this pulsation of the flow affects the profile of the amplitude of velocity across a pipe. The result is an increase in flowrate under a constant pressure gradient for some non-Newtonian fluids, or the energy for pumping could be minimised at constant flowrate by suitably dividing it between that required to provide a steady pressure gradient and that required to produce an oscillating component. This is known as "split-energy pumping" and may have many potential industrial applications.

Most experimental verification of the effect has involved aqueous polymer solutions which in general exhibit viscoelastic behaviour. However, the simplified theory of Barnes et al (50) does not include viscoelastic effects but nevertheless predicts pressure gradient reductions. It appears that the presence of viscoelastic properties enhances the drag reduction effect. It has been hypothesised (49) that for the flow of concentrated slurries the thickness of a particle-depleted layer at the pipe wall may be increased by the oscillating component of the pressure gradient, and the central core of slurry then moves as a plug on a lubricating layer of liquid, leading to a possible reduction in power consumption. The problem in the exploitation of this phenomenon is the dampening of the oscillatory component of the applied pressure gradient at large distances downstream from the point at which oscillations to the flow are imposed.

Round et al (51-53) pointed out that flowing systems, typically fluids in pipelines, all have some sort of pulsations inherent in them. These are usually regarded as a nuisance which should be eliminated as simply and as economically as possible, and pulsation dampers are often used to achieve this. However, situations in which these pulsations could be exploited mean that the capital costs of both a pulsation damper and a device for imparting pulsations to the flow could be saved.

Benefits can be gained for either "settling" or "non-settling" suspensions. Round (51) noted that the pulsation of "settling" suspensions in turbulent flow resulted in:

(1) a decrease in the terminal settling velocity;
(2) a decrease in the energy required to suspend particles.

This enhancement of suspension occurs particularly for flow in inclined pipes where there exists a component of the flow opposing the sedimentation direction of the particles, but even in horizontal pipes it has been observed by Chan et al (54) that

particle suspension is easier to achieve in a pulsating flow than in a steady flow. Round (51) considers the application of pulsations with most potential is to settling suspensions and, in fact, particles could be transported in a purely symmetric oscillatory flow without a net steady flow component by suitable adjustment of the pipe internal geometry. This has been done using baffles in a pipe.

Without pipe internals, purely oscillatory flow in a regular manner does not produce a net flow. However, asymmetric oscillations which do produce a net flow may be considered to consist of steady flow plus oscillatory components. Colamussi and Merli (55) have indicated the advantages of asymmetric pulsing in a solid-liquid system by means of flow interruptions. Such a system has been successfully applied to a dredging operation which transported a mixture of gravel, soil and water.

Vibration or Oscillation of Pipe

Oscillation of the pipe rather than the slurry obviously has less practical appeal but a handful of studies has provided information on reduction in head loss. Deysarkar and Turner (56) transported a normally stiff paste of fine iron ore containing 16% by weight water through a mechanically vibrated tube. The paste appeared to liquefy and flowed through the tube under a "very modest" pressure head. If no slip of the paste was assumed to exist at the tube wall, then the calculated viscosity for different tube diameters could be correlated with the peak cyclic acceleration given by Af^2, where A is the peak-to-peak amplitude and f is the angular frequency.

Manero and Mena (57,58) passed Newtonian, non-elastic and non-Newtonian, elastic aqueous polymer solutions through relatively small diameter tubes which were oscillated using a variable speed motor with an eccentric shaft. No change in flowrate under a constant pressure gradient was observed for the Newtonian liquids but an increase in flowrate was found for elastic fluids, the percentage increase being greater at small flowrates.

It is difficult to envisage that it would be technically feasible or economic to oscillate a pipeline along its axis over any significant distance and hence the exploitation of such a phenomenon is probably limited to short pipe lengths. Since in-factory pipelines often have vibrating equipment attached to them, it may well be that the effect is, sometimes unknowingly, being exploited. Frictional energy losses resulting from pipe flow within chemical plant are relatively minor compared with other energy requirements, and hence it appears doubtful that it would ever be worthwhile to install dedicated vibrational equipment, unless it is used specifically to prevent some pipework blocking.

USE OF MODIFIED PIPE CROSS-SECTION GEOMETRIES FOR SETTLING SLURRIES

Helical Ribs

Pipes containing helical ribs attached to the inner pipe wall surface have been shown in several studies (59-66) to have significant advantages for some slurry pipeline transport requirements. For rapidly-settling slurries, head losses can be lower than for smooth pipes at low velocities. Charles (60) used ribs with pitch-to-diameter ratios of 1.79, 3.3 and 5.2 for a 51 mm diameter pipe (see Fig. 7). At high velocities the ribs increase the head loss. It seems feasible though to operate ribbed pipes at lower velocities with reduced risk of blockage compared with plain pipes at similar velocities and lower operating velocities will aid in reducing energy consumption.

The pitch to diameter ratio is related to the rib angle, θ, measured from the pipe axis by:

$$\tan \theta = \frac{\pi D}{p} \qquad (2)$$

Charles (60) suggested that a p/D ratio of 5 should prove useful in many applications but, in a later study, Singh and Charles (64) found that the optimum p/D ratio seems to be about 8(θ = 31.4°) with a rib height of 10 to 15% of the pipe diameter.

Both Charles et al (61) and Schriek et al (63) have looked at how power consumption can be reduced using helical ribbed pipes. Because the advantage of the ribbed pipe is only realised at velocities below the critical deposit velocity in the smooth pipe, in order to provide the same solids flowrate, either the diameter of the ribbed pipe must be larger than the diameter of the smooth pipe if the delivered concentration is the same in both cases, or the slurry flowing through the ribbed pipe must be more concentrated if the diameters of the pipes are the same in both cases. Using two somewhat arbitrary comparisons, Charles et al (61) were able to show that the power consumption for a 13% slurry flowing through a 48 mm pipe (with p/D = 5.15) at a velocity of 1 m s^{-1} was some 60% of that required for the same slurry concentration passing through a 36 mm dia smooth pipe at 1.68 m s^{-1}. A p/D ratio of 3.32 gave a similar power consumption, while a p/D ratio of 1.80 gave a higher power consumption.

In the second comparison, the slurry concentrations must be different if the pipe diameter is to remain approximately the same, again for a fixed solids throughput. The power consumption for a 5% slurry through 50 mm pipe at 2.1 m s^{-1} is greater than that for a 13% slurry flowing through a 48 mm ribbed pipe at 1 m s^{-1} at either p/D = 5.15 or 3.3, but is slightly less for p/D = 1.80.

Schriek et al (63) obtained head loss and energy consumption data for sand slurry flow in a 150 mm diameter pipe which showed similar trends to the data of Charles et al using a 48 mm diameter pipe. Energy consumption calculations showed that for p/D < 4, ribbed pipes are not very attractive in reducing energy requirements, whereas in the range 5 < p/D < 10, the minimum energy consumption is relatively insensitive to p/D.

It is probable that there are numerous situations where ribbed pipes would be desirable. Capital cost of a ribbed pipe is obviously going to be larger but it is technically relatively straight-forward to provide a plastic pipe with ribs. In the Schriek et al study the 150 mm diameter ribbed pipe was manufactured by extruding over a mandrel which could be rotated at various speeds to alter the p/D ratio. Thus, the ribs were an integral part of the pipe itself. However, extruded plastic pipe can withstand only low pressures, but

it may be possible to develop a liner with the ribs for installation into steel pipe.

Since slurry pipelines can be operated at lower velocities when helical ribs are incorporated, pipe wall wear rates can be expected to be lower, but erosion of the ribs could be a potential problem and this must be evaluated as far as possible in any feasibility study owing to the high capital cost investment in a ribbed pipe. Smith et al (66) measured wear rates for a ribbed pipe (p/D = 4) and smooth pipe using a weight loss method. For the same velocity higher erosion rates were observed with the ribbed pipe but when comparison is made at the corresponding deposit velocities for the ribbed and smooth pipes, wear rates are similar. Also, Smith et al point out that in analysing wear rate data, it is the maximum local penetration rate which determines pipe durability, and so the distribution of wear rates must be considered. Although the flows in ribbed and smooth pipes are most similar at the lowest velocites, one might expect the wear to be more evenly distributed in the ribbed pipe.

Unfortunately there currently exists no systematic design approach for helically-ribbed pipe and each system must be evaluated both by desk study and probably pilot-scale testwork to investigate such variables as the viscosity of the slurry, the size distribution of the solids and the density ratio of solid to suspending medium. However, there exists quite a large volume of information for sand slurries and starting points, such as a p/D ratio lying in the range 5 to 8, can be used. Schriek et al (63) state that it is arbitrarily assumed that one rib for every 50 mm of pipe diameter will produce adequate agitation, but this seems a reasonable "rule-of-thumb" since their data for three ribs in a 150 mm diameter pipe and one rib in a 50 mm diameter pipe are in "reasonably good agreement".

Despite the lack of information concerning the effects of a number of variables, such as the optimisation of rib shape and size, ribbed pipe has special advantages for relatively coarse material, and for intermittent operation when unavoidable shutdowns are anticipated. The ribs could aid in the restarting of such lines over uneven terrain and on steep gradients and there appears to be no reason why ribs should not be installed only in sections of a line where blockage is of concern.

Use of Segmented Pipe

A number of studies have indicated that a circular cross-section to a pipe is not the most favourable for minimising head losses in slurry pipe flow and experiments on non-circular cross-section pipes have been undertaken. Although the results of these studies appear attractive from an operating cost saving point-of-view, it is probable that few, if any, industrial installations exploit this saving owing to technical and cost disincentives in manufacturing non-circular geometry pipework.

Wang and Seman (67) using six pipe geometries showed without doubt that pipe geometry and particularly the base area of the pipe significantly affect the slurry flow. The investigation revealed that the head loss at the minimum carrying velocity in a wide-base rectangular pipe was approximately 20% less than that in a circular pipe of the same cross-sectional area. All non-circular cross-sections, however, suffer from the disadvantage that points of stress intensification occur in the pipe wall and thicker walled pipes become necessary. To avoid this, Führböter (68) proposed that a segment plate be placed inside the pipe which would at the same time act as a wear plate. By equalizing the pressure above and below the segment plate by means of, for instance, a few small holes in the plate the advantage of uniform stress distribution in a circular pipe is retained.

Führböter (68) calculated that transport in a segmented pipe is particularly beneficial when transporting large solids and predicted that the best position of a segment plate is at approximately one third of the internal pipe diameter above the bottom level of the pipe as shown in Fig. 8, that is at h/D = 1/3.

In order to verify this theoretical finding Sauermann (69,70) carried out tests in pipes of 100 mm nominal diameter. Segment plates were placed at three different heights approximately at, above and below the level h/D = 1/3. Using both sand and anthracite slurries, Sauermann showed that a segmented pipe could reduce the power consumption and that in both cases, least power consumption was required when h/D = 0.313 (see Fig. 9 for anthracite slurry). For the highest solids concentrations, reductions in power consumption were of the order of one-sixth.

CONCLUSIONS

This paper has described techniques which have been proven technically on the laboratory scale, and sometimes on the pilot-scale, for reducing the head loss for slurry flow in horizontal pipes. Some of the techniques are also applicable to vertical pipe flow. Tables 1 and 2 summarise these techniques in order to indicate whether they apply to essentially "non-settling" slurries, settling slurries or both and whether the head loss reduction effect occurs in the laminar or turbulent flow regimes. A brief summary of the believed mechanism for head loss reduction is also included.

The purpose of this paper has been to advertise many of the phenomena causing head loss reductions. Few industrial installations exploit this emerging technology, but there may be many slurry pipeline systems, currently in existence or planned for the future, whose economics could appear substantially better if one of these head loss reduction techniques were to be adopted.

REFERENCES

A. Effect of Changing Particle Size

1. Wagstaff, I. and Chaffey, C.E.: J. Coll. Interf. Sci., 59, pp 53-62 (1977).

2. Chaffey, C.E. and Wagstaff, I.: J. Coll. Interf. Sci., 59, pp 63-75 (1977).

3. Papir, Y.S. and Krieger, I.M.: J. Coll. Interf. Sci., 34, p 126 (1970).

4. Woods, M.E. and Krieger, I.M.: J. Coll. Interf. Sci., 34, p 91 (1970).

5. Chaffey, C.E.: Colloid and Polymer Sci., 255, p 691 (1977).

6. Krieger, I.M.: Trans. Soc. Rheol., 7, p 101 (1963).

B. Effect of Changing Particle Size Distribution

7. Farris, R.J.: Trans. Soc. Rheol., 12, pp 281-301 (1968).

8. Eveson, G.F., Ward, S.G. and Whitmore, R.L.: Disc. Farad. Soc., 11, pp 11-14 (1951).

9. Mooney, M.: J. Coll. Sci., 6, pp 162-170 (1951).

10. Brinkman, H.C.: J. Chem. Phys., 20, p 571 (1952).

11. Roscoe, R.: Brit. J. Appl. Phys., 3, pp 267-269 (1952).

12. Lee, J.W. and Lee, K.J.: Proc. 3rd Congress of Fluid Mechanics Fundamentals II, pp 32-37 (1983).

13. Chong, J.S., Christiansen, E.B. and Baer, A.D.: J. Appl. Polym. Sci., 15, pp 2007-2021 (1971).

14. Skvara, F. and Vancourova, M.: Silikaty, 17, pp 9-18 (1973).

15. Parkinson, C., Matsumoto, S. and Sherman, P.: J. Coll. Interf. Sci., 33, pp 150-160 (1970).

16. Goto, H. and Kuno, H.: J. Rheol., 28, pp 197-205 (1984).

17. Ferrini, F., Battarra, V., Donati, E. and Piccini, C.: Proc. Hydrotransport 9, paper B2. Organised by BHRA Fluid Engineering, Cranfield, Beds (Oct 1984).

18. Klose, R.B.: Proc. Hydrotransport 9, paper C3. Organised by BHRA Fluid Engineering, Cranfield, Beds (Oct 1984).

C. Effect of Changing Particle Shape, Including High Aspect Ratio Fibres

19. Clarke, B.: Trans. I.Chem.E., 45, p 251 (1967).

20. Kerekes, R.J.E. and Douglas, W.J.M.: Can. J. Chem. Eng., 50, pp 228-231 (1972).

21. Vaseleski, R.C. and Metzner, A.B.: A.I.Ch.E.J., 20, pp 301-306 (1974).

22. Kale, D.D. and Metzner, A.B.: A.I.Ch.E.J., 20, pp 1218-1219 (1974).

23. Bobkowicz, A.J. and Gauvin, W.H.: Can. J. Chem. Eng., 43, pp 87-91 (1965).

24. Robertson, A.A. and Mason, S.G.: TAPPI, 40, p 326 (1957).

D. Addition of Soluble Ionic Compounds

25. Horsley, R.R. and Reizes, J.A.: Proc. Hydrotransport 7 Conference. Paper D3. Organised by BHRA Fluid Engineering (Nov 1980).

26. Horsley, R.R.: Proc. Hydrotransport 8 Conference. Paper H1. Organised by BHRA Fluid Engineering (Aug 1982).

27. Horsley, R.R.: J. Pipelines, 3, pp 87-96 (1982).

28. Sikorski, C.F., Lehman, R.L. and Shepherd, J.A.: Proc. 7th Slurry Transport Association Conference, pp 163-173 (1982).

29. Shook, C.A. and Nurkowski, J.: Can. J. Chem. Eng., 55, pp 510-515 (1977).

30. Teckchandani, N. and Shook, C.A.: J. Pipelines, 2, pp 43-47 (1982).

E. Addition of Soaps

31. Zakin, J.L., Poreh, M., Brosh, A. and Warshavsky, M.: Chem. Eng. Prog. Symp. Ser. No. 11, 67, pp 85-89 (1971).

F. Addition of Soluble High Molecular Weight Polymer
(see Ref. 31)

32. Radin, I., Zakin, J.L. and Patterson, G.R.: A.I.Ch.E.J., 21, pp 358-371 (1975).

33. Golda, J.: Proc. 9th Slurry Transport Association Conference, pp 55-61 (1984).

34. Pollert, J.: Proc. 2nd Conference on Drag Reduction. Paper B3. Organised by BHRA Fluid Engineering, Cranfield, Beds (1977).

35. Kolar, V.: Proc. Hydrotransport 1 Conference, Paper F2. Organised by BHRA Fluid Engineering, Cranfield, Beds (1970).

36. Sifferman, T.R. and Greenkorn, R.A.: Soc. Pet. Eng. J., 21 (6), pp 663-669 (1981).

G. Air Injection into Flowing Slurry

37. Ward, H.C. and Dallavalle, J.M.: Chem. Eng. Prog. Symp. Series No. 10, pp 1-14 (1954).

38. Cheng. D.C-H., Jones, P.E., Keen, E.F., Laws, K.G. and French, R.J.: Proc. Pneumotransport 1 Conference. Paper A1. Organised by BHRA Fluid Engineering, Cranfield (Sept 1971).

39. Carleton, A.J., Cheng, D.C-H. and French, R.J.: Proc. Pneumotransport 2 Conference. Paper F2. Organised by BHRA Fluid Engineering, Cranfield (Sept 1973).

40. Heywood, N.I. and Richardson, J.F.: Proc. Hydrotransport 5 Conference. Paper C1. Organised by BHRA Fluid Engineering, Cranfield (May 1978).

41. Farooqi, S.I., Heywood, N.I. and Richardson, J.F.: Trans. I.Chem.E., 58, pp 16-27 (1980).

42. Heywood, N.I. and Charles, M.E.: Proc. Hydrotransport 7 Conference. Paper E1. Organised by BHRA Fluid Engineering (Nov 1980).

43. Farooqi, S.I. and Richardson, J.F.: Trans. I.Chem.E., 60, pp 323-333 (1982).

44. Chhabra, R.P, Farooqi, S.I., Khatib, Z. and Richardson, J.F.: J. Pipelines, 2, pp 169-185 (1982).

45. Chhabra, R.P., Farooqi, S.I., Richardson, J.F. and Wardle, A.P.: Chem. Eng. Res. Des., 61, pp 56-61 (1983).

46. Khatib, Z. and Richardson, J.F.: Chem. Eng. Res. Des., 62, pp 139-154 (1984).

47. Dziubinski, M., and Richardson, J.F.: J. Pipelines, 5, pp 107-111 (1985).

48. 'MIXOKRET', Putzmeister Ltd, Commercial Brochure, DF 520-2GB.

H. Oscillation of Slurry Flowrate or Pressure Gradient

49. Edwards, M.F. and Wilkinson, W.L.: Trans. I.Chem.E., 49, p 785 (1971).

50. Barnes, H.A., Townsend, P. and Walters, K.: Nature, 244, p 585 (1969).

51. Round, G.F.: Proc. Hydrotransport 3 Conference. Paper B3. Organised by BHRA Fluid Engineering, Cranfield (1974).

52. Round, G.F., Latto, B. and Lau, K-Y.: Proc. Hydrotransport 4 Conference. Paper D1. Organised by BHRA Fluid Engineering, Cranfield (1976).

53. Round, G.F., Hameed, A. and Latto, B.: Proc. 6th Slurry Transport Association Conference, pp 121-130 (1981).

54 Chan, K.W., Baird, M.H.I. and Round, G.F.: Proc. Roy. Soc. (London), 330, pp 537-559 (1972).

55. Colamussi, A. and Merli, V.: Dock and Harbour Authority, pp 471-482 (March 1972).

I. Vibration or Oscillation of Pipe

56. Deysarkar, A.K. and Turner, G.A.: J. Rheol., 25 (1), pp 41-54 (1981).

57. Manero, O. and Mena, B.: Rheol. Acta., 16, pp 573-576 (1977).

58. Manero, O., Mena, B. and Valenzeula, R.: Rheol. Acta., 17, pp 693-697 (1978).

J. Use of Spiral Ribs

59. Wolfe, S.E.: CIM Bulletin, 60 (658), pp 221-223 (1967).

60. Charles, M.E.: Proc. Hydrotransport 1 Conference, Paper A3. Organised by BHRA Fluid Engineering, Cranfield (1970).

61. Charles, M.E. et al: Can. J. Chem. Eng., 49, pp 737-741 (1971).

62. Kuzuhara, S.: Proc. Hydrotransport 3 Conference, Paper F2. Organised by BHRA Fluid Engineering, Cranfield (1974).

63. Schriek, W. et al: Saskatchewan Res. Council Report no. E72-4, CIM Bulletin, 67 (750), p 84 (1974).

64. Singh, V.P. and Charles, M.E.: Can. J. Chem. Eng., <u>54</u>, pp 249-254 (1976).

65. Kuzuhara, S. and Shakouchi, T.: Proc. Hydrotransport 5 Conference, Paper H3. Organised by BHRA Fluid Engineering, Cranfield (May 1978).

66. Smith, L.G., Shook, C.A., Haas, D.B. and Husband, W.H.W.: Proc. Hydrotransport 5 Conference, Paper B2. Organised by BHRA Fluid Engineering, Cranfield (May 1978).

K. Use of Segmented Pipe

67. Wang, R.C. and Seman, J.J.: US Dept. of Interior, Bureau of Mines Report of Investigation, 7725 (1973).

68. Führböter, A.: Mitteilungen des Franzius - Instituts für Grund und Wasserbau der Technischen Hochschule, Hannover, Heft 19 (1961).

69. Sauermann, H.B.: Proc. Hydrotransport 5 Conference, Paper A4. Organised by BHRA Fluid Engineering, Cranfield (May 1978).

70. Sauermann, H.B.: Proc. 7th Slurry Transport Association Conference, pp 57-60 (1982).

TABLE 1. – Summary of Techniques to Reduce Head Loss for Pipeline Flow of "Non-Settling" Slurries in the Laminar Regime

Technique	Requirement for Suspension Flow Property	Essential Mechanism
Increase in particle size	Newtonian or non-Newtonian.	Reduction in Brownian motion contribution to viscosity and/or reduction in increase in effective volume fraction of solids arising from reduced importance of adsorbed layer effect.
Broader particle size distribution	Newtonian or non-Newtonian.	Wider particle size distribution at fixed solids concentration leads to lower suspension viscosity. (Alternatively higher solids loading is achievable at a fixed suspension viscosity and hence fixed head loss.)
Reduce particle angularity	Newtonian or non-Newtonian.	Reduction in viscosity arising from reduced surface friction contacts between particles at high solids concentrations.
Addition of soluble ionic compounds as dispersants	Usually non-Newtonian, shear-thinning.	Changes pH and/or zeta potential on fine particle surfaces, causing particle dispersion and hence reduces suspension viscosity.
Air injection into flowing slurry	Must be non-Newtonian, shear-thinning with or without yield stress.	Presence of air reduces wetted pipe wall area on which suspension shear stress acts so reducing drag and head loss. Increase in slurry velocity at fixed slurry flowrate tends to increase head loss marginally owing to shear-thinning behaviour.
Oscillation of slurry flow or applied pressure gradient	Non-Newtonian, shear-thinning. Viscoelastic property may enhance effect.	Probable creation of annulus adjacent to pipe wall which is depleted in particles and hence lower wall shear stress occurs.
Pipe vibration	Non-Newtonian, shear-thinning.	Possible creation of slip layer near to wall depleted in solids. May also be thixotropic structure breakdown effect occurring.
Periodic injection of water (or other suspending medium) along pipeline length.	Newtonian or non-Newtonian but probably latter with high yield stress.	Water injection gives local reduced solids concentration at pipe wall and hence lower head loss.

TABLE 2. – Summary of Techniques to Reduce Head Loss for Pipeline Flow of Settling Slurries in the Turbulent Flow Regime

Technique	Essential Mechanism
Fibre addition	Two mechanisms possible: with 10 to 1000 ppm of fibres turbulent drag reduction occurs probably because of dampening of turbulent eddies. With fibre concentrations up to 5%, fibres form a central plug with water annulus. In both cases, technique is also applicable to "non-settling" slurries.
Addition of soaps	Probable reduction in inter-particle friction.
Addition of soluble, long-chain polymers	Toms' effect occurs in turbulently-flowing suspending medium. Mechanism still unclear but probably dampening of turbulent eddies.
Oscillation of slurry flowrate	Reduction in deposit velocity arising from more effective particle suspension in oscillating flow.
Use of helical ribs attached to inner pipe wall	Ribs reduce deposit velocity by continuously picking up particles tending to settle. Lower velocity means lower head loss.
Use of segmented pipe	Mechanism unclear but deposit velocity reduced and hence pipeline can be operated at lower slurry flowrates and hence lower head losses are incurred.

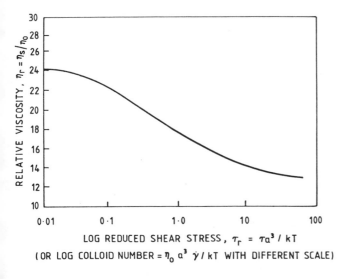

Fig. 1. Effect of particle size on suspension viscosity when Brownian motion effects are important. (After Krieger (6) and Papir and Krieger (3).)

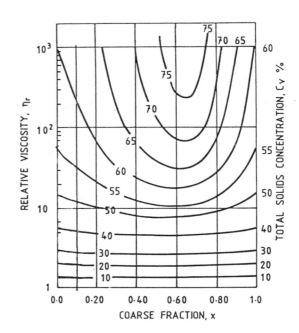

Fig. 2. Predictions of suspension viscosity for a mixture of coarse and fine particle fractions. (From Farris (7).)

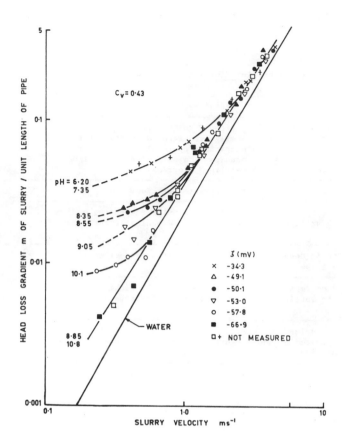

Fig. 3. Effect of pH and zeta potential on 43% concentration by volume of a $d_{50} = 17$ μm sand slurry in pipe flow. (From Horsley and Reizes (25).)

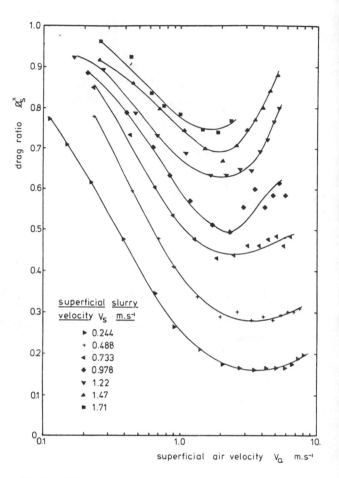

Fig. 5. Effect of air injection on the head loss of 22% by volume flocculated kaolin slurry flowing in a 42 mm dia pipe. (From Heywood and Richardson (40).)

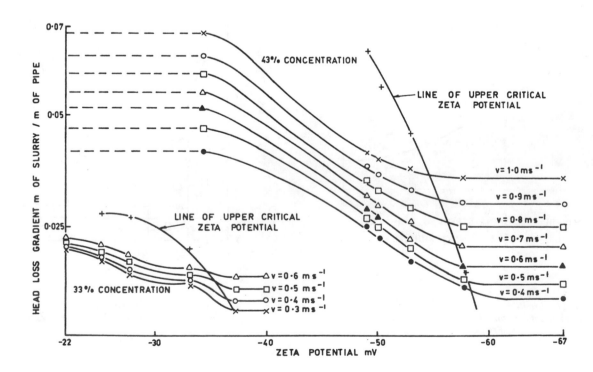

Fig. 4. Effect of zeta potential on head loss gradient for 33% and 43% concentrations by volume of a $d_{50} = 17$ μm sand slurry. (From Horsley and Reizes (25).)

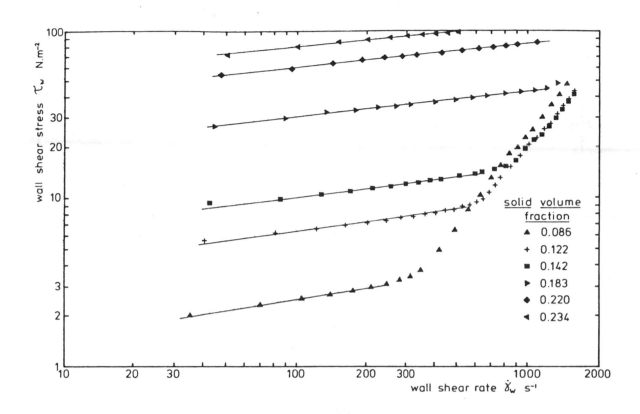

*Fig. 6.   Flow curves for flocculated kaolin slurries obtained using a
42 mm dia pipe. (From Heywood and Richardson (40).)*

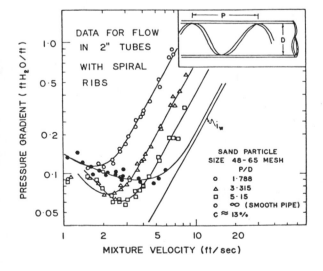

*Fig. 7.   Reduction in head losses obtained in a 48 mm dia pipe for a
13% settling sand slurry. (From Charles (60).)*

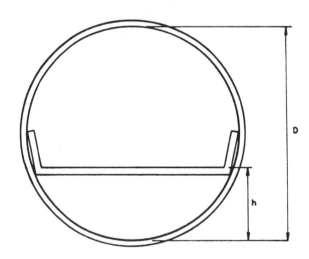

Fig. 8.   Design of segmented pipe used by Sauermann (69).

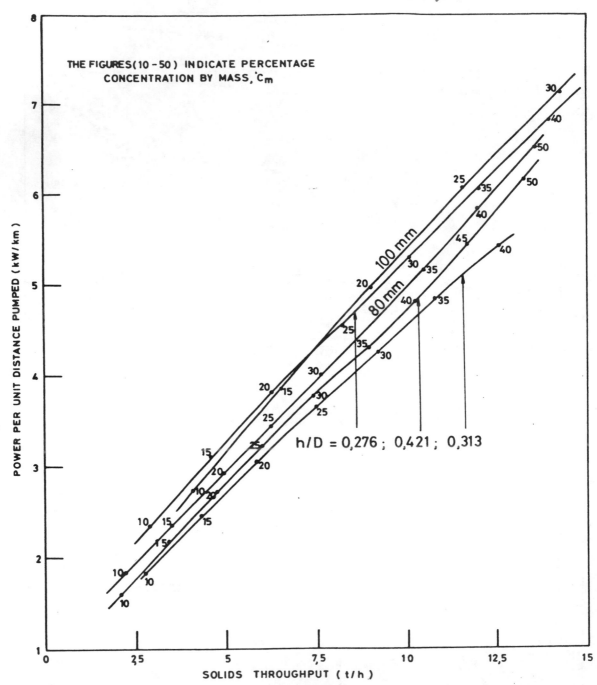

Fig. 9. Power reductions for anthracite slurry using a segmented
pipe at three alternative positions in 80 mm and 100 mm dia pipe.
(From Sauermann (69).)